Topics in Applied Physics Volume 30

Topics in Applied Physics Founded by Helmut K. V. Lotsch

Excimer Lasers

Edited by Ch. K. Rhodes

With Contributions by
Ch. A. Brau   H. Egger   A. Gallagher   P. W. Hoff
K. Hohla   D. L. Huestis   M. Krauss   G. Marowsky
M. V. McCusker   F. H. Mies   H. Pummer
Ch. K. Rhodes   F. K. Tittel

Second Enlarged Edition

With 100 Figures

Springer-Verlag
Berlin Heidelberg GmbH 1984

Dr. *Charles K. Rhodes*

University of Illinois at Chicago, College of Liberal Arts and Sciences, Department of Physics
P.O. Box 4348, Chicago, IL 60680, USA

ISBN 978-3-540-13013-0   ISBN 978-3-540-35671-4 (eBook)
DOI 10.1007/978-3-540-35671-4

Library of Congress Cataloging in Publication Data. Main entry under title: Excimer lasers. (Topics in applied physics; v. 30). Includes bibliographical references and index. 1. Excimer lasers. I. Rhodes, Charles K. II. Brau, Charles A., 1938–. III. Series. TA1695.E93 1984   621.36'63   83-20088

© by Springer-Verlag Berlin Heidelberg 1979 and 1984
Originally published by Springer-Verlag Berlin Heidelberg New York Tokyo 1984

2153/3130-543210

Preface to the Second Edition

The second edition of this Topics volume attempts to incorporate several developments that have occurred since the publication of the first edition. Several specific areas are prominent in this revision. These include the considerable knowledge on trimer species that has been obtained, the very significant advances made in spectral brightness of rare-gas-halogen systems, an expanded discussion of applications, a section on the mercury halides, and an account of refinements in the kinetic modeling and techniques of excitation for rare-gas-halogen media.

Naturally, the preparation of this new edition required the cooperation of many individuals, always generously given. As in the first edition, it is the editor's final responsibility to orchestrate the wide range of contributions, a task I hope has been achieved to the benefit and satisfaction of the reader.

The editor wishes to acknowledge and express his sincere gratitude to the group of persons whose work made possible the publication of this volume. Certainly the contributors who provided the new material must be thanked for interrupting their busy lives to do the thoughtful and careful writing necessary for their contributions. Very considerable thanks must also go to Mrs. Peggy Hrynko for much valuable assistance and, particularly, to Ms. Rhonda De Witt for extremely careful typing of the manuscript. Also, the efforts of Mr. Scott Stirton in the development of a considerably expanded index have been unusually valuable. Finally, the patience of Dr. Helmut Lotsch of Springer-Verlag is greatly appreciated, since his effort is central in reaching the goal of publication.

Chicago, September 1983   *Ch. K. Rhodes*

Preface to the First Edition

The development of excimer laser systems marked a significant turning point in the development of coherent sources. The progress of the last few years has been largely predicated upon the combined knowledge of several disciplines including atomic and molecular physics, optical technology, and pulsed power technology, the latter mainly associated with electron beam devices.

The purpose of this volume is to provide a comprehensive view of this marvelously exciting field that will be of value to both active researchers and neophytes alike. Since a clear understanding of both theory and experiment is necessary to achieve this goal, these issues are presented as an integrated whole.

The preparation of this work involved the dedicated cooperation of many authors dispersed both geographically and intellectually. Naturally, the editor has the responsibility to integrate and balance a diverse range of opinions to the satisfaction of all, a job not always readily accomplished. We hope that this has been performed to the satisfaction of the reader.

The editor wishes to express his gratitude to the many persons whose efforts made this book possible. Most important are the authors whose work constitutes the backbone and substance of this volume and whose normal professional lives are very busy ones, indeed. I can say that they worked diligently and with good humor in the preparation of their contributions. Their aid and advice has been invaluable to the editor. Considerable thanks must go to Mrs. Carol Peckham and Mrs. Janice Cox by whom the carefully prepared typed manuscript was prepared. Special notice is reserved for Ms. Nancy Garvey for a careful proofreading of the final manuscript. Finally, the patient assistance of Dr. Helmut Lotsch of Springer-Verlag is greatly appreciated as his efforts have been central to the finished publication in your hands.

Menlo Park, Calif.   *Ch. K. Rhodes*
November 1978

Contents

Contributors

Brau, Charles A.
  University of California, Los Alamos National Laboratory,
  Los Alamos, NM 87545, USA

Egger, Hans
  University of Illinois at Chicago, Department of Physics,
  P.O. Box 4348, Chicago, IL 60680, USA

Gallagher, Alan
  JILA, University of Colorado and National Bureau of Standards,
  Boulder, CO 80302, USA

Hoff, Paul W.
  Division of Laser Fusion, Department of Energy,
  Washington, DC 20545, USA

Hohla, Kristian
  Lambda Physik, Wagenstieg 8,
  D-3400 Göttingen, Fed. Rep. of Germany

Huestis, David L.
  Molecular Physics Laboratory, SRI International,
  Menlo Park, CA 94025, USA

Krauss, Morris
  United States Department of Commerce, National Bureau of Standards,
  Washington, DC 20234, USA

Marowsky, Gerd
  Max-Planck-Institut für Biophysikalische Chemie, Postfach 968,
  D-3400 Göttingen, Fed. Rep. of Germany

McCusker, Michael V.
  Spectra Physics Inc., 1250 West Middlefield Road,
  Mountain View, CA 94043, USA

Mies, Frederick H.
   United States Department of Commerce, National Bureau of Standards,
   Washington, DC 20234, USA

Pummer, Herbert
   University of Illinois at Chicago, Department of Physics,
   P.O. Box 4348, Chicago, IL 60680, USA

Rhodes, Charles K.
   University of Illinois at Chicago, College of Liberal Arts and Sciences,
   P.O. Box 4348, Chicago, IL 60680, USA

Tittel, Frank K.
   Rice University, Department of Electrical Engineering,
   Houston, TX 77251, USA

1. Introduction

P. W. Hoff and Ch. K. Rhodes

With 2 Figures

The concept of bound-free excimer system as a laser medium was initially enunciated [1.1] in 1960. The first successful laboratory demonstrations, however, were not accomplished until approximately 1972. It was the new application of an available technology, namely, electron beams, which enabled a sufficiently high excitation rate to achieve the optical amplification required. It is now apparent that these developments comprise a rapidly maturing hybrid technology which will have a considerable impact on future scientific progress and applications.

The physical mechanisms underlying this broad class of laser systems comprise the central topic of this book. This involves the examination of an extensive array of molecular systems from both theoretical and experimental viewpoints, the task reserved for the following chapters.

The organization of this volume is straightforward. Chapter 2 provides the main theoretical framework needed for an understanding of the currently known excimer molecules. This includes a description of valence, ion-pair, and Rydberg systems as well as an analysis of the emission processes and calculations of the optical cross section. Chapter 3 leads directly to the experimental findings in describing three classes of related excimers, namely, the rare gas dimers, the rare gas oxides, and those arising in excited diatomic halogen molecules. The rare gas halogen systems appear in Chap. 4, which also includes a considerable discussion on the technologies applied for excitation. Metal vapor excimers are analyzed in Chap. 5, including some relatively exotic molecules such TlHg. Chapter 6 closes the discussion of different excimer species with an analysis of triatomic rare-gas-halogen systems. Chapter 7 provides a detailed description of high-spectral-brightness rare-gas-halogen lasers, and the production of picosecond pulses with these devices. Finally, Chap. 8 reviews applications of excimer lasers, purely scientific and applied, current as well as proposed, which are viewed as relevant examples.

Excimers are not simple systems, a fact noted by *Finkelnburg* [1.2] in his early discussion on continuous spectra. Certainly one source of interpretive difficulties with continuous spectra stems directly from the absence of two basic tools generally used in the analysis of molecular electronic transitions; viz., vibrational and rotational progressions. Among the simplest systems representing the prototype of molecular bound-free transitions, and for which there exists a considerable literature concerning its properties, is He_2. The complex

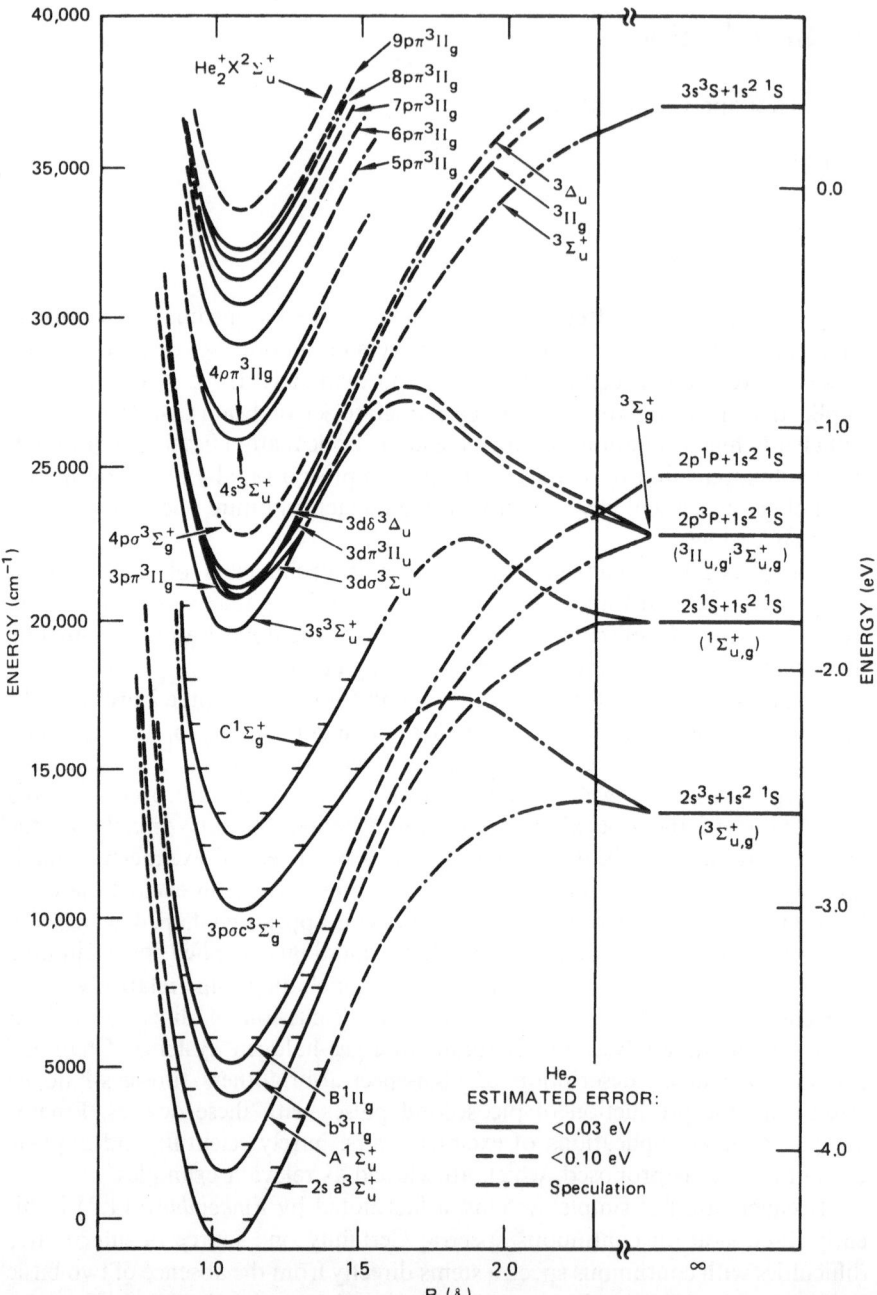

Fig. 1.1. Potential-energy curves (rotationless) for selected electronic states of He₂. Energy in cm⁻¹ is based on $N=0$, $v=0$ of the $a^3\Sigma_u^+$ state, while energy in electron volts is based on the lowest level in $X^2\Sigma_u^+$ of He₂⁺. When practical, the observed vibrational levels are indicated by horizontal lines at the edges of their appropriate curves. The repulsive ground $^1\Sigma_g^+$ state, which lies at $\sim -22.4\,\mathrm{eV}$, is not shown. (Printed with permission)

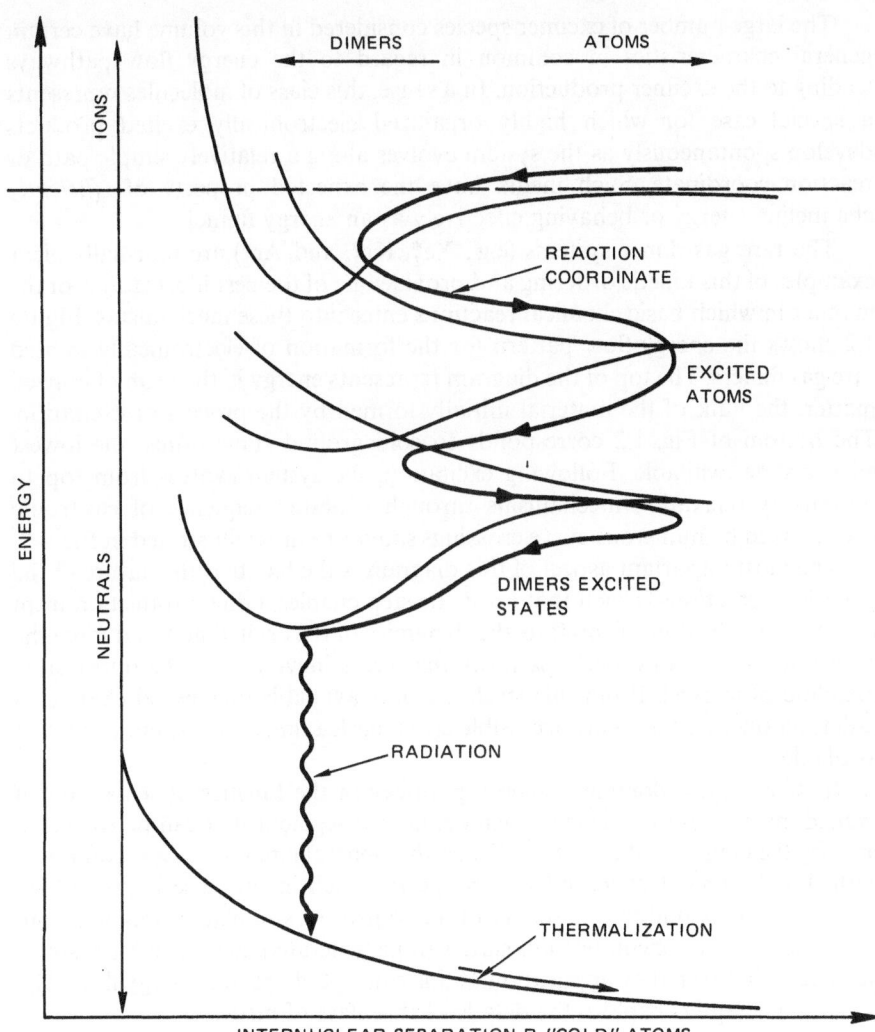

DIMERS ⟷ ATOMS

IONS

NEUTRALS

ENERGY

REACTION
COORDINATE

EXCITED
ATOMS

DIMERS EXCITED
STATES

RADIATION

THERMALIZATION

INTERNUCLEAR SEPARATION R "COLD" ATOMS

Fig. 1.2. Dimer chemical pathway

nature of these molecular systems is immediately illuminated by the fact that over 60 electronic states are currently known even for this elementary system [1.3, 4].

The molecular curves of *Ginter* and *Battino* shown in Fig. 1.1 illustrate this complexity for the simplest of excimer systems, He_2^*. We immediately note the families of closely nested curves for the excited states as well as the level crossings linking specific levels. These crossings, as we shall see in subsequent chapters, provide important kinetic pathways for the population of the excimer states.

The large number of excimer species considered in this volume have certain general characteristics in common in regard to the energy flow pathways leading to the excimer production. In a sense, this class of molecules represents a special case for which highly organized electronically excited products develop spontaneously as the system evolves along a relatively simple path or reaction coordinate. Such media have the expected property of *efficiently* channeling energy or behaving effectively as an energy funnel.

The rare gas dimer systems (e.g., Xe_2^*, Kr_2^*, and Ar_2^*) are unusually clear examples of this kinetic ordering and provide one of the best illustrations of the manner in which basic chemical reactions enter into these mechanisms. Figure 1.2 shows the energy flow pattern for the formation of electronically excited rare gas dimers. The top of the diagram represents energy in the form of ionized matter, the bulk of the material initially formed by the process of excitation. The bottom of Fig. 1.2 corresponds to cold ground state atoms, the lowest energy state available. Following excitation, the system evolves from top to bottom by relaxation mechanisms through a nested sequence of electronic states linked by numerous curve crossings similar to those illustrated in Fig. 1.1.

The most important aspect of this diagram is the fact that the nature of the particle interactions is such that no single step enables a direct transition from the top to the bottom. Therefore, the dynamics of the excited material force the energy to flow in an orderly pathway that leads inevitably to the upper state manifold of interest. From this level, the only available downward channel is radiation since no thermally accessible crossings leading to the ground state are available.

In summary, a dominant energy pathway in the kinetics of relaxation of excited material is established which contains a segment that can be traversed only by the emission of radiation. When this general circumstance is combined with the attainment of *inversion*, a property which in this case is present on account of the bound-free character of the transition, stimulated emission from this ordered and nonequilibrium state can be generated. Several laser systems have been discovered by application of this concept of ordered energy flow. The remaining chapters explore the details of this class of media.

References

1.1 F.G.Houtermans: Helv. Phys. Acta. **33**, 933 (1960); also see
    C.K.Rhodes: IEEE J. QE-**10**, 153 (1974)
1.2 W.Finkelnburg: *Kontinuierliche Spektren* (Springer, Berlin 1938)
1.3 M.L.Ginter, R.Battino: J. Chem. Phys. **52**, 4469 (1970) (Here, as well as in references cited herein, the molecular properties of the He_2 systems are discussed)
1.4 W.A.Fitzsimmons: In *Atomic Physics 3*, ed. by S.J.Smith, G.K.Walters (Plenum Press, New York 1973) p. 477 (description of experimental data on He_2)

2. Electronic Structure and Radiative Transitions of Excimer Systems

M. Krauss and F. H. Mies

With 16 Figures

The interaction of the ground states of two closed-shell systems (atom or molecule) is usually repulsive except for a weak long-range van der Waals or electrostatic interaction. But the interaction of an excited state with the ground state of the same fragments can form a strong chemical bond. The bound excited state has been called an excimer or exciplex state and the bound to continuum transition defines an excimer emission system [2.1]. In the case of degenerate open-shell fragments (e.g., P-states, etc.) more than one molecular state will correlate with the degenerate asymptote. Some or all of the curves may be essentially repulsive, but there may also be a bound ground state. The excited state of the open-shell fragment interacting with a ground state fragment can also yield strongly bound molecular states which can be defined as excimer states. The analysis of excimer emission systems in this chapter assumes that the lower repulsive state always arises from ground state fragments which may, however, be degenerate. Rare gas atoms interacting in their ground state provides an example of a single repulsive curve arising from a non-degenerate asymptote. The interaction of two hydrogen atoms in their ground ^2S states will, on the other hand, yield two states a $^1\Sigma_g^+$ and a $^3\Sigma_u^+$ which are, respectively, strongly bound and repulsive. By generalizing the definition of an excimer emission system to include degenerate asymptotic ground states, we would include almost all possible chemical systems. In order to keep the review within limits, we have restricted consideration to the interactions of either rare gas or Group II atoms with other atomic species for which excimer emission systems have been found and were originally suggested as laser candidates [2.2]. Only two triatomic classes of excimers have been included which are again based on rare gas or Group II molecule fragments.

The most important intrinsic characteristic of the laser transitions is the small signal gain cross section which is proportional to

$$\sigma \sim \lambda^2 A g(\varepsilon).$$

In addition to the wavelength dependence, λ, there are two factors, the Einstein coefficient A for spontaneous emission from the upper excimer state and the continuum line shape factor g which is the Franck-Condon factor for a bound to continuum transition. Both the transition probabilities and energy curve parameters of the excimer molecule which determines the Franck-Condon

Fig. 2.1a–c. Excimer radiative transitions to "repulsive" curves that are **a** steeply repulsive, **b** flat, and **c** long-range region of bound curve

factor are little known experimentally. Analysis of excimer transitions has often been based on theoretical interpretations which require that the electronic structure of the ground and excited states is understood at least qualitatively correctly. Even the rough energy ordering of the higher excited states is lacking which precludes guesses as to absorption possibilities from the excimer excited state.

Ground state "repulsive" energy curves can be classified in two extreme categories, a) very repulsive, or b) flat, in the neighborhood of the emission intensity maximum as seen in Fig. 2.1. The details of the electronic interactions that lead to these curves will be described below for the specific cases. The interaction of open-shell atoms with either closed-shell (e.g., $s^2 p^6$) or closed sub-shell (e.g., s^2) systems is usually repulsive because the considerable energy required to excite or transfer electrons from closed-shells inhibits bond formation. In the ground states the atomic electrons tend to remain localized and the interaction of the atoms is then repulsive and proportional to the charge overlap. For open-shell systems certain orientations of the charge tend to minimize the overlap and the repulsive interaction can be quite flat or else the charge distribution can be oriented to maximize the repulsion and yield a very repulsive curve. A simple example to illustrate the orientation effect is the interaction of a singly occupied p orbital in an alkali atom with a closed shell rare gas atom. If the p orbital is oriented along the line connecting the atom centers, the charge of the $p\sigma$ electron will repulsively overlap the charge of the rare gas atom at long distances. When the p electron is oriented perpendicular, the $p\pi$ electron charge has a node along the line of centers which diminishes the overlap and the repulsion.

The flat curves arise when repulsive ground state configurations mix with bound excited configurations and obtain some bound character. This characterization is useful only when the mixing is weak and the state is still properly described as repulsive. The bound state curves will all have minima due to either a weak van der Waals dispersion interaction or a combination of the dispersion and configuration mixing interactions. The equilibrium internuclear separations, R_e, are invariably to much larger distances than the excimer R_e for the pure dispersion. But the other shallow wells that result can have all possible relations to the excited state R_e depending on the type and strength of the configuration mixing. As will be seen in the last section, the different types of ground state curves lead to very different Franck-Condon vibrational envelopes for the excimer transition. Continuum transitions are also possible to the large separation regions of strongly bound ground states as illustrated in Fig. 2.1 c. Examples of this type of transition occur in the case of $H_2(1\Sigma_u^+ \rightarrow {}^1\Sigma_g^+)$ and the laser transitions of the halogen and group II-halide diatomic molecules. From low v' vibrational levels of the excited state, the transitions are predominantly to the high v'' bound levels and such transitions are not usually considered in the excimer category.

The bonding of the excited state can be classified into three extreme cases. There are bound covalent, Rydberg, and charge transfer excimer states. Covalent bonding occurs when a pair of electrons shares a bonding orbital that is distributed over the two atoms. As we shall see the valence excimer states of the Group II homonuclear molecules provide examples of this type of binding. The Rydberg orbitals are diffuse and have relatively little molecular bonding or anti-bonding characteristics. The basic bonding behavior of the Rydberg molecular state is, therefore, determined by the bonding characteristics of the limiting positive ion curve. Rare gas excimer states provide the example of Rydberg excimers. Charge transfer excited states are well known in the case of the rare gas halides. It is likely that such states are very widespread. They result when the ionization potential of the electron donor is relatively high such that the ion-pair asymptote is so high in energy relative to the neutral atom asymptote that the Coulombic attraction is insufficient to make the ion-pair state the ground state. Nonetheless, the ion-pair excited state is low enough in energy to be isolated from perturbations of other excited states and predissociations. The electrostatic interactions determine the binding in the ion-pair excited state just as it does for the ground state in the alkali halides. Mixing between the ion-pair and covalent states are important in determining details of the energy curves of both the ground and excited state curves.

The review is organized around the different categories of excimer systems. Each category provides an example of one or more of the types of electronic bonding. The ground and excited state electronic structure will be analyzed separately and then the observed spectroscopic transitions coupling these states will be reviewed for the typical cases of rare gas dimer, rare gas-halide, rare gas-group VI, and Group II–Group II excimers. Much briefer discussions will be given of other cases, since no new concepts would be involved and/or there is very little theoretical or experimental information available. In the final section the quantal theory of stimulated emission and absorption in an excimer system will be reviewed. The Franck-Condon factors for the bound-continuum transitions to both the repulsive and flat ground states will be discussed in detail.

2.1 Rare Gas Dimers: A Rydberg Example

Since there are no valence excited states or negative ion (bound or low-lying resonance) states of a rare gas atom, the dimer excited states are entirely Rydberg in character. *Mulliken* has given a qualitative analysis of the diatomic rare gas molecules with particular consideration to He_2 [2.3] and Xe_2 [2.4]. The bound Rydberg curves behave analogously to the limiting bound ion curve in the vicinity of the equilibrium distance. But there are repulsive curves that also correlate to the ground ionic asymptotes. In addition to families of bound Rydberg configurations, there will thus be families of repulsive Rydberg configurations and the final energy states can be complicated mixtures of these configurations. The qualitative behavior of the Rydberg states can be deduced from that of the dimer ions.

2.1.1 Dimer Ions

For He_2^+, in addition to the bound $^2\Sigma_u^+$ state, there is a repulsive $^2\Sigma_g^+$ state since there are only two valence orbitals, one bonding and the other anti-bonding, into which the three electrons are distributed. Other homopolar ions involving p-electrons have four valence orbitals, the σ_g, π_u, π_g, σ_u, in ascending order of the orbital energy, into which eleven electrons are distributed. If the strongly anti-bonding σ_u orbital is only singly occupied, the ionic $^2\Sigma_u^+$ ground state is found to be quite bound. The π_g orbital, on the other hand, is weakly anti-bonding and if it is singly occupied the $^2\Pi_g$ energy curve is quite flat and attractive at large distances but is essentially a repulsive curve at shorter distances. In the case of both the $^2\Pi_u$ and $^2\Sigma_g^+$ energy curves bonding orbitals are singly occupied and the curves are very repulsive. These expectations are borne out by a number of ab initio calculations [2.5, 6]. Representative energy curves are shown in Fig. 2.2.

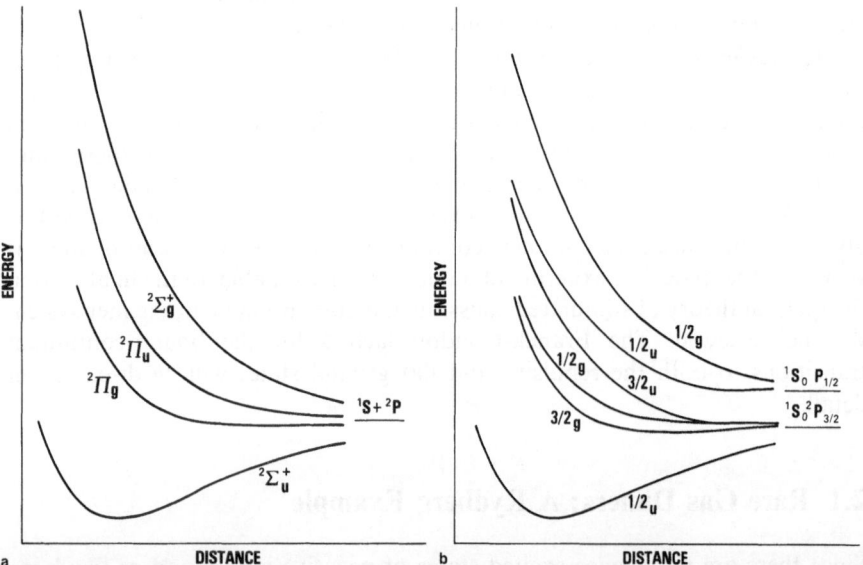

Fig. 2.2a and b. Energy curves of a homonuclear diatomic positive ion **a** without and **b** with spin-orbit coupling

Table 2.1. Comparison of the dissociation energy and equilibrium internuclear separation of the ground states of the neutral and positive ion rare gas dimer and the lowest energy triplet Rydberg dimer. The data on the Ne ion and Rydberg dimers are least accurate but are included to illustrate the qualitative behavior

	D_e [eV]			R_e [Å]		
	R_2	R_2^+	R_2^*	R_2	R_2^+	R_2^*
He	$0.9 \cdot 10^{-3\,a}$	2.47^e	2.55^g	2.96^a	1.08^e	1.045^j
Ne	$3.6 \cdot 10^{-3\,b}$	1.16^e	0.50^e	3.10^b	1.75^e	1.79^e
Ar	$12 \cdot 10^{-3\,c}$	1.23^f	0.68^h	3.76^c	2.48^i	2.43^h
Kr	$17 \cdot 10^{-3\,d}$	1.15^f		4.00^d	2.79^i	
Xe	$24 \cdot 10^{-3\,d}$	1.03^f		4.36^d	3.27^i	

[a] J.M. Farrar, Y.T. Lee: J. Chem. Phys. **56**, 5801 (1972).
[b] J.M. Farrar, Y.T. Lee, V.V. Goldman, M.L. Klein: Chem. Phys. Lett. **19**, 359 (1973).
[c] J.A. Barker, R.A. Fisher, R.O. Watts: Mol. Phys. **21**, 657 (1971).
[d] J.A. Barker, R.O. Watts, J.K. Lee, T.P. Schafer, Y.T. Lee: J. Chem. Phys. **61**, 3081 (1974).
[e] [2.5].
[f] [2.7].
[g] [2.3].
[h] [2.15].
[i] [2.8].
[j] M.L. Ginter: J. Chem. Phys. **42**, 561 (1965).

Spin-orbit coupling will be quite significant in these systems as we go from Ne to Xe. Explicit calculations have been made for a model spin-orbit interaction for Ne_2^+ and Ar_2^+ [2.6]. *Mulliken* notes that as we go to Xe_2^+ a Hund's case (c) representation is more appropriate near R_e. This will have a significant effect on the transition moments [2.6]. Qualitatively, it is easy to see that there is only one strong case (a) transition from the ground state. A simple molecular orbital analysis of the transition $^2\Sigma_u^+ \to {}^2\Sigma_g^+$ shows that the transition moment is proportional to R since exciting a σ_g molecular orbital to a σ_u is equivalent to transferring the electron from one atom to the other. But σ to π excitations will be weak due to the inherently small overlap of the orbitals. Spin-orbit coupling will permit the $^2\Sigma_u^+ \to {}^2\Pi_g$ transition to borrow intensity from the strong $^2\Sigma_u^+ \to {}^2\Sigma_g^+$ and $^2\Pi_u \to {}^2\Pi_g$ transitions.

Recent photoionization data [2.7] have provided reliable dissociation energies for Ar_2^+, Kr_2^+, and Xe_2^+ which are noted in Table 2.1. Ab initio calculations exist for He_2^+ and less accurate calculations have been reported for Ne_2^+ [2.5]. Recently, configuration interaction calculations have been completed on Ar_2^+, Kr_2^+, and Xe_2^+ with a model spin-orbit interaction [2.8]. The calculated dissociation energies are 1.19, 1.05, and 0.79 eV, respectively for Ar_2^+, Kr_2^+, and Xe_2^+. *Wadt* notes that the most likely source of error is the lack of polarization of the core which becomes more important as the atom gets larger. The dipole polarizability will have a significant contribution from core excitations for the heavier atoms.

Analogous photoionization studies [2.9] have also provided experimental values of 0.37, 0.14, and 0.59 eV for the XeKr, XeAr, and KrAr ions, respectively. These values are considerably less than the values for the homonuclear molecules indicating the importance of exchange contributions to the bonding. Theoretical calculations considering these species are few and are mostly concerned with the scattering behavior. The ground state of HeNe$^+$, $1^2\Sigma^+$, has been calculated [2.10] to be bound by 0.6 eV with an R_e at 1.49 Å.

2.1.2 Ground State

The ground state interatomic potentials have been described accurately for all the homonuclear systems using beam scattering, spectroscopy, gas transport properties, and solid state data. There is considerable research still arguing small differences, but the potentials are very accurately determined in the bound region [2.11] with perhaps the least testing of these potentials in the repulsive regions of interest in excimer transitions. The D_e and R_e are given in Table 2.1 for all the homonuclear pairs. The heteronuclear ground state potentials are also known experimentally for most of the pairs [2.12].

These ground state systems are van der Waals molecules. The energy of interaction can be described approximately as the sum of a repulsive interaction of the overlapping of the atomic charge distributions and the attractive dispersion energy due to the correlation of the interacting fragments [2.13]. Due to the weakness of the attractive interaction the equilibrium internuclear separation in these systems will be to much larger distances than the R_e of the Rydberg excited states.

2.1.3 Rydberg States

The complexity of even the lowest Rydberg levels is illustrated by the estimated potential curves of Xe$_2$ presented by *Mulliken* [2.4]. *Cohen* and *Schneider* [2.5] calculated all the $3s$ Rydberg states arising from the $^2\Sigma^+_{u,g}$ and $^2\Pi_{u,g}$ states of Ne$_2^+$ and obtained curves that show the essential correctness of the Σ^+ curves predicted by *Mulliken*. In contrast to *Mulliken* their Π_g curves are repulsive and dominated by the $^2\Pi_g$ ion core. This is surprising since the $3p\pi$ orbital energy should not be much higher than the $3p\sigma$ orbital energy. As *Mulliken* indicated, the Π_g energy curves should result from the configuration mixing between $(^2\Pi_g)$ $3s$ and $(^2\Sigma^+_u)$ $3p\pi$ configurations. The long range behavior is still a considerable research problem. Figure 2.3 shows the qualitative curves and the essential isolation of the case (c)1_u, O$_u^-$, and O$_u^+$ states which arise from the $^3\Sigma^+_u$ and $^1\Sigma^+_u$ states. These states are the three most important to the understanding of the excimer radiation in the rare gas dimer. In Ne$_2$ *Schneider* and *Cohen* calculated lifetimes for the $1_u(^3P_2)$ and O$_u^+(^3P_1)$ to be 11.9 µs and 2.8 ns, respectively, but the spin-orbit interaction increases as we go to Xe$_2$ and the analogous lifetimes are determined [2.14] to be 99 ns and 5.5 ns with $1_u(^3P_2)$ not very metastable anymore.

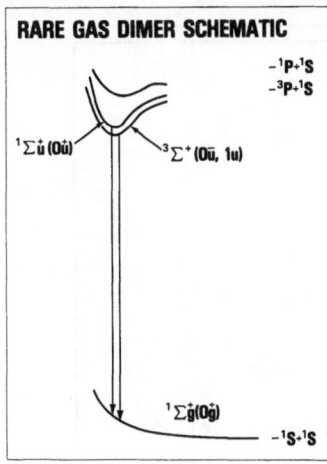

RARE GAS DIMER SCHEMATIC

$-{}^1P \cdot {}^1S$
$-{}^3P \cdot {}^1S$

$^1\Sigma_u^+(O_u^+)$ $^3\Sigma^+(O_u^+, 1_u)$

$^1\Sigma_g^+(O_g^+)$

$-{}^1S \cdot {}^1S$

Fig. 2.3. Schematic energy curves of the lowest Rydberg states of a homonuclear rare gas molecule and the excimer radiative transitions to the repulsive ground state

The transition moments for the Ne$_2$ $^1\Sigma_u^+$ and $^1\Pi_u$ states are calculated to decrease slowly from the asymptotic value. But the transitions can still be described qualitatively in terms of a one-electron excitation from the ground state atom to the Rydberg electron. Such a matrix element is about 0.375 a.u. which is sufficiently large to yield a radiative lifetime for the singlet state of about 1 ns. *Schneider* and *Cohen* also found that the transition moments can vary significantly with distance due to the spin-orbit mixing so there should be considerable variation in lifetime with vibrational quantum number.

An accurate calculation of the $^3\Sigma_g^+$ and $^3\Sigma_u^+$ excimer states of Ar$_2$ has been reported by *Saxon* and *Liu* [2.15]. The bound $^3\Sigma_u^+$ curve has a well depth of 0.68 eV with an equilibrium internuclear separation of 2.43 Å while the $^3\Sigma_g^+$ state is found to be metastable. Elastic scattering calculations using these curves are found to be in agreement with experimental differential cross section data [2.15]. The energy curve of the lowest Rydberg state is found to be considerably less bound than the limiting dimer ion curve for both Ne$_2$ and Ar$_2$, as seen in Table 2.1 but not in the case of He$_2$ [2.3]. In the neighborhood of R_e, the energy curves of the Rydberg state and ion are very similar. *Stevens* [2.16] found that the curves superimpose on the left-hand limb and near the R_e, as shown in Fig. 2.4. As the atoms separate, the Rydberg electron is localized and tends to screen the ion. This diminishes the energy required to separate the atoms since the ion induced-dipole energy will exceed the long-range dispersion interaction. In the case of Ar$_2$ the induction energy at 5.5 bohr is more than 0.3 eV and accounts for the reduction in the Rydberg energy. The Ne$_2$ calculations are not as accurate but still reflect the large qualitative change in bond energies. For He$_2$ the dipole polarizability is so small that this effect is much reduced.

The data on the heteropolar excited states are quite sparse. But the emission spectra that have seen would also indicate shallower excimer wells than

Ar_2^* $^1\Sigma_u^+$ POTENTIAL ENERGY CURVE

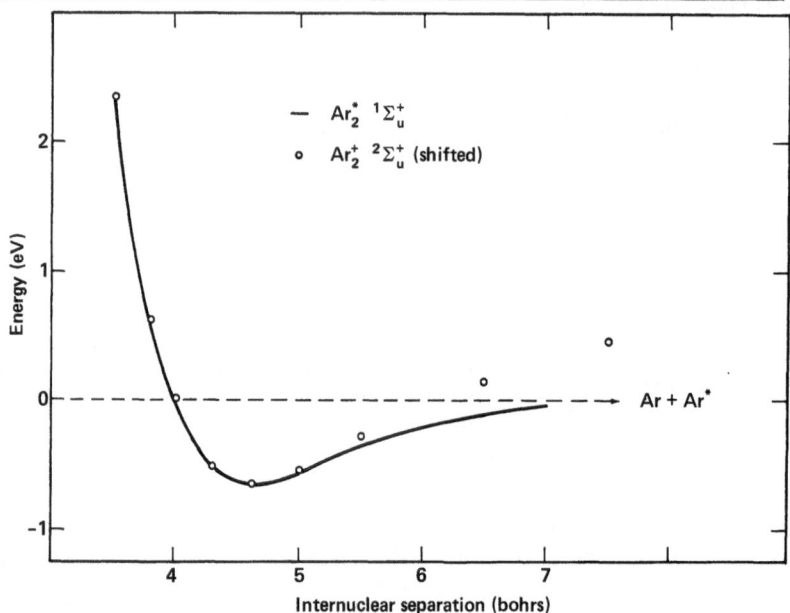

Fig. 2.4. Comparison of the potential energy curves of Ar_2^+ and Ar_2^*

the ion states. *Cheshnovsky* et al. [2.17] and *Verkhovtseva* et al. [2.17] found only KrAr and XeKr excimer states. The exchange interactions are apparently weaker as the ionization potentials of a rare gas pair get very different.

The binding energies of only $KrAr^+$ and $XeKr^+$ are found to be appreciable. Since the Rydberg bonding energies are diminished by screening, the theoretical expectation that only KrAr and XeKr excimers will be appreciably bound is borne out by the observations.

2.1.4 Observed Spectroscopy

The molecular emission spectra in the vacuum ultraviolet of the rare gases are characterized by two types of continua. The "first continuum" is an oscillatory structure to the red of the atomic line and the "second continuum" is a roughly Gaussian structure at much longer wavelengths than the atomic line. The oscillatory structure in the "first continuum" is attributed to continuum-continuum, quasi bound continuum, or high-bound level-continuum transitions while the "second continuum" is due to the lower vibrational bound levels emitting to the ground state repulsive curve. The O_u^- and 1_u states can be considered degenerate with all the oscillator strength in the 1_u. O_u^+ states are probably maintained in a Boltzmann equilibrium by collisions and both contribute to the bound (low v)-continuum spectra. Whether the "first continuum" is due to high v in O_u^+ or continuum states of 1_u or O_u^+ is a subject of continuing research. In He_2 the 60 nm bands originate in quasi-bound [2.18] or continuum levels [2.19] of the $A^1\Sigma_u^+$ state since the spin-orbit coupling is so weak that the $^3\Sigma_u^+$ state is metastable under the experimental conditions. The emission spectra have been analyzed in the case of Ar_2 to obtain an upper state potential curve which was assigned to the O_u^+ state [2.20]. The absorption spectra from the bound van der Waals molecules has been used to determine the spectroscopic properties of the ground state but, as yet, has not proved fruitful in providing excimer state information [2.21].

Absorption of the excimer radiation by the excited states leads to ionization. Cross sections have been deduced from calculated values for the excited atoms. Much theoretical effort is being expended in calculating photoionization cross sections [2.22], but few molecular data are available to date.

2.2 Rare Gas-Halides: Ion-Pair Excimers

2.2.1 The Ground State

In the absence of the charge transfer mixing with the excited state, both the Σ and Π valence states are expected to be repulsive from the exchange repulsion of the F valence electrons with the closed shell of the rare gas. The first calculations on the KrF [2.23] and XeF [2.24] ground state found no binding even though configuration interaction sufficient to include charge transfer was included. The exchange repulsion in the ground $^2\Sigma^+$ states is not directly comparable to the analogous mixed rare gas interaction. The $^2\Pi$ states would be analogous but the smaller $^2\Sigma^+$ repulsion is due mostly to the smaller overlap of only one F $p\sigma$ electron with the closed shell of the rare gas. In both KrF and XeF the charge transfer mixing does significantly reduce the repulsion and push the repulsion turning point to a sufficiently small distance that the van der Waals contribution is appreciable. A calculation [2.25] of the XeF $^2\Sigma^+$ state finds that the well depth of about $1200\,cm^{-1}$ that is experimentally determined [2.26, 27] can be accounted for by a combination of charge transfer and van der Waals contributions. The $^2\Sigma^+$ ground states of the rare gas halides are, therefore, a kind of electron donor complex as defined by *Mulliken* and *Person* [2.28]. The ionic-covalent mixing in the rare gas halides is essentially reversed in the analogous mixing in the alkali halides [2.29]. The repulsive valence curves are lower in energy than the ionic and since the configuration mixing is dominated by only these two configurations, the valence curve is lowered while the energy of the ionic curve is raised an equal amount.

The ionic-covalent mixing is proportional to the overlap of the orbitals between which the electron is transferred. This overlap is much larger for transfer between σ orbitals than between π orbitals. The covalent $^2\Pi$ curves are

Table 2.2. Dissociation energies and equilibrium internuclear separations of the ground and ion-pair excited states of the rare-gas fluorides. The dissociation energy in the ion-pair state is relative to the separated ions

	RF			
	Ne[a]	Ar	Kr[c]	Xe[e]
D_e [eV]				
X				~0.14
B	6.41		5.30 (5.54±0.03[d])	5.30
C	6.35		5.24	
D			5.26	
R_e [Å]				
X				~2.4
B	2.00	~2.4[b]	2.51 (2.27[d])	2.49
C	1.99		2.44	
D			2.47	

[a] N.W. Winter, C.F. Bender, T.N. Rescigno: J. Chem. Phys. (to be published). Spectroscopic constants for states calculated neglecting the spin-orbit coupling.
[b] [2.39].
[c] [2.32].
[d] J.Tellinghuisen, A.K. Hays, J.M. Hoffman, G.C. Tisone: J. Chem. Phys. **65**, 4473 (1976).
[e] [2.26].

inherently more repulsive and also have much less configuration mixing or charge transfer. The $^2\Pi$ states wells are determined predominantly by the van der Waals interaction.

Taking all of this into account, one predicts that the XeF $^2\Sigma^+$ ground state will be the most stable charge transfer rare gas halide diatomic complex. The XeF R_e is about 0.4 Å shorter than the R_e of the ionic B state. On the other hand, the XeCl ground state R_e exceeds the B state R_e by about 0.2 Å and the $D_e(X^2\Sigma^+)$ is only about 255 cm^{-1} [2.30]. KrF emission spectra is bound-to-free in the gas phase and the well depth has not been measured. Charge transfer is certainly significant in the ground state of KrF as we shall see in analyzing the ionic states but it cannot overcome the large exchange repulsion in this system. The dissociation energies and equilibrium separations of the ground and excited states of the rare-gas fluoride molecules are given in Table 2.2.

2.2.2 The Ion-Pair Excited States

Ion-pair states can be modelled by an electrostatic model of the *Rittner* type [2.31]. The ionic states correlate diabatically to ionic fragments $R^+(^2P)$ $+ X^-(^1S)$ and analogous to the neutral ground state yield $^2\Sigma^+$ and $^2\Pi$ curves.

The configuration mixing between the neutral and ionic configurations for KrF will explain the crossing between the Σ and Π curves found by *Dunning* and *Hay* [2.32]. A Rittner type model with configuration mixing has been developed and applied to KrF [2.33]. The model potential for the rare gas halides differs from those applied to the alkali halides in two ways. There is an anisotropy inherent in the interaction energy of a degenerate atom and the configuration mixing is explicitly included. Although the anisotropy in the dipole polarizability [2.34] and van der Waals constants [2.35] can be quite large for a single valence p electron, it is much smaller for a p^5 hole configuration. There is an appreciable anisotropy, though, in the ion-quadrupole interaction and the exchange repulsion, both of which lead to the $^2\Sigma^+$ being more bound than the $^2\Pi$. The extent to which appreciable overlap modifies the electrostatic interactions is unknown and a Rittner type of analysis is necessarily very empirical.

The ion-pair and valence configuration mixing will increase as R decreases and tend to repel the excited $^2\Sigma$ state relative to the $^2\Pi$ state. The experimental evidence on the relative positions of the excited $^2\Sigma^+$ and $^2\Pi$ states is not clear since transitions from the C state in the gas phase have not been analyzed [2.36]. Matrix isolation studies indicate the C state is, in fact, lower in energy than the B for XeF [2.37]. The matrix shifts of the excitation energy are quite large and different for different states. They are to the red for the rare gas halides and to the blue in the alkali halides [2.38], reflecting the extent to which ion-pair states cluster in the matrix. It is possible that such a differential effect can alter the ordering of the levels.

The ab initio calculations indicate that the ArF* states [2.39] do not cross near R_e but on the left-hand limbs and that the splitting between the curves is about 0.25 eV near R_e. As the rare gas atom increases in size, the fluoride configuration mixing will be substantial to larger distances and the repulsion of the $^2\Sigma^+$ state will be relatively the largest for the XeF system. Nonetheless, the most recent accurate ab initio results for the fluorides, chlorides, and XeBr find that the $^2\Sigma_{1/2}$ excited state is always lower in energy than the $^2\Pi_{3/2}$ [2.40, 41].

The theoretical transition probabilities are largest for all the systems for the $^2\Sigma^+ \rightarrow {}^2\Sigma^+$ transition. This is due to the large charge transfer component that mixes into the ground state and vice versa for the excited ionic state. Since the dipole moment for the ionic state is proportional to the internuclear distance, this term will dominate the transition moment as the ionic-covalent configuration mixing becomes appreciable near the R_e of the excited state. Spin-orbit coupling will mix Σ and Π contributions into both excited $\Omega = 1/2$ states to yield lifetimes of the order of 7 ns for KrF, for example in the $B_{1/2}$ state, and about twice that in the $D_{1/2}$ state. $^2\Pi \rightarrow {}^2\Pi$ transitions are also influenced by the charge transfer mixing but this is much less important in this case. Ab initio calculations for all dipole-allowed transition moments have been made [2.32, 40]. Since the charge transfer and spin-orbit effects are both dependent strongly on the distance, the transition probabilities will be a strong function of vibrational level.

2.2.3 Rydberg States

There are two sets of adiabatic correlations that describe the positive ion states of the rare gas halides. For $R^+(^2P) + X(^2P)$ we have $^{1,3}\Sigma^+(2)$, Σ^-, $\Pi(2)$, Δ and for $R(^1S) + X^+(^3P; ^1D; ^1S)$ we have, respectively, $^3\Sigma^-$, $^3\Pi$; $^1\Sigma^+$, $^1\Pi$, $^1\Delta$; $^1\Sigma^+$. Only one configuration, a $^1\Sigma^+$, is chemically bound where singly occupied $p\sigma$ orbitals on each atom overlap to form a σ bonding orbital. The bound $^1\Sigma^+$ state need not be the ground state depending on the rare gas and halogen ionization potentials. For ArF^+, KrF^+, and XeF^+, for example, the $^1\Sigma^+$ bound state is the ground state. For NeF^+ or HeF^+ the lowest states would resemble the rare gas oxide energy curves with the additional ion-induced multipole attraction. The Rydbergs for the $^3\Pi$ and $^3\Sigma^-$ ion states are the only significant states involved in the ion-pair Rydberg mixing in NeF on the right-hand limb while the $^1\Sigma^+$ Rydberg states would mix on the left-hand limb.

There are no quantitative studies of the configuration mixing between ion-pair and Rydberg configurations. For the Rydbergs of bound limiting ion curves, there will be curve crossings at both short and large distances with stronger couplings at the shorter distance. The equilibrium internuclear distances between the Rydberg states and ion-pair states are also so different that the absorption transitions from the ion-pair states will occur into the "long-range" region of the Rydberg states where strong configuration and spin-orbit mixing must occur in order to insure correct asymptotic behavior. The bonding behavior of Rydberg states interacting with open-shell atoms is also little understood.

The KrF calculation of *Dunning* and *Hay* [2.32] shows the lowest $^2\Sigma^+$ Rydberg state has a barrier near 2.6 Å. The $^2\Sigma^+$ curve that correlates with the $Kr(^1P) + F(^2P)$ asymptote would show a bound $5p\sigma$ Rydberg state at distances near 1.7 Å but, unfortunately, the excited energy curve behavior near the R_e of the ion-pair curves is not known. These $^2\Sigma^+$ and $^2\Pi$ Rydberg states will determine the self-absorption behavior of the KrF excimer radiation but the electronic structure and energy curves are not well understood.

2.2.4 Observed Spectra

Emissions have been reported from a number of rare gas halides including NeF, ArF, KrF, XeF, ArCl, KrCl, XeCl, ArBr, XeBr, and XeI. As seen from Table 2.3 only six of these, ArF, KrF, XeF, KrCl, XeCl, and XeBr, have been made to oscillate on at least one transition. The upper excimer states were immediately identified as ion-pair states while the lower states are essentially unbound covalent states [2.42]. The allowed electric dipole transitions are given schematically in Fig. 2.5. From observation and theoretical considerations [2.32] only three strong transitions are expected, $B–X$, $D–X$, and $C–A$. These transitions are not observed for systems in which the ion-pair states are no longer isolated but are embedded in a manifold of Rydberg states. Rydberg

Table 2.3. Rare-gas halide molecules that fluoresce (F) and lase (L) are indicated in the appropriate square. The ionization potentials of the atoms are given in parentheses

	F (17.42)	Cl (13.01)	Br (11.84)	I (10.45)
He (24.58)				
Ne (21.56)	F			
Ar (15.75)	L, F	F	F	
Kr (14.00)	L, F	L, F		
Xe (12.13)	L, F	L, F	L, F	F

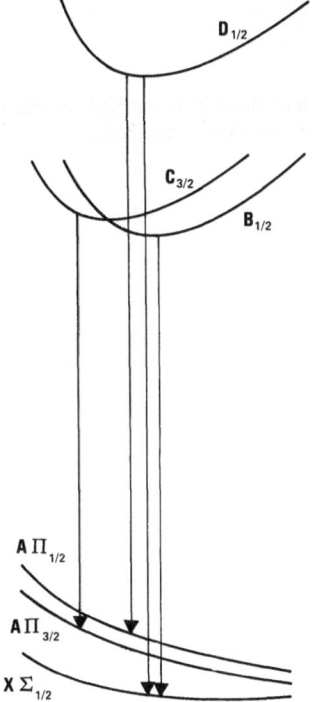

Fig. 2.5. Schematic of the radiative transitions for the ion-pair rare-gas halide excimer states. Only the strong transitions are represented by the solid lines

states of the systems will be the lowest states when the ionization potential of the halogen is much lower than the rare gas as in HeI, while the ion-pair states will be lowest when the relative ionization potentials are reversed as in XeF. Although the production mechanisms of the excimer states will vary, the probability that oscillation can be achieved and the efficiency of fluorescence is diminished if many predissociating Rydberg levels mix with the ion-pair energy curves [2.43]. Those systems which have exhibited stimulated emission have a dominant ion-pair structure and relatively few Rydberg asymptotes below the

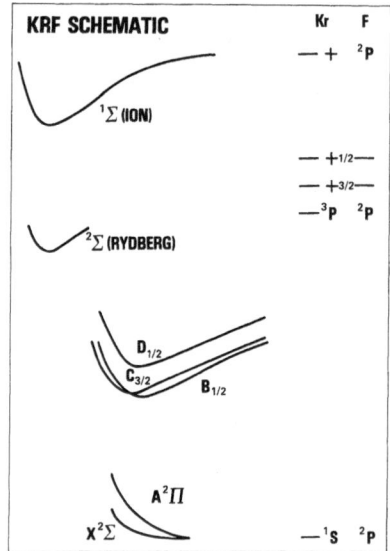

Fig. 2.6. Energy level diagram of the KrF molecule including estimated Rydberg levels [2.32]

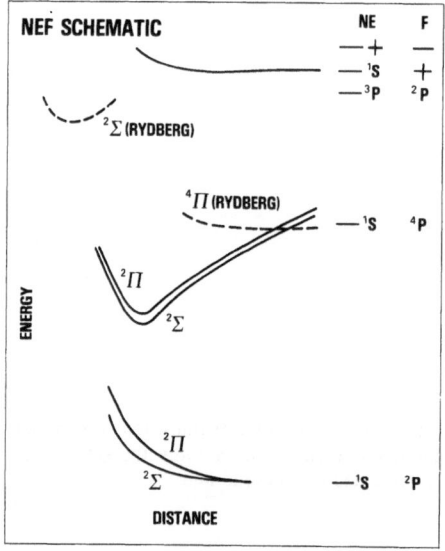

Fig. 2.7. Energy level diagram of the NeF molecule including estimated Rydberg levels

ion-pair limit. Competition between emission from an excited ion-pair state and the production of excited halogen atoms by predissociation may be assessed from diatomic halogen emissions produced by energy transfer.

Two examples will illustrate the general rare gas halide electronic structure, KrF and NeF. The schematic for KrF in Fig. 2.6 shows that only one Rydberg asymptote (ignoring fine structure splitting) lies below the ion-pair asymptote.

Rydberg curve crossings on the right-hand limb of the ion-pair state will occur at very large distances (~ 50 bohr) and consequently the ion-pair and Rydberg curves will mix very weakly. On the other hand the schematic for NeF in Fig. 2.7 shows an infinity of Rydberg asymptotes which intersect the ion-pair curves. Nonetheless, in the case of NeF the well depth for the ion-pair state is sufficiently large, that the energy minimum is still well below the lowest Rydberg asymptote. The ion-pair states can be produced in vibrational levels safe from predissociation or deactivated into such levels which can then radiate.

2.3 Rare Gas+Group VI: Valence and Ion-Pair Excimers

The electronic states arising from the interaction of a rare gas atom and a Group VI atom can be considered in three categories. In the first category are the low-lying valence states that arise from the interaction of the ^3P, ^1D, and ^1S valence states with the ground state rare gas atom. Second are the electronic states that are dominantly ion-pair with an ionicity, R^+X^-. Third are the Rydberg states that correlate to Rydberg asymptotes of either a rare gas or Group VI atom. In Table 2.4 the molecular states that correlate adiabatically to representative fragments in these categories are given.

2.3.1 Valence States

Most of the available experimental and theoretical information is on the valence states. The recent interest has developed because of the laser possibilities of the auroral, ^1S–^1D, and transauroral, ^1S–^3P, transitions [2.44]. Although the emission spectra of the collision-induced auroral line were investigated some time ago [2.45], it is only recently that detailed information has become available on the spectroscopic properties and collisional lifetimes.

Table 2.4. Adiabatic molecular state correlations for the interaction of a rare-gas and Group VI atom. Valence, ion-pair, and Rydberg correlations are included

Atomic states		Molecular states
R	X	
^1S	^3P	$^3\Pi, \Sigma^-$
^1S	^1D	$^1\Sigma^+, \Pi, \Delta$
^1S	^1S	$^1\Sigma^+$
^1S	^5S [ns]	$^5\Sigma^-$
^1S	^3S [ns]	$^3\Sigma^-$
^3P [ns]	^3P	$^{1,3,5}\Sigma^+, \Sigma^-(2), \Pi(2), \Delta$
$^2P(+)$	$^2P(-)$	$^{1,3}\Sigma^+(2), \Sigma^-, \Pi(2), \Delta$

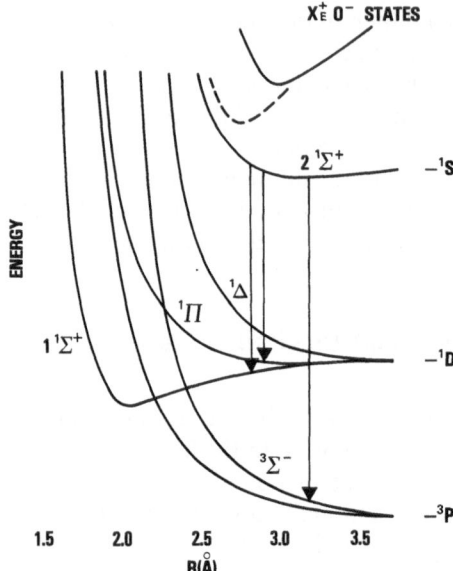

Xₑ⁺ 0⁻ STATES

Fig. 2.8. Energy curves of XeO including estimated ion-pair and Rydberg curves [2.48]

This has followed the observation of laser oscillations on the auroral bands of ArO, KrO, and XeO [2.46]. The experimental evidence on the collisional deactivation of the O(^1S) atom by a rare gas atom is that the process occurs entirely by radiative transitions [2.47]. This is in agreement with theoretical calculations which show that the energy curves from the O(^1S) and O(^1D) asymptotes do not cross in thermally accessible regions [2.48]. The theoretical calculations also show that, with the exception of XeO, the transitions are mostly continuum to continuum with only a small bound to continuum excimer contribution.

The schematic energy curves for XeO are shown in Fig. 2.8. Transitions have been observed near both the auroral and transauroral line for XeO. The theoretical assignments are clear for the auroral bands which are $2^1\Sigma^+ - 1^1\Sigma^+$ and $2^1\Sigma^+ - 1^1\Pi$ as shown in Fig. 2.8. Photoselection experiments with XeO on rare gas matrices show that the green auroral bands are essentially due to $\Sigma - \Sigma$ and $\Sigma - \Pi$ transitions [2.49].

The $1^1\Sigma^+$ energy curve is crossed on the right-hand limb by both the $^3\Pi$ and $^3\Sigma^-$ energy curves in the ab initio calculation. The deactivation of O(^1D) by the rare gases is known to be fast [2.50] and recent spectroscopic evidence in the matrix [2.49] finds linewidth behavior that can be interpreted in terms of a curve crossing on the right-hand limb.

The absolute accuracy of the theoretical calculations is not sufficient that the spectroscopic properties can be relied upon. Dispersion effects are important and not completely included. The $2^1\Sigma^+$ and $1^1\Pi$ states are calculated to have very small wells that are much smaller than the experimental values. *Cooper* et al. [2.45] had determined that $2^1\Sigma^+$ well depth is about 508 cm^{-1}, but *Maki* et al. [2.51] found the D_e is nearer $690\,cm^{-1}$ and the R_e is about $2.9\,\text{Å}$. *Goodman* et al. [2.49] have revised slightly the absolute vibrational numbering proposed by *Huestis* et al. [2.52] and estimated the $^1\Pi$ well depth is $505\,cm^{-1}$. The dissociation energy calculated by *Dunning* and *Hay* for the $1^1\Sigma^+$ state is $0.70\,eV$, nearly twice the estimates of *Huestis* et al. [2.52]. But *Goodman* et al. [2.49] have shown that the absolute vibrational numbering of the $1^1\Sigma^+$ state has not yet been determined, and the preliminary spectroscopic evidence is that the ab initio results are more realistic.

The qualitative features of the interaction energy are understood. The interaction between a closed shell atom and an open shell atom is basically repulsive because of overlap effects and the repulsion increases with the number of electrons that occupy the $p\sigma$ orbital of the open shell atoms. The attractive interactions are of two types; first, the dispersion and electrostatic interactions, and second, the charge transfer contribution that arises from configuration interaction between the ionic and covalent configurations. As the ionization potential of the rare gas decreases, the charge transfer contribution increases. The charge transfer mixing manifests itself most obviously in the bound $1^1\Sigma^+$ states but occurs in all the states and makes a significant contribution to the transition probabilities and predissociation probabilities to the extent the charge transfer modifies the spin-orbit coupling.

The transition probability between the singlet valence states is entirely induced by the collision. The charge transfer mixing dominates the $2^1\Sigma^+ - 1^1\Sigma^+$ transition at shorter distances but when the atomic overlap is small the quadrupole-induced dipole transition moment becomes significant. The electrostatic interaction is even more important for the $2^1\Sigma^+ - 1^1\Pi$ transition [2.53]. Accurate calculation of the transition moment is made difficult by the necessity of calculating both effects accurately which has not yet been done.

2.3.2 Ion-Pair States

The charge transfer contribution to specific states occurs through specific ion-pair configurations. The ion-pair configurations arising from $R^+(^2P)$ and $O^-(^2P)$ are both singlets and triplets, but since the charge is localized, the splitting between the different spin states is relatively small compared to the spin-orbit coupling for Xe^+ and even Kr^+. Different spin states of the same ion-pair configuration are, therefore, strongly coupled into a Hund's case (c) representation. This provides enhanced transition probabilities among the different multiplicity valence states which are mixed with appropriate ion-pair configurations. As the size and spin-orbit coupling for the Group VI atom increases from oxygen to tellurium, the spin-orbit effects within the valence manifold will compete with ion-pair contributions.

The ion-pair states, themselves, have been observed in the gas phase as the upper states in emission for Ar, Kr, and Xe [2.42] and in laser excited fluorescence [2.49] and absorption [2.54] in a matrix. The matrix fluorescence, in fact, indicates two different upper states. As seen in Table 2.4 there are a number of ion-pair states expected with little, as yet, known of the energy splittings and couplings among these states. As in the case of the rare gas halides, configuration mixing with the lower valence states will perturb the ion-pair level structure determined by the electrostatic and single-configuration repulsive interactions. It is likely that the $^3\Sigma^+$ state is lowest and weakly coupled to other states.

2.3.3 Rydberg States

The Rydberg level structure of rare gas oxides will be complicated because the ionic level structure is complicated. Xe has a lower ionization potential than O and the dimer ion will have a $^2\Pi$ ground state analogous to IO. The dissociation energy is probably about 2 eV with an R_e of 1.9 Å or nearly one Ångstrom shorter than the R_e for the ion-pair states. The Rydberg states for the $^2\Pi$ limit would in fact be predissociated by the left-hand repulsive limbs of all the valence states.

For Kr through He the O atom has a lower ionization potential than the rare gas and there will be two low-lying ion bound levels, the $^2\Pi$ and a $^4\Sigma^-$. The $^4\Sigma^-$ state is primarily bound by the ion-induced dipole attraction but the proximity of an excited $^4\Sigma^-$ curve can permit some stabilization of the lower curve by configuration mixing. In ArO$^+$ the $^4\Sigma^-$ curve has been reported to have $D_e = 0.68$ eV and $R_e = 2.02$ Å [2.55]. The $^2\Pi$ curve for ArO$^+$ is more likely to have an R_e of 1.5 Å but the deeper well is probably not enough to offset the difference in the ionization potentials. Rydbergs of the $^4\Sigma^-$ curve are still likely to intersect the ion-pair curves of ArO. The $^2\Pi$ Rydbergs will connect diabatically to R*+O asymptotes and will play an important role in both energy transfer and ion-pair molecular production [2.56].

2.4 Rare Gas+Group IV: Analogous to Group VI

There is very little information on the interaction of a rare gas atom and an excited state of a Group IV atom with the exception of deactivation data. The valence energy curve behavior and ordering is essentially inverted relative to those Group VI since there are two less *p* electrons. The $^3\Sigma^-$ curve is lower in energy than the $^3\Pi$ and the $^1\Delta$ state can be the lowest energy state arising from the ^1D asymptote. The extent of binding in the $^1\Delta$ or $1^1\Sigma^+$ states is uncertain because the ion-pair mixing cannot be estimated. Superficially, the smaller electron affinities for the Group IV atoms relative to those in Group VI [2.57] would indicate a smaller charge transfer contribution. But the ion-pair energy curves for the rare-gas Group IV molecules will be relatively more bound. Since the Rydberg states are also relatively lower in energy, the mixing between ion-pair and Rydberg configurations would have to be considered separately for each atom pair.

The relatively rapid deactivation of $C(^1D_2)$ by Xe indicates that a curve crossing occurs in a thermally accessible region. But the rates are much lower for deactivating the $^1D_2$ states of Pb or Sn [2.58]. The large spin-orbit coupling has an effect on the adiabatic behavior of the energy curves. There is no calculation of this spin-orbit effect on the energy transfer and it is not possible at this time to interpret the deactivation rates in terms of adiabatic energy curves.

2.5 Rare Gas+Alkali: An Effective One-Electron Interaction

The difference between the ionization potentials and excitation energies of the rare gas and alkali atom is so great that the interaction energy can be described in terms of a one-electron potential model. Extensive model potential calculations have been reported by *Pascal* and *Vandeplanque* for all alkali-rare gas atom pairs [2.59]. The attractive van der Waals interaction is represented by the dipolar coupling of alkali valence electron with a polarizable rare gas atom. The repulsion is described by a one-electron pseudo-potential that represents the exchange repulsion arising from the overlap of the valence electron with the rare gas atom [2.60]. Taking explicit account of the configuration mixing among the closely spaced molecular states, curves have been calculated for excited states about 3 eV above the ground state. The potential parameters are fitted to the ground state data.

The accuracy of the calculated excited states is still not known. There are some differences with the few ab initio calculations, especially for the LiHe and NaHe systems. The character of the interaction behavior of the $X^2\Sigma^+$, $A^2\Pi$, and $B^2\Sigma^+$ states can be seen in the plots of the charge density as a function of the internuclear distance [2.61]. Repulsive overlap is obviously greater for the directed $p\sigma$ orbital than for the s orbital. At moderate distances this repulsion is almost entirely the exchange repulsion of an undistorted atomic orbital overlapping a closed shell. The dipole moment function of the LiHe ground state calculated by both the Hartree-Fock method or by a perturbation treatment that considers only the overlap moment due to exchange are in good agreement [2.62]. The $A^2\Pi$ state binding for LiHe and NaHe was found to be largely due to the unscreening of the Li^+ or Na^+ core for a $p\pi$ approach of the valence electron on the rare gas atom. The Hartree-Fock energy curves for the A state and positive ion were similar near their equilibrium separation. The binding is determined by the ion-induced dipole attraction. The model potential does not account for the binding in the A states of LiHe or NaHe, since there is no provision for any attractive interaction but the dispersion interaction. However, for the larger rare gas atom, the overlaps are now larger and no attractive well is found for the A states at the Hartree-Fock level of accuracy [2.63].

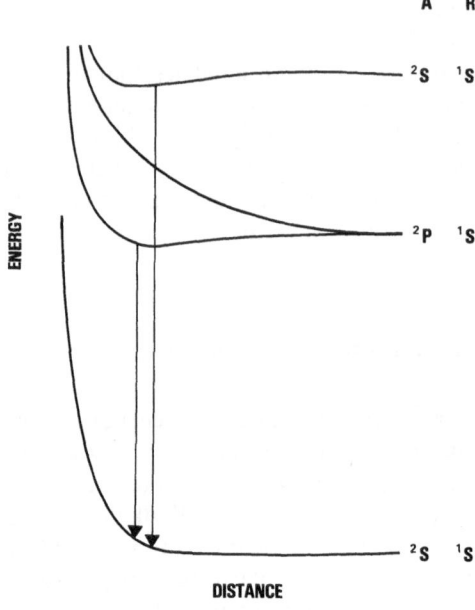

Fig. 2.9. Ground and lower excited states of the interaction of an alkali atom with a rare gas atom [2.59]

It should also be noted that *Pascale* and *Vandeplanque* [2.59] have presented energy curves with asymptotes close to 40,000 cm^{-1}. For any light alkali interacting with a Xe atom, the ion-pair states will have T_e in that range. Configuration mixing would be weak at the energies considered, but the method does not make allowance for such mixing in the model hamiltonian.

The energy curves of the interaction of a rare gas atom with the ground and first excited atomic state of an alkali is given schematically in Fig. 2.9. Excimer transitions occur from the $A^2\Pi$ state to the ground $X^2\Sigma^+$ state. Such transitions have been observed both in absorption and emission. Analyzing the temperature and pressure dependence of the emission lineshape, *Gallagher* et al. [2.64] have deduced potential curves for the Cs [2.64], Rb [2.65], Na [2.66], and Li [2.67] rare gas molecules. In the neighborhood of the van der Waals minimum the X potential curves are reasonably well known [2.68]. Comparisons of the potential curves with both model potential, ab initio, and perturbation calculations have been made with inconclusive results. For excited states that correlate adiabatically to $(n+1)^2S$ or n^2D asymptotes, the curve structure and transition probability behavior are sufficiently complicated that emission spectra from such excited states are yet to be definitely assigned [2.69], although it is probable that transitions are observed to states correlating to upper D and S alkali atomic levels.

The transition moment of the atomic resonance transition of the alkali atoms is very large, of the order of 2.6 a.u. The electrostatic and dispersion contributions to the moment are relatively small and the moment is not calculated to change much at internuclear distances accessible to low energy collisions. Collision-induced transition probabilities to states that correlate to ^2S or ^2D asymptotes have not been reported as yet.

2.6 Rare Gas+Group II: Closed Shell and Sub-Shell

The ground states of rare gas and Group II atoms are van der Waals molecules like those of interacting rare gas atoms. The repulsive curves that result when the dispersion is neglected have been reported for BeNe, MgNe, MgAr, ZnNe, and ZnAr by *Bender* [2.70]. These calculations also show that the excited states arising from the $^1S + {}^3P$ and $^1S{}^1P$ asymptote behave in a similar fashion to the states that arise from a ^2P alkali atom colliding with a rare gas atom. The excited singlet states are somewhat less repulsive than this simple picture would indicate which can be ascribed to interaction with Rydberg states. The calculations show that the moments for the allowed dipole transitions vary only slightly as a function of the internuclear distance from the asymptotic atomic value.

The binding in the Rydberg states of these systems is likely to be appreciable since the dissociation energies of the limiting ion states are found to be large. In the case of RBe$^+$, the ground state dissociation energies have been obtained for Ar, Kr, and Xe [2.71]. These dissociation energies are much larger than the values for the RLi$^+$ ground state which are also dominated by the electrostatic ion induced-dipole interaction. The dissociation energy for the Π state in ArBe$^+$ is deduced to be 12,767 cm^{-1}. The explanation for this strong bonding must consider the highly polarizable nature of a single s or p electron outside a closed shell. A hybridized sp or pd orbital would minimize the exchange repulsion of the interaction. In the case of the Π states, the atomic cores can more readily penetrate along the node of the π orbitals.

Absorption and emission data have been observed in the gas phase for Hg complexes associated with the ^3P and ^1P states [2.72, 73]. In the gas phase there is also considerable information on the quenching of the ^3P_J excited states of Group II atoms indicating that the interaction with a rare gas atom is weak [2.74]. Since most of the gas phase spectroscopic data relates to the interaction of Hg with rare gas atoms and especially Xe, the electronic structure of this system will be discussed in some detail.

Gutcheck et al. [2.75] have reported detailed emission spectra for an e-beam excited mixture of Hg and Xe. Excimer transitions from the molecular states arising from the Hg(^3P_J) states are tentatively assigned. In addition, satellite bands are observed which are similar to those observed in optical absorption and emission [2.73]. From an analysis of the temperature dependence of the intensity, the well depth of the upper state has been deduced to range from 500–1500 cm^{-1} with the evidence most convincing for the higher value [2.72]. The band to the red of the 2536 Å resonance line that peaks around 2700 Å is

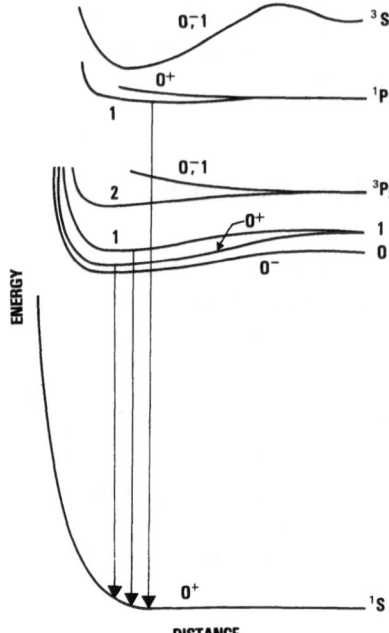

Fig. 2.10. Schematic of the interaction energy curves for Xe and Hg

assigned to the $^3O^+ \rightarrow {}^3O^+$ transition as shown in Fig. 2.10. *Gutcheck* et al. [2.75] assume that the $^3P_1$ is rapidly deactivated to $^3P_0$ but the $^3O^-$ level that correlates with this asymptote is probably in collisional equilibrium with the $^3O^+$ state that lies only slightly higher in energy. Both of these states arise through spin orbit coupling from a $^3\Pi$ state in which the overlap of the $p\pi$ orbital on the Hg with the rare gas atom is much reduced over a $p\sigma$ orientation.

There are considerable data [2.76] indicating that the $^3P_2$ levels are not readily deactivated. The Ω states correlating to the $^3P_2$ asymptote are split energetically from the ^{3}1 state more than the $^3O^+$ and $^3O^-$ states are, especially on the left-hand limb where the latter states nearly merge. The 2100 Å band can then be assigned ^{3}1$\rightarrow {}^3O^+$, as shown in Fig. 2.10.

The relatively narrow lines observed for the $7^3S_0 \rightarrow 6^3P_1$ collision broadened transition are surprising. Several classes of Rydberg states of the Group II-rare gas system are expected to be bound since relatively strong ionic bonds are expected. A molecular transition in HgAr$^+$ has been observed by *Bridge* [2.77] and related to the atomic Hg$^+$ transition $5d^96s^2(^2D_{5/2}) \rightarrow 5d^{10}6s(^2S_{1/2})$. The ground and excited state dissociation and equilibrium distance were deduced to be $D''_e \sim 1630\,\mathrm{cm}^{-1}$, $r''_0 = 2.868\,\text{Å}$ and $D'_e \sim 1820\,\mathrm{cm}^{-1}$, $r^1_0 = 2.964\,\text{Å}$. Although the spectroscopic analysis [2.78] indicates a spin-orbit behavior analogous to the $5d^96s^2$ configuration for the excited state, the molecular orbital analysis would suggest that configuration mixing with an excited $^2\Pi$ state is responsible since the doubly occupied $6s^2$

subshell would interact repulsively with the rare gas atom very much like the neutral ground state. However, a $5d^{10}6p$ configuration would permit the closer approach of the rare gas atom to the ion and yield a strong attractive electrostatic interaction.

The Rydbergs to both the $X^2\Sigma^+$ and $A^2\Pi$ ion states will be substantially bound in the region of the equilibrium distance. But the $7^3S - 6^3P$ emission does not show the upper state to be bound. The long range properties of Rydberg state ground state interactions have been examined in a model calculation for the alkali-rare gas interaction but for the Group II system only the calculation of van der Waals coefficients has been made [2.79]. The possibility of small or even repulsive van der Waals interactions for excited states has not been carefully investigated to date for either system. In the case of the 7^3S interaction, the long range behavior could be repulsive even though the molecular $^3\Sigma^+$ state is bound near R_e.

But the complexity of the Rydberg behavior may not be the most important feature in understanding the absorptions up from the populated ^3P states. Ion-pair states must be considered here as they have for all other excimers. In this case we are concerned about ion-pair molecular states with a negative ion asymptote that is stable or a shape resonance, i.e., the $R^+ + Hg^-$ asymptote. From the correlation scheme we note that a number of curves result and such states will fall above the lowest Rydberg state arising from the 7^3S asymptote and strongly perturb the Rydberg series to the ground state ion. As usual predictions of the absorption spectrum from metastable excited states will be very difficult.

2.7 Rare Gas + Group III and V

The spectra of Tl and rare gases have been studied and energy curves for the systems were deduced [2.80]. A schematic curve is shown in Fig. 2.11 and the emission transitions from the $B^2\Sigma_{1/2}$ state indicated. *Cheron* et al. noted that the bonding behavior of the p valence orbital is similar to that of the first excited state of an alkali atom. The Π type system will be attractive while the Σ system is very repulsive. For the case of Tl the spin-orbit splitting is large and the molecular states that correlate to the $^2P_{1/2}$ and $^2P_{3/2}$ atomic asymptotes are dominated by Hund's case (c) coupling near the ground state equilibrium separation. The $X\,1/2$ state is mostly Π and will be attractive while the $A\,1/2$ is mostly Σ and is repulsive. Since the $X\,3/2$ is entirely Π, this state is found to be the most attractive valence state.

The $B^2\Sigma_{1/2}$ state is presumed to be an attractive Rydberg of the RTl^+, $^1\Sigma^+$, ion. The low ionization potential of Tl would preclude configuration mixing from either ion-pair or TlR^+ Rydberg states.

Only data on the rate of deactivation of the ^2P and ^2D excited states of Group V atoms are available [2.81]. The rates would indicate that there are no

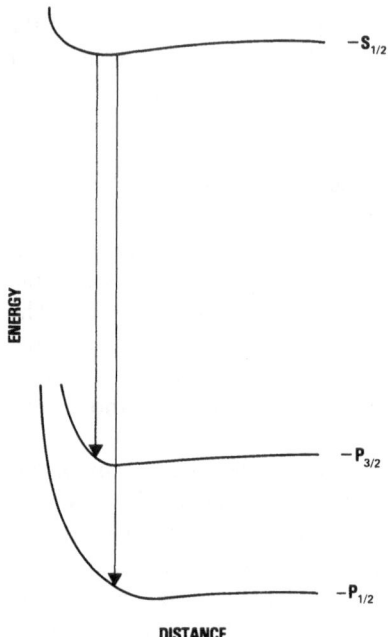

ENERGY

$-S_{1/2}$

$-P_{3/2}$

$-P_{1/2}$

DISTANCE

Fig. 2.11. Schematic energy curves of Tl + Xe [2.80]

curve crossings at thermal energies with the exception of Xe. The states that cross are likely to be the $^4\Sigma^-(^4S)$ with the $^2\Pi(^2D)$. The energetics would suggest that ion-pair mixing is smaller in these states than in the analogous Group VI states.

2.8 Group II + Group II: Covalent Excimers

The energy curves of the valence states of a Group IIb homonuclear molecule is given schematically in Fig. 2.12 as based on calculations of the Zn_2 [2.82] and Mg_2 [2.83] systems. These curves differ in a number of significant ways from those proposed much earlier [2.84] upon which the assignment of the excimer emission spectra was originally based. The most significant is the determination of the $^3\Pi_g$ state and its appropriate Hund's case (c) Ω components as the lowest energy state. Until recently [2.85], only the ungerade curves were considered as upper states of the observed excimer emission spectra. The ground states of the Group IIb homonuclear systems are not well determined and contribute to the difficulty of assigning the emission spectra. Even in the case of the Group IIa molecules where the ground state potential curves are well known in the bound region [2.86], the repulsive part of the curve is not accurately enough known to calculate the Franck-Condon envelopes from different excited bound states.

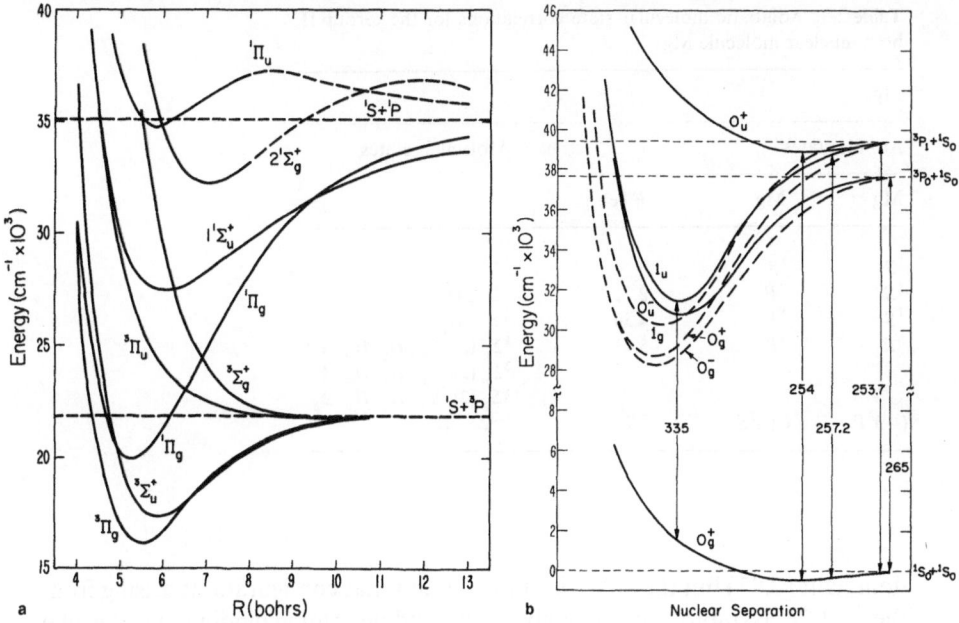

Fig. 2.12a and b. Energy level curves of **a** Mg_2 and **b** Hg_2: Only a portion of the energy curves arising from the $^3P_0$ and $^3P_1$ states are shown for Hg_2. The Hg_2 curve has been adjusted to agree with the experimental observations [2.93]

2.8.1 Valence and Ion Pair States

The ground state of the Group II homonuclear molecule can be considered a van der Waals molecule. The excimer bound states of greatest interest arise from the $^3P + {}^1S$ asymptote. In both the Mg_2 and Zn_2 cases the $^3\Pi_g$ state is lower in energy than the $^3\Sigma_u^+$ state. Model calculations of Hg_2 using either Mg_2 or Zn_2 curves as a basis find that the O_g^- energy curve is lowest and about 0.3 eV below the lowest ungerade curve, the O_u^-.

The lifetime of the 1_u state is calculated to be about 1.3 µs in either model calculation. The spin orbit coupling that makes the $^3P_1$ state dipole allowed is reduced by molecular formation which splits the 1 states apart. The transition probability to the O_u^+ state which goes adiabatically to the $^3P_1 + {}^1S$ asymptote will increase as the molecular states that mix come closer in energy. Transition moments to the singlet states from the ground state will be large and comparable to the large atomic resonance transition. The adiabatic correlations for Mg_2 are given in Table 2.5 and show that understanding these systems requires an understanding of more than just the covalent states arising from the $^1S + {}^3P$ and $^1S + {}^1P$ asymptotes. Ion pair configuration mixing is dominant only for the $^1\Pi_u$ in Mg_2, but is significant for other singlet states and for higher excited states. For Zn_2 the ion-pair states were interpreted as

Table 2.5. Adiabatic molecular state correlations for the Group II homonuclear molecule Mg_2

Mg₂			
Atomic states			Molecular states
Mg	Mg	E [eV]	
^1S	^1S	0	$^1\Sigma^+$
^1S	^3P	2.71	$^3\Sigma^+, \Pi_{g,u}$
^1S	^1P	4.34	$^1\Sigma^+, \Pi_{g,u}$
^3P	^3P	5.43	$^1\Sigma_g^+(2), \Sigma_u^-, \Pi_g, \Pi_u, \Delta_g;$
			$^3\Sigma_u^+(2), \Sigma_g^-, \Pi_g, \Pi_u, \Delta_u;$
			$^5\Sigma_g^+(2), \Sigma_u^-, \Pi_g, \Pi_u, \Delta_g$
$(-)^2P$	$(+)^2S$	∼7.9	$^{1,3}\Sigma^+, \Pi_{g,u}$

dominant [2.87] but the analysis of Mg_2 found that configurations arising from the $^3P + {}^3P$ asymptote are strongly bound and determine binding in states like the $2^1\Sigma_g^+$ excited state. The Zn_2 calculation also found strongly bound $^3\Sigma_g^-$ and $^1\Delta_g$ states that correlate with the $^3P + {}^3P$ asymptote. For the Group IIa atoms, Ca, Sr, Ba, and Rn, the ^3D and ^1D levels interpose between the ^3P and ^1P excited states and further complicate the molecular energy curve structure.

   Very little is known about the heteromolecules such as HgCd. The emission spectra have been reported [2.88] and analyzed in terms of a pair of bound excimer states. In contrast to the homonuclear molecules, the radiative transitions are likely to be from the $^3\Pi$ state, split and made allowed by spin-orbit mixing. Transitions are possible from the 1 and O^+ states to the ground O^+ state. The dissociation energy of the $^3\Pi$ state should be significantly lower in HgCd than for the dissociation energy deduced for the $^3\Pi_g$ state in Hg_2.

   The complexity of the excited state structure is suggested by the relatively large number of asymptotes below the $Cd(^1S) + Hg(^1P)$ asymptote and the relatively low-lying $^3P + {}^3P$ and ion-pair asymptotes. The cross sections for Hg–Cd and Hg–Zn collisions where the optically excited $Hg(^3P_1)$ transfers its excitation to the $^3P_1$ of Cd or Zn are large, but the interpretation of the experiments are controversial [2.89, 90]. The transfer has been assigned to an atom-atom collision [2.91] which would imply a curve crossing connecting the two asymptotes. There is insufficient information on these systems to speculate as to whether such a crossing occurs simply through two triplet curves or is mediated through a low-lying singlet. It has also been suggested that metasta-ble diatomic Hg_2 is formed first which then transfers its energy to the Cd or Zn ground state atoms [2.92]. At the present time the experimental evidence is too sparse and our theoretical understanding too limited to warrant drawing even schematic energy curves.

2.8.2 Rydberg States

In addition to the complicated valence configuration mixing, there is also the need to consider Rydberg configurations for the singlet and repulsive triplet states. There are two low-lying bound positive ion states in Mg_2^+ and the relatively low-lying $^2\Sigma_g^+$ repulsive curve. The Rydberg manifolds overlap the same energy region as the singlet excited states. Especially on the left-hand limbs of these singlet curves the electronic structure can be well represented by a Rydberg function. The detailed mixing of Rydberg and valence configurations is still a considerable research question and will be of even more importance if transitions up from the excited excimer states are considered. At the present time the energy curve behavior and electronic structure in the region of 6.5–7.5 eV excitation energy for the Hg_2 molecule can only be guessed. The valence energy results would suggest a weak O_g^-, $O_g^+ \rightarrow 1_u$ transition is possible at the lower part of this energy range as seen from model energy curves obtained by both *Hay* et al. [2.82] and *Mies* et al. [2.91].

Since the bond energy of Hg_2^+ is about 1 eV, the $7s$ and even the $7p$ Rydbergs of the $^2\Sigma_u^+$ ion can be important as upper levels for absorptions from the lowest excimer states. The level structure will be complicated by repulsive curves at about 8.5 eV arising from the $^1S + {}^3P(d^9s^2P)$ asymptote and the triplet and singlet ion-pair curves not already involved in perturbing the valence curves arising from the $^1S + {}^1P$ asymptote. It is interesting to speculate on the possibility that transient spectra observed by *Schlie* et al. [2.92] could in part arise from transitions from one of the Rydberg or ion-pair excited states to the lower valence excimer states.

2.8.3 Observed Spectra

Even with the improved energy curves the definitive assignment of excimer features in the Hg_2 spectra has proved to be very difficult. Two prominent emission bands are observed at 335 and 485 nm. By analysis of the temperature and pressure dependence of these bands, *Smith* et al. [2.93] confirmed that the 335 nm system is diatomic in origin, but they deduced the 485 nm system to originate from a triatomic complex. The theoretical results strongly suggest that the 335 nm band should be assigned to the $1_u(^3\Sigma_u^+) \rightarrow O_g^+(X^1\Sigma_g^+)$ transition as was originally suggested in early analyses but the O_u^- or gerade levels are not involved in the 485 nm band system.

Significant populations are expected in all the bound states below the 1_u shown in Fig. 2.12 for any photon pumped arrangement for which fluorescence from the 1_u is significant. Absorptions from the lowest O_g^-, O_g^+, 1_g, O_u^-, and 1_u states then can all be significant. Little experimental and theoretical work has yet been done in probing absorptions from the populated excited states except for studies studying gain as the fluorescing wavelengths [2.94]. *Drullinger* et al. [2.95] have recently reported an absorption study at 1.06 μm which tried to pump the $O_g^+ \to 1_u$ transition. However, the model calculations would predict a transition energy of 0.4 eV and the Franck-Condon factors may be small at about 1.1 eV. They obtained a null result, but *Mosburg* and *Wilke* [2.96] have reported pumping the 1_u state with an HF laser.

An emission band at 470 nm has been observed for Cd_2 [2.97]. This system is probably at $1_u \to O_g^-$ transition as found for Hg_2. It should be noted that *Hamada* [2.98] also found a continuum emission with a maximum about 453 nm.

In the case of the Group IIa molecules, the triplet spectra have not been observed but the strong $X^1\Sigma_g^+ \to A^1\Sigma_u^+$ transition has been observed for Mg_2 [2.99] and possible for Ca_2 [2.100]. Absorption is also observed to the blue of the atomic resonance which is consistent with the theoretically predicted maxima in the $^1\Pi_u(1_u)$ curve [2.101]. *Andrews* [2.102] has preliminary evidence of absorptions to two states in Ca_2 which can be interpreted as $^1\Sigma_u^+$. These observations will require the reinterpretation of the absorption data of *Balfour* and *Whitlock*. As in the case of homonuclear dissociating to a $^1S + {}^1P$ asymptote, the states dissociating to the $^1S + {}^1D$ asymptote will also have a long-range electrostatic interaction, but in this case it is a quadrupole-quadrupole coupling and the long-range behavior will go as R^{-5}.

2.9 Group II + Alkali

The ground states of alkali-Hg molecules have been deduced from scattering data [2.103]. A pseudo-potential calculation by *Duren* [2.104] was parametrized to fit the ground state for NaHg. This calculation assumes that the attractive interaction is determined entirely by the van der Waals interaction and ignores configuration mixing with bound excited states of Hg and ion-pair states. *Duren* noted that an attempt to fit the Li–Hg failed and noted this is indicative of perturbations due to the excited configurations. This also suggests that the pseudo-potential model is inapplicable to the excited states as well.

The likely effect of the excited configurations on both the ground and excited states can be gauged from the rather large bond energies determined for positive ions like XeF^+ or a molecule such as AgCu. The $^2\Pi$ state that correlates with $Hg(^3P) + Na(^2S)$ will be substantially bound and likely to mix with the $^2\Pi$ arising from $Hg(^1S) + Na(^2P)$. In addition, the $Hg^-(^2P) + Na^+(^1S)$ asymptote is only at about 5.3 eV and will yield both $^2\Sigma^+$ and $^2\Pi$ curves with excitation energies between 2–2.5 eV and also significantly perturb states arising from most of Rydberg asymptotes. There is so little explicit information on the energy curves of other intersections of a column of the periodic table with the Group II atoms that it is best to avoid speculation and just note that the complexity described for alkali-Group II systems is expected for any other including Group II–Group III systems.

2.10 Polyatomic Excimers:
Ion-Pair Clusters and Triatomic Metals

Many types of polyatomic excimers can be envisaged. Only two have played a significant role to date in emission spectra or kinetics significant in a possible lasing medium. Broad-band emission spectra have been observed in electron beam excitation of Ar, Kr, and F_2 mixtures [2.105, 106]. These emissions have been assigned to a triatomic excimer R_2F^* where R is either Ar or Kr. In addition, the 485 nm band in the excited Hg spectrum has been assigned [2.93] to an Hg_3^* upper state rather than the Hg_2^* that presumed for so long.

Using a Rittner-type analysis [2.105] or by arguing from analogy of the structure of dialkali halides [2.106, 107] the equilibrium geometry of the upper state is predicted to be an isosceles triangle for the R_2X^*. For this geometry the homonuclear diatomic positive ion retains much of the electronic structure of the isolated species. This is confirmed in recent ab initio calculations of Ar_2F [2.108] and Ne_2F [2.109]. For a linear geometry, on the other hand, there is a transfer of charge and the most stable electronic structure of the excited state is better described as $Ar(Ar^+F^-)$ or $(Ar^+F^-)Ar$ in analogy with the structures proposed for the stable RX_2 molecules [2.110].

At the present time emission is found from Ar_2F^* to peak at 290 nm and for Kr_2F^* at 415 nm. Searches for Xe_2F^* emission have been without success. Both the ab initio and the Rittner analysis predict long lifetimes for the dipole allowed transitions in distinction to the diatomic rare gas halide excimer states. At high pressures this allows these states to be deactivated before they radiate. It should be noted that R_2X^* is only the first possible clustering and that R_3X^* states should also be stable to the extent that R_3^+ molecule ions are stable [2.111, 112].

The absorption spectrum of the R_2X^+ molecules will mimic that of the homonuclear positive ions, R_2^+. The absorption cross sections should remain large for the triatomic transitions that derive from the $\sigma_u - \sigma_g$ transition either directly in the ultraviolet or through spin-orbit coupling in the visible transition. The u→u transitions in the homonuclear system will also become allowed for the triatomic system. In fact the ab initio calculations of Ar_2F show that transitions from the $3^3B_2(\Pi_u)$ are relatively strong. A qualitative examination of the wave function shows the overlap of the Ar_2^+ and F^- orbitals is large for this state which ensures large mixing of ionic and covalent $^3B_2$ states. The polyatomic transition probabilities can be deduced from the fragment overlaps but the analogy to the diatomic is limited.

Adiabatic correlation diagrams (ACD) show that the R_2X^* systems correlate only to higher energy excited states. The ACD for Kr_2F^* is shown in Fig. 2.13. On the other hand, RX_2^* systems are shown to correlate to lower energy asymptotes and, at the least, would be susceptible to predissociation. It is likely that ionic triatomic excited states play a role in the kinetic production of excited atoms or diatomic fragments. A recent example would be the role

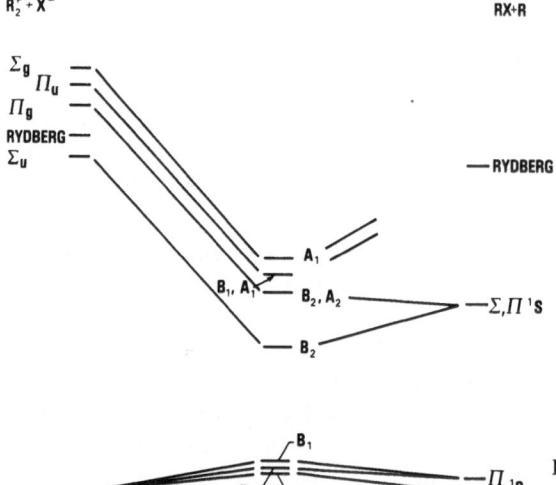

Fig. 2.13. Adiabatic correlation diagram for the Kr_2F system

Xe_2O^* plays in the production of $O(^1S)$. There is direct evidence for such a species in the photoproduction of $O(^1S)$ in a matrix by absorption into a diatomic XeO^* ion-pair state [2.49].

The electronic structure of triatomic metallic excimer states has not been determined as yet. The kinetic analysis indicates this excited system is only about 0.3 eV bound relative to the lowest Hg_2^* state. Slight bonding for a linear triatomic could be understood by analogy with the diatomic excimers. One electron in the most anti-bonding combination of s orbitals is excited into a bonding combination of p orbitals. This model would suggest that other group II triatomic excimers are stable. A broad emission band found by *Drullinger* [2.97] could be attributed to Cd_3.

2.11 Excimer Emission and Gain Cross-Sections

The attractive upper state potential $V_U(R)$ in an excimer transition defines a set of electronic-rotational-vibrational bound states $|U; J', M', v'\rangle$ with discrete total energies $E^U_{J',v'}$. Above the dissociation limit, $\varepsilon'' > 0$, of the lower state potential, $V_L(R)$, energy normalized continuum states exist, $|L; J'', M'', \varepsilon''\rangle$, which dissociate into ground state atomic fragments $X + M$ that separate with relative kinetic energy ε''. Emission from the discrete upper levels, which are $2J' + 1$ degenerate, into the continuous, infinitely degenerate, lower states yields a continuous distribution of emitted photons, $hv_f \equiv hck_f$,

$$\sum_{M'} XM(U; J', M', v) \rightarrow \sum_{J'', M''} X \cdots M(L; J'', M'', \varepsilon'') + hv_f.$$

Energy conservation requires that

$$E^U_{J',v'} = \varepsilon'' + h\nu_f. \tag{2.1}$$

The resultant spectral distribution is determined by the "continuum line shape" [2.113].

$$g^c_{v'}(\varepsilon'') = \frac{\langle v'|t_{U\to L}(R)|\varepsilon''\rangle^2}{\langle v'|t^2_{U\to L}(R)|v'\rangle} \cong \langle v'|\varepsilon''\rangle^2. \tag{2.2}$$

The quantity $g^c_{v'}(\varepsilon'')d\varepsilon''$ is the probability[1] that molecules in upper vibrational level $|v'\rangle$ undergo spontaneous emission into a band of energy, $d\varepsilon''$, centered about ε''.

The integrated line shape equals unity if the entire emission is continuous, and it is less than one if there are any contributions from bound state levels $|v''\rangle$ in the V_L potential. The electronic transition moment $t_{U\to L}(R)$ is an implicit function of the interatomic distance R. If this R-dependence is negligible over the amplitude of the vibration the line shape reduces to a simple Franck-Condon overlap integral $\langle v'|\varepsilon''\rangle^2$.

The excimer gain cross section for molecules initially in state J', v' is simply related to the line shape [2.113],

$$\sigma^{U\to L}_{J',v'}(h\nu_f) = \lambda^2_f \frac{A^{U\to L}_{v'}}{8\pi} h g^c_{v'}(\varepsilon'')|_{\varepsilon'' = E^U_{J',v'} - h\nu_f}. \tag{2.3}$$

The continuum line shape, which has units of reciprocal energy, replaces the usual Lorentzian line shape for bound-bound transitions, and is evaluated at the kinetic energy dictated by energy conservation in (2.1). The cross section is proportional to the total spontaneous emission rate from the upper state $|U, J', M', v'\rangle$ into the entire manifold of final states associated with the lower state $|L\rangle$, i.e.,

$$A^{U\to L}_{v'} \equiv \frac{64\pi^4}{3h} \lambda^{-3}_f \langle v'|t^2_{U\to L}(R)|v'\rangle. \tag{2.4}$$

The rate is only an insensitive function of the rotational quantum number J' and this index has been suppressed. The dominant effect of J' is manifested in the concentration of upper state molecules, $\varrho_{J',v'}$, and secondarily in the choice of $\varepsilon'' = E^U_{J',v'} - h\nu_f$, although the matrix element $\langle v'|\varepsilon''\rangle^2$ is generally unaffected by such small changes in ε''.

The critical parameter in determining excimer gain is $g^c_{v'}(\varepsilon'')$ which is a measure of the extent to which the total upper state emission is distributed

---

1 These probabilities $g^c_{v'}$ must be weighted by the factor $(h\nu_f)^3$ associated with each final state in order to reproduce the spontaneous emission spectrum.

among the final continuum states. This depends primarily on the nature of the lower state potential $V_L(R)$ in the vicinity of the upper state equilibrium position, R'_e, which we may characterize with two parameters

$$\varepsilon_0 \equiv V_L(R'_e) \tag{2.5}$$

and

$$\varepsilon_0 \alpha \equiv -\partial V_L/\partial R|R'_e. \tag{2.6}$$

For a steeply repulsive potential ε_0 closely approximates the kinetic energy ε'' associated with the peak in the $v'=0$ emission. The parameter α is a measure of the potential gradient, and the emission probability is *roughly* dispersed over a range of kinetic energies $\varepsilon'' = \varepsilon_0 \pm \varepsilon_0 \alpha |\Delta R_{v'}|$ where $|\Delta R_{v'}|$ is the "maximum" amplitude of the vibrational state $|v'\rangle$. For large $\varepsilon_0 \alpha$ the range of ε'' is large, and the value of $g_{v'}^c$ at any prescribed kinetic energy is small, i.e., $O(1/2\varepsilon_0 \alpha |\Delta R_{v'}|)$ or less. However, if ε_0 or $\varepsilon_0 \alpha$ are small, then one might expect a more narrow band of final kinetic energies to be accessible, and therefore a larger line shape factor, and higher gain.

These two limiting cases are expressible in analytic form if the upper state potential $V_U(R)$ is harmonic, and $t_{U \to L}$ is constant. The amplitude of a molecule, with reduced mass μ_{xm}, in vibrational state v', with vibrational energy $E_{v'} = h\nu_{osc}(v'+1/2)$, is approximately limited to the range

$$k_{osc}|R - R'_e| \le 2v' + 1$$

where the wave number k_{osc} which characterizes the oscillator is

$$k_{osc}^2 = \frac{2\mu_{xm}}{\hbar^2}(h\nu_{osc}/2). \tag{2.7}$$

The two limiting cases of a steeply repulsive (a), and a weak, slowly varying (b) lower state potential, are determined by the magnitude of the reduced parameter

$$\delta_{v'} = \left(\frac{h\nu_{osc}}{\varepsilon_0}\right)\left(\frac{k_{osc}}{\alpha}\right)\frac{1}{\sqrt{2v'+1}}. \tag{2.8}$$

Case (a) applies when $\delta v' \ll 1$, and a *crude* estimate of the line shape is obtained from evaluating $(2\varepsilon_0 \alpha |\Delta R_{v'}|)^{-1} = (1/2\varepsilon_0)(k_{osc}/\alpha)(2v'+1)^{-1/2}$. A more precise estimate is given in the next section. Case (b) applies when $\delta_{v'} \gg 1$, and the line shape becomes proportional to $1/h\nu_{osc}$.

2.11.1 Cross-Section for Steeply Repulsive Potentials: $\delta_{v'} \ll 1$

When $\delta_{v'} \ll 1$ the "reflection method" [2.114] for evaluating $\langle v'|\varepsilon''\rangle^2$ is valid. Generally $V_L(R)$ is well approximated by an exponential in the vicinity of R'_e,

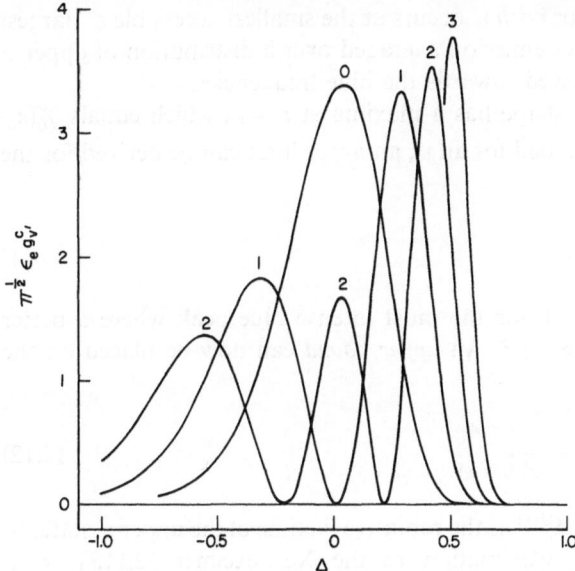

Fig. 2.14. Line shape for excimer transitions to steep repulsion potential from upper vibrational levels $v' = 0$, 1, 2, and 3. $\varDelta = (\varepsilon_0 - \varepsilon'')/\varepsilon_0$

$V_L(R) = \varepsilon_0 \exp[-\alpha(R - R'_e)]$, and if the upper state is harmonic, i.e., $lv' > = k_{osc}^{1/2} R_{v'}(X)$, where $X = k_{osc}(R - R'_e)$ and $R_{v'}(X)$ is a normalized harmonic oscillator wave function $\int R_{v'} R_{v'} dx = 1$, the line shape is expressible as,

$$g^c_{v'}(\varepsilon'') = \frac{S}{\varepsilon''} R^2_{v'}(-S \ln \varepsilon'' / \varepsilon_0), \qquad (2.9)$$

where $S = k_{osc}/\alpha$. This approximation is justified, and easily improved, by more rigorous *WKB* stationary phase methods [2.115] devised by *Jablonski* [2.116] but the result (2.9) is quite adequate for $\delta_{v'} \ll 1$.

Some general properties of the line shape for steep potentials can now be summarized. (An extensive discussion of the reflection method has been given by *Mulliken* [2.117].)

(I) The excimer emission extends over a range of kinetic energies which is large compared to $E_{v'}$; i.e., $\varepsilon_0 \exp[-\sqrt{2v' + 1/S}] \leqq \varepsilon'' \leqq \varepsilon_0 \exp[+\sqrt{2v' + 1/S}]$, or more approximately,

$$-hv_{osc}/\delta_{v'} \leqq \varepsilon'' - \varepsilon_0 \leqq hv_{osc}/\delta_{v'}. \qquad (2.10)$$

(II) The line shape factor has $v' + 1$ maxima since the vibrational function $R^2_{v'}$ has v' nodes. This feature, together with the range defined by (2.10) is demonstrated in Fig. 2.14, for a representation case, with $S = 3.32$, which is meant to approximate the Xe_2 excimer [2.108].

(III) The largest peak for *each* v' occurs at the smallest accessible ε'' (largest $h\nu_f$), and the overall excimer emission averaged over a distribution of upper v' states will generally be skewed towards the blue frequencies.

(IV) For $v'=0$ the line shape has a maxima at $\varepsilon''=\varepsilon_0$ which equals $g_0^c(\varepsilon_0)$ $=S/\sqrt{\pi}\varepsilon_0$. Since $R_{v'}^2$ is bounded for all v', an *upper* limit can be derived for the value of $g_{v'}^c(\varepsilon'')$,

$$g_{v'}^c(\varepsilon'') \lesssim \frac{S}{\varepsilon''}\frac{2}{3} \quad \text{for all } v'. \tag{2.11}$$

This applies particularly well for the most intense blue peak where a better estimate for $v'>0$ is $g_{v'}^c \cong (S/\varepsilon'')/3$. An *upper bound* can now be placed on the gain cross section (2.3)

$$\sigma_{J',v'}^{U \to L}(h\nu_f) \lesssim \frac{\lambda_f^2}{12\pi}\frac{S}{(\varepsilon''/hc)(\tau_{v'}c)}\bigg|_{\varepsilon''=E_{J',v'}^U - h\nu_f}, \tag{2.12}$$

where $\lambda_f=(c/\nu_f)$ and $\tau_{v'}=1/A_{v'}^{U \to L}$ is the radiative lifetime of the upper v' state. In the application of this approximation to the Xe_2 excimer [2.113]; with $\delta_0=0.05$, $S=3.32$, $\varepsilon_0/hc=5000\,\text{cm}^{-1}$, $\lambda_f=1700\,\text{Å}$ and $\tau_{v'}=10^{-8}\,\text{s}$, the bound yields $\sigma_{v'} \lesssim 1.7 \times 10^{-17}\,\text{cm}^{-2}$.

2.11.2 Cross Section for Weak Lower State Potentials: $\delta_{v'} \gg 1$

If the lower state potential is slowly varying, i.e., $\alpha \to 0$, or small, i.e., $\varepsilon_0 \to 0$, in the vicinity of R_e', then $\delta_{v'}$ in (2.8) is no longer small, and the reflection method is invalid. This occurs in many classes of excimer systems where R_e' is displaced to large distances relative to the repulsive inner core of V_L. In the limit

$$(\varepsilon''-\varepsilon_0)\delta_{v'} \gg h\nu_{osc} \tag{2.13}$$

an alternate expression for the line shape can be developed [2.118] if the upper state is harmonic,

$$g_{v'}^c(\varepsilon'') \cong \frac{4}{h\nu_{osc}}R_{v'}^2(K)\frac{\sin^2}{K}\left(KX_L+\frac{\pi v'}{2}\right), \tag{2.14}$$

where

$$K^2 = \frac{\varepsilon''-\varepsilon_0}{h\nu_{osc}/2}$$

and $X_L>2v'+1$ is a measure of the maximum amplitude the upper state may experience without penetrating the "core" of the lower state potential, $V_L(R_c)=\varepsilon''$,

$$X_L \cong \int_{R_c}^{R_e'} d(k_{osc}R)\sqrt{\frac{\varepsilon''-V_L(R)}{\varepsilon''-V_L(R_e')}}.$$

This result is exact for a perfectly flat potential, $\delta_{v'} = \infty$, and gives an analytic interpretation of the interesting numerical results obtained by *Golde* [2.119] for a potential $V_L = 0$ with an infinite barrier at R_c, i.e., $K^2 = 2\varepsilon''/hv_{osc}$, and $X_L = k_{osc}(R'_e - R_c)$.

The general properties of this line shape expression are quite distinct from the strongly repulsive case.

(I) *If* $\varepsilon_0 > 0$, such that the lower state potential is negligible, or repulsive, in the region of R'_e, the energy parameter K has a threshold $K_0 = 0$ when $\varepsilon'' = \varepsilon_0$, and extends to infinity. An integration of $g^c_{v'}(\varepsilon'')$ over ε'', assuming $\overline{\sin^2} = 1/2$ over a small increment of $d\varepsilon''$, yields unity, and the entire emission from v' occurs in the continuum. Since $R_{v'}(K)$ rapidly vanishes for $K > 2v' + 1$ the emission occurs in an energy region determined by $E_{v'}$

$$\varepsilon_0 \leq \varepsilon'' \leq hv_{osc}(v' + 1/2) + \varepsilon_0$$

and, if $E^U_{0,v'} = E^U_{0,0} + E_{v'}$, the excimer radiation $hv_f = E^U_{0,v'} - \varepsilon''$ is constrained to the region

$$E^U_{0,0} - \varepsilon_0 \lesssim hv_f \lesssim E^U_{0,v'} - \varepsilon_0.$$

(II) *If* $\varepsilon_0 < 0$, such that R'_e sits above an attractive lower state potential, the energy parameter K has a finite threshold

$$K^2_0 = \frac{-2\varepsilon_0}{hv_{osc}}$$

and the bounds on ε'' are reduced to

$$0 \leq \varepsilon'' \leq hv_{osc}(v' + 1/2) + \varepsilon_0.$$

The reduction is a measure of the contribution that *bound states* $|v''\rangle$ make to the emission. This can be determined by integrating (2.14) from K_0 to infinity. As K_0 approaches $\sqrt{(2v' + 1)}$, and therefore ε_0 approaches $-E_{v'}$, the integrated line shape vanishes. The excimer radiation is limited to the region

$$E^U_{0,0} - \varepsilon_0 \leq hv_f \leq E^U_{0,v'}.$$

(III) The line shape is a complicated function of *two* oscillatory terms, the rapidly varying $\sin^2(KX_L + \pi v'/2)$ term, and the more slowly varying vibrational function $R^2_{v'}(K)$ which has $v'/2$ nodes in the range $0 \leq K \leq \sqrt{2v' + 1}$ for even v', and $(v' + 1)/2$ nodes for odd v' states. Some representative line shapes are shown in Fig. 2.15 for $\varepsilon_0 \geq 0$ and $X_L = 10$. For $\varepsilon_0 < 0$ the blue portion of the spectrum for each v', from threshold to $-\varepsilon_0$ above threshold will appear as a discrete spectrum.

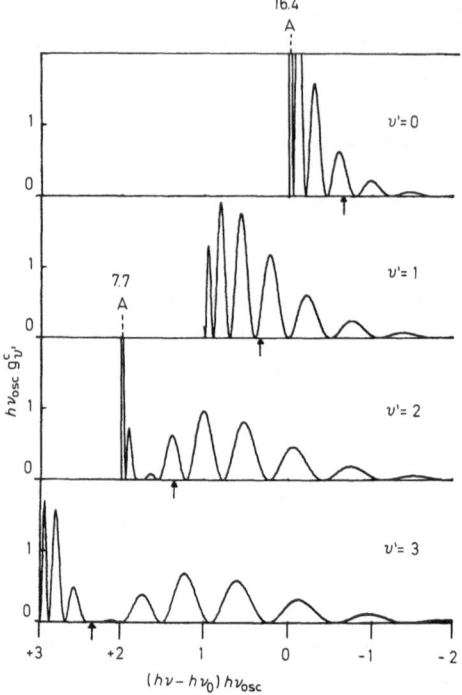

Fig. 2.15. Line shape for excimer transitions to a slowly varying potential from upper vibrational levels $v'=0$, 1, 2, and 3. $(h\nu - h\nu_0)/h\nu_{osc} = v' + (\varepsilon_0 - \varepsilon'')/h\nu_{osc}$, and $h\nu_0 = E_{0,0}^U - \varepsilon_0$

(IV) The quantitative aspects of the line shape at the blue end of the spectrum *must not* be taken too seriously, since we are dealing with continuum wave functions $|\varepsilon''\rangle$ very close to threshold, and very probably violating the conditions imposed by (2.13) beyond any hope of quantitative accuracy. However, the expectation of exceptionally high gain cross-sections at the blue wavelength limit, especially for $\varepsilon_0 \to 0$ and symmetric vibrational levels, is undoubtedly valid. An approximate upper bound to the line shape can be obtained from (2.14)

$$g_{v'}^c(\text{threshold}) < \frac{6}{\pi} \frac{X_L}{2v'+1} \frac{1}{h\nu_{osc}}.$$

(V) The results at the red end of the spectrum, where $\varepsilon'' \to E_{v'}$, are much more reliable and quantitatively accurate. This is demonstrated in Fig. 2.16, where spectra for $v'=100$ and $v'=200$ are presented for $X_L=21$. These are in excellent agreement with the accurate calculations and experimental observations *Tellinghuisen* [2.120] had made of the McLennan bands of I_2, and lend substantial confidence to the applicability of the simple analytic expression. The energy spacing of the maxima due to the $\sin^2(KX_L)$ oscillations are approximately uniform in the region, i.e., $\Delta\varepsilon'' = h\nu_{osc}\pi\sqrt{2v'+1}/X_L$, while the

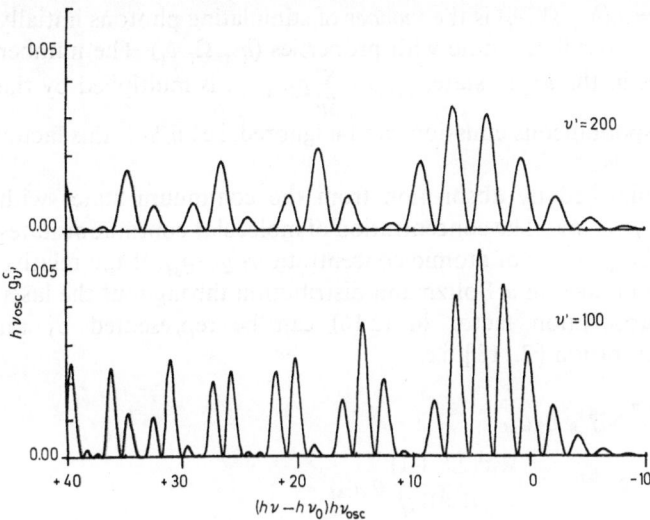

Fig. 2.16. Model calculations for high vibrational level excimer transitions representative of the McLennan bands of I_2. $(h\nu - h\nu_0)/h\nu_{\text{osc}} = \nu' + (\varepsilon_0 - \varepsilon'')/h\nu_{\text{osc}}$, and $h\nu_0 = E_{0,0}^U - \varepsilon_0$

underlying "vibrational" modulations become more closely spaced as ε'' decreases. An upper bound to the line shape in the vicinity of the largest vibrational peak is obtainable from (2.14),

$$g_{\nu'}^c(\varepsilon'' \simeq E_{\nu'}) \lesssim \frac{2.7}{\sqrt{2\nu'+1}} \frac{1}{h\nu_{\text{osc}}}. \tag{2.15}$$

These estimates of $g_{\nu'}^c$ may be introduced into (2.3) to obtain bounds on the gain cross sections.

2.12 Total Excimer Gain, Including Final State Absorption

The cross section (2.3) is used to determine the total gain per unit distance of travel, experienced by photons with energy $h\nu_f = hck_f$, and polarization $\hat{e}_f$, and propagating in the direction of the wave vector $\boldsymbol{k}_f$ with polar coordinates $\Omega_f = (\theta_f, \phi_f)$.

$$\text{Gain } (h\nu_f, \Omega_f, \hat{e}_f) \equiv \frac{\cdot \partial n_f}{c \partial t n_f} = \sum_{J',v'} \sigma_{J',v'}^{U \to L}(h\nu_f) \left[\left(\frac{n_f + 1}{n_f} \right) \varrho_{J',v'} \right.$$

$$\left. - e^{+h\nu_f/kT} \varrho_{J',v'}^{(\text{eq})} \right]. \tag{2.16}$$

The quantity $n_f = n_f(h\nu_f, \Omega_f, \hat{e}_f)$ is the *number* of stimulating photons initially present in the photon oscillator state with properties $(h\nu_f, \Omega_f, \hat{e}_f)$. The number density of molecules in the *upper* state, $\varrho_{J',v'} = \sum_{M'} \varrho_{J',M',v'}$, is multiplied by the factor $(n_f + 1)/n_f$. If spontaneous emission can be ignored, i.e., $n_f \gg 1$, this factor is unity.

The gain is diminished by absorption from the continuum states with kinetic energy $\varepsilon'' = E^U_{J',v'} - h\nu_f$. The concentration of molecular continuum states is proportional to the product of atomic concentrations $\varrho_X \cdot \varrho_M$. If the relative kinetic energy is maintained in a Boltzmann distribution throughout the laser amplification the absorption factor in (2.16) can be represented by an "equilibrium" concentration [2.113], i.e.,

$$e^{h\nu_f/kT} \varrho^{(eq)}_{J',v'} = e^{h\nu_f/kT} K^{eq}_{(J',v')} \varrho_X \varrho_M$$

$$= e^{h\nu_f/kT} e^{-E^U_{J',v'}/kT} \frac{W^{el}_U (2J'+1)}{Q_{tr} W^{el}_X W^{el}_M} \varrho_X \varrho_M$$

$$= e^{-\varepsilon''/kT} \frac{W^{el}_U (2J'+1)}{Q_{tr} W^{el}_X W^{el}_M} \varrho_X \varrho_M , \tag{2.17}$$

where $Q_{tr} = (2\mu_{xm} \pi kT/h^2)^{3/2}$ is the translational partition function, and W^{el} is the electronic degeneracy of the indicated state. Inversion in an excimer system can always be achieved by maintaining the ground state fragments in low concentration. However, one of the more attractive features of excimer emission is that, even in the presence of large atomic concentrations, by *choosing* a photon energy $h\nu_f$ from the continuum distribution sufficiently displaced to the red, the Boltzmann factor $\exp(-\varepsilon''/kT)$ in (2.17) may be made small enough to maintain inversion.

Notice that $\varrho^{(eq)}_{J',v'}$ as defined by (2.17) is just the concentration of *upper* state molecules that *would* exist in thermal equilibrium with the absorbing atomic fragments. Therefore, it is easily seen that (2.16) satisfies detailed balance. When the radiation field is in thermal equilibrium, i.e., $n_f^{(eq)} = 1/[\exp(h\nu_f/kT) - 1]$, the gain expression vanishes.

References

2.1   B.Stevens, E.Hutton: Nature **186**, 1045 (1960)
       T.Förster: *The Exciplex*, ed. by M.Gordon, W.R.Ware (Academic Press, New York 1975)
2.2   F.G.Houtermans: Helv. Phys. Acta '33, 933 (1960)
2.3   R.S.Mulliken: Phys. Rev. **136**, A962 (1965)
       M.L.Ginter, C.M.Brown: J. Chem. Phys. **56**, 672 (1972)
2.4   R.S.Mulliken: J. Chem. Phys. **52**, 5170 (1970)
2.5   B.Liu: Phys. Rev. Lett. **27**, 1251 (1971)
       T.L.Gilbert, A.C.Wahl: J. Chem. Phys. **55**, 5247 (1971)
       J.S.Cohen, B.Schneider: J. Chem. Phys. **61**, 3230 (1974)

2.6   B.Schneider, J.S.Cohen: J. Chem. Phys. **61**, 3240 (1974)
       W.J.Stevens, M.Gardner, A.Karo, P.S.Julienne: J. Chem. Phys. (to be published)
2.7   C.Y.Ng, D.J.Trevor, B.H.Mahan, Y.T.Lee: J. Chem. Phys. **66**, 446 (1977)
2.8   W.R.Wadt: J. Chem. Phys. (to be published)
2.9   C.Y.Ng, P.W.Tiedemann, B.H.Mahan, Y.T.Lee: J. Chem. Phys. **66**, 5737 (1977)
2.10   R.J.Blint: Phys. Rev. **A14**, 971 (1976)
2.11   J.A.Barker: J. Chem. Phys. **63**, 2767 (1975)
2.12   J.M.Parson, T.P.Schafer, F.P.Tully, P.E.Siska, Y.C.Wong, Y.T.Lee: J. Chem. Phys. **53**, 2123 (1970)
       J.M.Parson, T.P.Schafer, P.E.Siska, F.P.Tully, Y.C.Wong, Y.T.Lee: J. Chem. Phys. **53**, 3755 (1970)
2.13   W.J.Stevens, A.C.Wahl, M.A.Gardner, A.M.Karo: J. Chem. Phys. **60**, 2195 (1974)
       P.Toennies: Chem. Phys. Lett. **20**, 238 (1973)
       B.Liu, D.McLean: J. Chem. Phys. **59**, 4557 (1973)
2.14   J.W.Keto, R.E.Gleason,Jr., G.K.Walters: Phys. Rev. Lett. **33**, 1365 (1974)
       P.K.Leichner, K.F.Palmer, J.D.Cook, M.Thieneman: Phys. Rev. **A13**, 1787 (1976)
2.15   R.P.Saxon, B.Liu: J. Chem. Phys. **64**, 3291 (1976)
       K.T.Gillen, R.P.Saxon, D.C.Lorents, G.E.Ice, R.E.Olson: J. Chem. Phys. **64**, 1925 (1976)
2.16   W.J.Stevens: Private communication
2.17   O.Cheshnovsky, A.Gedanken, B.Raz, J.Jortner: Chem. Phys. Lett. **22**, 23 (1973)
       E.T.Verkhovtseva, A.E.Ovechkin, Ya.M.Fagel: Chem. Phys. Lett. **30**, 120 (1975)
2.18   K.M.Sando: Mol. Phys. **23**, 413 (1972)
2.19   F.H.Mies, A.L.Smith: J. Chem. Phys. **45**, 994 (1966)
2.20   R.C.Michaelson, A.L.Smith: J. Chem. Phys. **61**, 2566 (1974)
2.21   Y.Tanaka, K.Yoshimo, D.E.Freeman: J. Chem. Phys. **59**, 5160 (1973)
       E.A.Colburn, A.E.Douglas: J. Chem. Phys. **65**, 1741 (1976)
2.22   A.U.Hazi, T.N.Rescigno: Phys. Rev. (to be published)
2.23   B.Liu, H.F.Schaefer,III: J. Chem. Phys. **55**, 2369 (1971)
2.24   D.H.Liskow, H.F.Schaefer,III, P.S.Bagus, B.Liu: J. Am. Chem. Soc. **95**, 4056 (1973)
2.25   M.Krauss, B.Liu: Chem. Phys. Lett. **44**, 257 (1976)
2.26   J.Tellinghuisen, G.C.Tisone, J.M.Hoffman, A.K.Hays: J. Chem. Phys. **64**, 4796 (1976)
2.27   J.Goodman, L.E.Brus: J. Chem. Phys. **65**, 3808 (1976)
       B.S.Ault, L.Andrews: J. Chem. Phys. **65**, 4192 (1976)
2.28   R.S.Mulliken, W.B.Person: *Molecular Complexes* (Wiley-Interscience, New York 1969)
2.29   J.J.Ewing, R.Milstein, R.S.Berry: J. Chem. Phys. **54**, 1752 (1971)
       R.Grice, D.R.Herschbach: Mol. Phys. **27**, 159 (1974)
2.30   J.Tellinghuisen, J.M.Hoffman, G.C.Tisone, A.K.Hays: J. Chem. Phys. **64**, 2484 (1976)
2.31   E.S.Rittner: J. Chem. Phys. **19**, 1030 (1951)
       P.Brumer, M.Karplus: J. Chem. Phys. **58**, 3903 (1973)
2.32   T.H.Dunning,Jr., P.J.Hay: Appl. Phys. Lett. **28**, 649 (1976)
       P.J.Hay, T.H.Dunning,Jr.: J. Chem. Phys. **66**, 1306 (1977)
2.33   M.Krauss: J. Chem. Phys. (to be published)
2.34   W.J.Stevens, F.P.Billingsley,II: Phys. Rev. **A8**, 2236 (1973)
       H.J.Werner, W.Meyer: Phys. Rev. **A13**, 13 (1976)
       E.A.Reisch, W.Meyer: Phys. Rev. **A14**, 915 (1976)
2.35   A.Khadjavi, A.Lurio, W.Hopper: Phys. Rev. **167**, 128 (1968)
       G.D.Mahan: J. Chem. Phys. **50**, 2755 (1969)
2.36   J.E.Velazco, J.H.Kolts, D.W.Setser: J. Chem. Phys. **65**, 3468 (1976)
2.37   B.S.Ault, L.Andrews: J. Chem. Phys. **64**, 3075 (1976)
2.38   M.Oppenheimer, R.S.Berry,Jr.: J. Chem. Phys. **54**, 5058 (1971)
2.39   Unpublished calculations of H.H.Michels, private communication; D.Neumann, M.Krauss
2.40   T.Dunning,Jr.: Rare-Gas Fluorides
       P.J.Hay: XeCl and XeBr, 7th Winter Colloqium, High Power Visible Lasers, Utah (February 1977)

44   *M. Krauss* and *F. H. Mies*

2.41 N.W.Winter: ArCl and KrCl, 32nd Symp. Mol. Spectrosc., Ohio State University (June 1977)
2.42 M.F.Golde, B.A.Thrush: Chem. Phys. Lett. **29**, 486 (1974)
      C.A.Brau, J.J.Ewing: J. Chem. Phys. **63**, 4640 (1974)
      J.J.Ewing, C.A.Brau: Phys. Rev. A**12**, 129 (1975)
      J.E.Velazco, D.W.Setser: J. Chem. Phys. **62**, 1990 (1975)
2.43 J.J.Ewing, C.A.Brau: In *Tunable Lasers and Applications*, ed. by A.Mooradian, T.Jaeger, P.Stokseth, Springer Series in Optical Sciences, Vol. 3 (Springer, Berlin, Heidelberg, New York 1976) p. 21
2.44 J.R.Murray, C.K.Rhodes: J. Appl. Phys. **47**, 5041 (1976)
2.45 C.D.Cooper, M.Lichtenstein: Phys. Rev. **109**, 2026 (1958)
      C.D.Cooper, G.C.Colb, E.L.Tolnas: J. Mol. Spectrosc. **7**, 223 (1961)
2.46 H.T.Powell, J.R.Murray, C.K.Rhodes: Appl. Phys. Lett. **25**, 730 (1974)
2.47 D.L.Cunningham, K.C.Clark: J. Chem. Phys. **61**, 1118 (1974)
      K.H.Welge, R.Atkinson: J. Chem. Phys. **64**, 531 (1975)
2.48 P.S.Julienne, M.Krauss, W.J.Stevens: Chem. Phys. Lett. **38**, 174 (1976)
      T.H.Dunning,Jr., P.J.Hay, R.C.Raffenetti: *Electronic Transition Lasers*, ed. by J.I.Steinfeld (MIT Press, Cambridge, Mass. 1976) p. 272
      T.H.Dunning,Jr., P.J.Hay: J. Chem. Phys. **66**, 3767 (1977)
2.49 J.Goodman, J.C.Tully, V.E.Bondybey, L.E.Brus: J. Chem. Phys. **66**, 4802 (1977)
2.50 R.J.Donovan, D.Husain: Chem. Rev. **70**, 489 (1970)
2.51 A.Maki, J.H.Simmons, J.Hougen: Private communication
2.52 D.L.Huestis, R.A.Gutcheck, R.M.Hill, M.V.McCusker, D.C.Lorents: Stanford Res. Inst. Rept. No. MP-75-18 (1975)
      D.C.Lorents, D.L.Huestis: In *Laser Spectroscopy*, ed. by S.Haroche, J.C.Pebay-Peyroula, T.W.Hänsch, S.E.Harris, Lecture Notes in Physics, Vol. 43 (Springer, Berlin, Heidelberg, New York 1975) p. 100
2.53 P.S.Julienne: J. Chem. Phys. (to be published)
2.54 B.S.Ault, L.Andrews: Chem. Phys. Lett. **43**, 350 (1976)
2.55 A.Ding, J.Karlau, J.Weise: Chem. Phys. Lett. **45**, 92 (1977)
2.56 D.L.King, L.G.Piper, D.W.Setser: J. Chem. Soc. Faraday Trans. II **73**, 177 (1977)
2.57 H.Hotop, W.C.Lineberger: J. Phys. Chem. Ref. Data **4**, 539 (1975)
2.58 J.J.Ewing, D.W.Trainor, S.Yatsiv: J. Chem. Phys. **61**, 4433 (1974)
      A.Brown, D.Husain: Int. J. Chem. Kinet. **7**, 77 (1975)
2.59 J.Pascale, J.Vandeplanque: J. Chem. Phys. **60**, 2278 (1974); Reports on *Molecular Terms of the Alkali-Rare Gas Atom Pairs:* I Li, II Na, III K, IV Rb; V Cs, (Centre d'Etudes Nucleaires de Saclay, France, March 1974)
2.60 W.E.Bayliss: J. Chem. Phys. **51**, 2665 (1969)
2.61 M.Krauss, P.Maldonado, A.C.Wahl: J. Chem. Phys. **54**, 4944 (1971)
2.62 A.J.Lacey, W.B.Brown: Mol. Phys. **27**, 1013 (1974)
2.63 T.Janis, A.C.Wahl: Private communication
2.64 R.Hedges, D.Drummond, A.Gallagher: Phys. Rev. A**6**, 1519 (1972)
2.65 D.L.Drummond, A.Gallagher: J. Chem. Phys. **60**, 3426 (1974)
2.66 G.York, R.Scheps, A.Gallagher: J. Chem. Phys. **63**, 1052 (1975)
2.67 R.Scheps, Ch.Ottinger, G.York, A.Gallagher: J. Chem. Phys. **63**, 2581 (1975)
2.68 U.Buck, H.Pauly: Z. Phys. **208**, 390 (1968)
      R.Duren, G.P.Raabe, C.Schlier: Z. Phys. **214**, 410 (1968)
      P.Dehrueraud, L.Wharton: J. Chem. Phys. **57**, 4821 (1972)
      D.J.Auerbach: J. Chem. Phys. **60**, 4116 (1974)
2.69 W.Happer, G.Moe, A.C.Tarn: Phys. Lett. **54**A, 405 (1975)
      A.Tarn, G.Moe, W.Park, W.Hopper: Phys. Rev. Lett. **35**, 85 (1975)
      B.Sayer, M.Ferray, J.Lozingot, J.Berlande: J. Phys. B**9**, L293 (1976)
2.70 C.F.Bender: 32rd Symp. on Mol. Spectrosc., Ohio State University (1977)

2.71   K.V.Subbaram, J.A.Coxon, W.E.Jones: Can. J. Phys. **53**, 2016 (1975)
       J.A.Coxon, W.E.Jones, K.V.Subbaram: Can. J. Phys. **53**, 2321 (1975)
       K.V.Subbaram, J.A.Coxon, W.E.Jones: Can. J. Phys. **54**, 1535 (1976)
2.72   H.E.Gunning, S.Penzes, H.S.Sandhu, O.P.Strausz: J. Am. Chem. Soc. **91**, 7684 (1969)
       O.P.Strausz, J.M.Campbell, S.DePaoli, H.S.Sandhu, H.E.Gunning: J. Am. Chem. Soc. **95**, 732 (1973)
2.73   J.T.Kielkopf, R.A.Miller: J. Chem. Phys. **61**, 3304 (1974)
2.74   W.H.Breckenridge, T.W.Broadbent: Chem. Phys. Lett. **29**, 421 (1974)
2.75   R.A.Gutcheck, R.M.Hill, D.L.Huestis, D.C.Lorents, M.V.McCusker: Stanford Res. Inst. Rept. No. MP-75-43 (1975)
2.76   F.J.Van Italie, L.F.Doemeny, R.M.Martin: J. Chem. Phys. **56**, 3689 (1972)
2.77   N.J.Bridge: J. Mol. Spectrosc. **42**, 370 (1972)
2.78   J.T.Hougen: J. Mol. Spectrosc. **42**, 381 (1972)
2.79   H.A.Hyman: J. Chem. Phys. **61**, 4063 (1974); Chem. Phys. Lett. **31**, 593 (1975)
2.80   B.Cheron, R.Scheps, A.Gallagher: J. Chem. Phys. **65**, 326 (1976)
2.81   J.J.Bevan, D.Husain: Int. J. Chem. Kinet. **7**, 63 (1975); J. Phys. Chem. **80**, 217 (1976)
2.82   P.J.Hay, T.H.Dunning, Jr., R.C.Raffenetti: J. Chem. Phys. **65**, 2679 (1976)
2.83   W.J.Stevens, M.Krauss: J. Chem. Phys. (to be published)
2.84   W.Finkelnburg, T.Peters: „Kontinuierliche Spektren", in Spektroskopie II. Spectroscopy II, ed. by S.Flugge, Handbuch der Physik, Vol. 28 (Springer, Berlin, Göttingen, Heidelberg 1957)
2.85   D.J.Eckstrom, R.A.Gutcheck, R.M.Hill, D.Huestis, D.C.Lorents: Stanford Res. Inst. Rept. No. MP-73-1 (1973)
2.86   K.C.Li, W.C.Stwalley: J. Chem. Phys. **59**, 4423 (1973)
2.87   H.S.Taylor, F.W.Bobrowicz, P.J.Hay, T.H.Dunning, Jr.: J. Chem. Phys. **65**, 1182 (1976)
       P.D.Burrow, J.A.Michejda, J.Comer: J. Phys. **89**, 3225 (1976)
2.88   M.W.McGeoch, G.R.Fournier, P.Ewart: J. Phys. B**9**, L121 (1976)
2.89   E.N.Morozov, M.L.Sosinskii: Opt. Spectrosc. **24**, 282 (1968)
       M.Czajkowski, L.Krause: Can. J. Phys. **52**, 2228 (1974)
2.90   E.K.Kraulinya, M.G.Arman: Opt. Spectrosc. **24**, 285 (1968)
2.91   F.H.Mies, W.J.Stevens, M.Krauss: Unpublished calculations
2.92   L.A.Schlie, B.D.Guenther, D.L.Drummond: Chem. Phys. Lett. **34**, 258 (1975)
2.93   E.W.Smith, R.E.Drullinger, M.M.Hessel, J.Cooper: J. Chem. Phys. **66**, 5667 (1977)
2.94   L.A.Schlie, B.D.Guenther, R.D.Rathge: Appl. Phys. Lett. **28**, 393 (1976)
2.95   R.E.Drullinger, M.M.Hessel, E.W.Smith: J. Chem. Phys. **66**, 5656 (1977)
2.96   E.R.Mosburg, M.Wilke: J. Chem. Phys. **66**, 5682 (1977)
2.97   R.E.Drullinger: Private communication
2.98   H.Hamada: Philos. Mag. **12**, 50 (1931)
2.99   W.J.Balfour, A.E.Douglas: Can. J. Phys. **48**, 901 (1970)
       C.R.Vidal, H.Scheingraber: J. Mol. Spectrosc. **65**, 46 (1977)
2.100 W.J.Balfour, R.F.Whitlock: Can. J. Phys. **53**, 472 (1975)
2.101 G.W.King, J.H.Van Vleck: Phys. Rev. **55**, 1165 (1939)
2.102 J.C.Miller, L.Andrews: Chem. Phys. Lett. (to be published)
2.103 V.Buck: Rev. Mod. Phys. **46**, 369 (1974)
2.104 R.Duren: Chem. Phys. Lett. **39**, 481 (1976)
2.105 Hawryluk, J.A.Mangano, J.H.Jacobs: 3rd Summer Colloquium on Electronic Transition Lasers (1976)
       J.A.Mangano, J.H.Jacob, A.Hawryluk: Appl. Phys. Lett. (to be published)
2.106 D.C.Lorents, R.M.Hill, D.L.Huestis, M.V.McCusker, H.H.Nakeno: 3rd Summer Colloquium on Electronic Transition Lasers (1976)
       H.H.Nakano, R.M.Hill, D.C.Lorents, D.L.Huestis, M.V.McCusker: Stanford Res. Inst. Rept. No. MP-76-99 (1976)
2.107 W.S.Struve: Mol. Phys. **25**, 777 (1973)
       P.K.Pearson, W.J.Hunt, C.F.Bender, H.F.Schaefer, III: J. Chem. Phys. **58**, 5358 (1973)

2.108 W. R. Wadt, P. J. Hay: Appl. Phys. Lett (to be published)
2.109 C. F. Bender, N. W. Winter: 32nd Symp. on Mol. Spectrosc., Ohio State University (1977)
2.110 P. S. Bagus, B. Liu, D. H. Liskow, H. F. Schaefer, III: J. Am. Chem. Soc. **97**, 7216 (1975)
2.111 P. L. Patterson: J. Chem. Phys. **48**, 3625 (1968)
        C. Vange, J. L. Whitten: Chem. Phys. Lett. **13**, 541 (1972)
        W. F. Liu, D. C. Conway: J. Chem. Phys. **62**, 3070 (1975)
        H. Helm: Phys. Rev. **A14**, 680 (1976)
        C. W. Werner, E. Zamir, E. V. George: Appl. Phys. Lett. **29**, 236 (1976)
2.112 M. Rokni, J. H. Jacob, J. A. Mangano, R. Brochu: Appl. Phys. Lett. **31**, 79 (1977)
2.113 F. H. Mies: Mol. Phys. **26**, 1233 (1973)
2.114 A. S. Coolidge, H. M. James, R. D. Present: J. Chem. Phys. **4**, 194 (1936)
2.115 F. H. Mies: J. Chem. Phys. **48**, 482 (1968)
2.116 A. Jablonski: Phys. Rev. **68**, 78 (1945)
2.117 R. S. Mulliken: J. Chem. Phys. **55**, 309 (1971)
2.118 F. H. Mies, P. S. Julienne: IEEE J. Quantum Electron. (to be published)
2.119 M. F. Golde: J. Mol. Spectrosc. **58**, 261 (1975)
2.120 J. Tellinghuisen: Chem. Phys. Lett. **29**, 359 (1974)

3. The Rare Gas Excimers

M. McCusker

With 13 Figures

In this chapter three types of excimer lasers will be discussed. These are the pure noble gas excimers, the rare gas oxides and the diatomic halogens. The closely related rare gas halide systems are examined in Chap. 4. The historical development of the noble gas excimer laser systems has already been discussed in Chap. 1, while the details of the potential curves relevant to the excimer systems appear in Chap. 2. Our discussion here, therefore, will be confined to a sketch of the present understanding of the kinetics and energy flow governing these systems, and also a brief summary of the more recent experimental efforts demonstrating the characteristics of these molecular systems.

The mixed gas systems such as xenon/oxygen or argon/iodine, have the gas of molecular additive in very dilute concentrations. An essential feature of these systems is the fact that the dominant channel for populating the upper laser level is via neutral energy transfer from the excited rare gas atoms and dimers. The rare gas energy donors (both atoms and dimers) are initially produced by excitation with an electron beam or an electron beam sustained discharge. Thus, the kinetics of the energy flow within pure rare systems are key elements of the behavior of mixed gas systems. Systems that are directly excited photolytically are exceptions, but even in these systems the light source is likely to be a rare gas laser with its internal kinetics.

Historically, a major factor in the development of noble gas excimer lasers was the knowledge that direct e-beam excitation would efficiently produce high densities of metastable noble gas atoms. However, in contrast to the high values of computed laser efficiency for the rare gas dimer systems, the actual laser efficiencies and powers that were initially demonstrated were rather low. In addition, since the emission wavelengths were in the vacuum ultraviolet, several technical problems connected with optics and beam transport were associated with these systems. In these circumstances, there was motivation for the investigation of methods that would selectively transfer the energy from the noble gas excimer to an appropriate acceptor molecule. If the energy transfer is sufficiently specific, a sizable population inversion may be generated in the acceptor species. An optimistic view then suggests that the additive can be selected to have properties that match the requirements of particular applications. For example, it may be desirable to have a long upper state lifetime applicable to high energy storage systems for fusion lasers. A wide variety of additives have been studied recently (O_2 and oxygen bearing compounds, N_2, CN, H_2O, mercury, the halogens and their compounds) and many, but not all, have been developed into successful lasers.

In Sect. 3.1 we shall briefly review the kinetics of the rare gas systems and explore the current status of the rare gas excimer laser development. We shall then consider the Group VI (oxygen, sulfur, selenium) laser development, and conclude with an outline of the recent progress of the diatomic halogen lasers (F_2, Cl_2, Br_2 and I_2).

3.1 Kinetics in the Pure Rare Gases

The kinetics of rare gas excimer lasers have been recently reviewed by *Lorents* [3.1], *Lorents* et al. [3.2], *Lorents* and *Olson* [3.3], *Fournier* [3.4], *Werner* et al. [3.5], *George* and *Rhodes* [3.6], *Rhodes* [3.7], *Hutchinson* [3.8] and recently again by *Werner* et al. [3.9]. While operation at high pressures allows for some simplification of the kinetic analysis (since collisions tend to vibrationally and electronically relax the population distributions to only a few states), these lasers, nonetheless, retain considerable complexity. The present understanding of these kinetics is incomplete and the relative importance of several collisional process remains controversial.

In general, noble gas excimer lasers operate at relatively high pressures (>2 atmospheres) and are energized by relatively high current, high voltage electron beams (~ 1 MeV, hundreds of $A\,cm^{-2}$). Under these conditions, the electron density in the generated plasma will be rather high ($>10^{14}\,cm^{-3}$). Laser action in the pure rare gases has received the most attention with particular emphasis on xenon at 172 nm. There have also been several efforts to explore the advantages of mixed rare gases, but this naturally tends to complicate the kinetics.

As was described in Chap. 2, the general molecular potential curves structures for all the heavier noble gases are similar. In particular, as *Mulliken* pointed out [3.10] the bound states of these excited molecules can be envisioned as a vertical sequence of Rydberg states beginning with the $^1\Sigma_u^+$ and $^3\Sigma_u^+$ excimer states. Many molecular properties of the neutrals such as well depths, vibrational frequency, and, in particular, equilibrium internuclear separation (R_0), can then be inferred from the corresponding molecular ion states which are known from scattering experiments [3.11, 12]. These higher lying bound levels are crossed by a large number of repulsive curves.

The mechanism for selectively populating the excimer levels can, in a simplified sense, be viewed as a sequence of collisional energy exchanges. Figure 3.1 portrays a schematic diagram of this sequence. The high energy electrons ionize or excite the host gas in the reactions

$$e^- + Ar \rightarrow Ar^+ + 2e^- \;, \tag{3.1}$$

$$e^- + Ar \rightarrow Ar^* + e^- \;. \tag{3.2}$$

We use argon as a specific case, but the conclusions also apply to krypton and xenon. The secondary electrons created will have a mean energy of five to seven

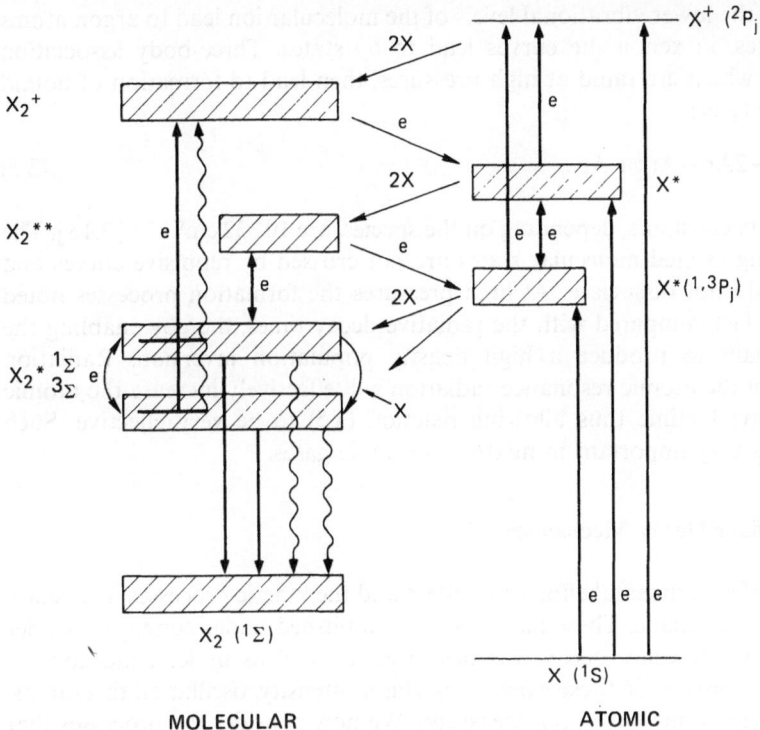

Fig. 3.1. Energy levels relevant to noble gas excimer production. The species X can refer to any noble gas except helium. Electron collisions are denoted by the letter e; photoemission and photoabsorption are indicated by wavy arrows

electron volts depending on the gas [3.11]. As discussed by *Lorents* and co-workers [3.12], the energy required for producing an ion pair is similar for all the heavier noble gases (approximately 24 eV).

At the high pressures relevant to the lasers (above a few atmospheres), three-body association to form the molecular ion is rapid through the process

$$Ar^+ + 2Ar \rightarrow Ar_2^+ + Ar \ . \tag{3.3}$$

The rate constant for this reaction is generally greater than $10^{-31} \, cm^6 \, s^{-1}$ [3.13]. Subsequently, a dissociative recombination reaction quickly forms an excited neutral in a dissociating state,

$$Ar_2^+ + e^- \rightarrow Ar_2 \rightarrow Ar^{**} + Ar \ . \tag{3.4}$$

Typical rate constants for this process are $10^{-6} \, cm^3 \, s^{-1}$ and decrease with increasing electron temperature [3.14]. It is assumed here that vibrational relaxation is rapid prior to the recombination. In argon, the repulsive curves that cross the lower vibrational levels of the molecular ion lead to argon atoms in $4p$ states, in xenon the curves lead to $6p$ states. Three-body association reactions, which are rapid at high pressures, then lead to formation of bound molecular levels

$$Ar^{**} + 2Ar \rightarrow Ar_2^* + Ar \ . \tag{3.5}$$

Typical rate constants, depending on the species, are $10^{-32} \, cm^6 \, s^{-1}$ [3.15]. The lowest lying excited molecular states are not crossed by repulsive curves and therefore do not dissociate. At high pressures the formation processes noted above are fast compared with the radiative decay times, thereby enabling the kinetic chain to produce a high density population inversion. Radiation trapping of the atomic resonance radiation will effectively increase the atomic excited state lifetime thus allowing reaction (3.5) to be very effective. Such trapping is very important in mixtures of noble gases.

3.1.1 Excimer Decay Mechanisms

A variety of experimental efforts to understand the kinetic processes in a xenon laser have been made. These have been accomplished under conditions under which the fluorescence decay was measured, as well as under conditions of stimulated emission. In these experiments laser intensity, oscillation thresholds, and frequency shifts have been measured. We now consider the processes that have been suggested to account for excimer fluorescence decay. The noble gases have two low-lying, but separated excimer states, both of which may radiate; these are the $^1\Sigma_u^+ (O_u^+)$ state that dissociates to $^3P_1$ and ^1S atoms and the $^3\Sigma_u^+ (1_u, O_u^-)$ state that dissociates to $^3P_2$ and ^1S (see Chap. 2). Although the radiative lifetimes of these two states are appreciably different, both contribute to the fluorescence observations. A summary of recent lifetime data taken from the review of *Lorents* [3.1] is shown in Table 3.1.

Several experiments have been made to examine the decay rate of the excimer emission at the relatively high excitation densities characteristic of lasers. The majority of these studies have concentrated on xenon. At all xenon pressures these experiments have, however, shown one rather than two decay times. At low pressure the decay rate is low and rises with increasing pressure asymptotically to a limiting value. A thorough comparison of several experiments has been made by *Lorents* and colleagues [3.2]. The data that were analyzed included those published by *Koehler* et al. [3.24], *Gerardo* and *Johnson* [3.25], *Wallace* et al. [3.26], and *Bradley* et al. [3.27]. For xenon, this asymptotic value is $6 \pm 1 \times 10^7 \, s^{-1}$; this agrees with the value that would be derived assuming that complete population equilibrium exists between the triplet and singlet excimer states. The inverse of the decay time then becomes

$$\frac{1}{\tau} = \frac{3}{4} \cdot \frac{1}{\tau_3} + \frac{1}{4} \cdot \frac{1}{\tau_1} \cong 5.2 \times 10^7 \, s^{-1} \tag{3.6}$$

Table 3.1. Radiative lifetimes of excimer states

Excimer	$^1\Sigma_u$ [ns]	$^3\Sigma_u$	Ref.
Ne_2^*	–	5.1	[3.16]
	2.8	11.9	[3.17]
	–	12 ± 6	[3.18]
		6.62	[3.19]
Ar_2^*	–	2.8	[3.20]
	–	3.7	[3.21]
	4.2 ± 0.13	3.2 ± 0.3	[3.22]
		4.0 ± 2.0	[3.18]
		3.22	[3.19]
Kr_2^*	–	1.7	[3.21]
	–	0.3	[3.23]
	–	0.36 ± 0.16	[3.18]
		0.35	[3.19]
Xe_2^*	5.5 ± 1.0	0.09 ± 0.05	[3.22]
	–	0.10 ± 0.05	[3.18]
	–	0.140 ± 0.45	[3.18]

where 3/4 and 1/4 are the fractional populations of the triplet and singlet populations, and τ_3 and τ_1 are the triplet and singlet lifetimes. Electrons are an important factor in establishing this equilibrium. In the presence of an electron beam, the relatively hot plasma electrons will rapidly mix the excimer states as well as their precursors in reactions of the type

$$e^- + Ar_2(^3\Sigma) \xrightarrow{k_e} e + Ar_2(^1\Sigma) . \qquad (3.7)$$

By assuming a reasonable rate constant $k_e \sim 2 \times 10^{-6}\,cm^{-3}\,s^{-1}$ and an electron density $10^{15}\,cm^{-3}$, we observe that process (3.7) can account for the equilibrium discussed above. In the high electron density environment implied by these conditions, super-elastic electron collisions can also enhance electronic relaxation in both the atomic and excimer manifolds. The super-elastic mechanism, however, can also deplete the excimer population inversion by causing transitions to the ground state. It is important to note that these super-elastic channels can provide a mechanism for electron heating, a factor which helps to maintain the $^1\Sigma_u^+$ population through reaction (3.7).

Atomic collisions can also influence the $^1\Sigma_u^+$ and $^3\Sigma_u^+$ populations. *Werner* et al. [3.5] discussed excimer relaxation by atomic collisions in the process

$$Ar_2^*(^1\Sigma) + Ar \rightleftarrows Ar_2^*(^3\Sigma) + Ar . \qquad (3.8)$$

However, since the measurements of *Keto* et al. [3.22, 28, 29] indicate that the cross section for this reaction is less than $6 \times 10^{-18}\,cm^2$, this reaction appears to be important only at extremely high densities. Therefore, it is concluded that the electrons are the dominant contributor to electronic state mixing.

Rapid electron mixing is vital to the efficient operation of a practical laser. The long lifetime of the triplet state leads to a relatively low value of the stimulated emission cross section in comparison to the contribution of the singlet manifold. However, as stated above, the excited state population will be distributed between the triplet and singlet excimers. Therefore, a requirement for efficient laser operation is the existence of a sufficient hot electron density to rapidly convert the triplet population into the much more strongly radiating singlet system.

Photoionization of the excimer states is a fundamental radiative loss process affecting laser performance. The excimer state can absorb the emitted radiation and be photoionized. The photoionization cross sections for the excimers have not been directly measured. Using a simple quantum defect approach, *Lorents* and *Huestis* [3.2] derived a value for the photoionization of any atom (or molecule) of

$$\sigma = \frac{8 \times 10^{-18}}{Z \left(\frac{\text{IP}}{13.6}\right)^{\frac{1}{2}} \varrho^3} \text{[cm}^2\text{]} \tag{3.9}$$

where ϱ is the relative energy of the photon ($h\nu/\text{IP}$), IP is the ionization potential, and Z is the ionic charge after photoionization. While this procedure may not always be accurate (for example, it fails for the alkali atoms at lower photon energies), it provides a useful estimate for the noble gas excimers. In particular, it predicts photoionization cross sections for xenon and argon of $2 \times 10^{-18} \text{cm}^2$ and $9 \times 10^{-19} \text{cm}^2$ at the excimer emission wavelengths. The important fact is that these are comparable to estimates of stimulated emission cross sections on the bound-free excimer transition. *Dunning* and *Stebbings* [3.30] have estimated an upper bound for photoabsorption of metastable noble gas atoms as $1.1 \times 10^{-18} \text{cm}^2$ for argon and $4.9 \times 10^{-19} \text{cm}^2$ for krypton in the near ultraviolet, *Hyman* [3.31] has made recent calculations of the excited atom photoionization, and emphasized that the s-states and p-states have markedly different photoionization cross section energy dependences.

At the high excited state densities reached by relatively high current electron beams ($>100 \text{A cm}^{-2}$) into high pressure gases, excimer–excimer annihilation will be particularly important (as will excited atom-excited atom collisions),

$$\text{Ar}_2^* + \text{Ar}_2^* \rightarrow \text{Ar}_2^+ + 2\text{Ar} + e^- \tag{3.10}$$

where the Ar_2^+ can subsequently recycle kinetically to form another excimer. *Gerardo* and *Johnson* [3.25], by analyzing fluorescence decay data, determined a rate constant of $3.5 \times 10^{-10} \text{cm}^3 \text{s}^{-1}$. In a separate experiment *Zamir* et al. [3.32] determined a value of $8 \times 10^{-11} \text{cm}^3 \text{s}^{-1}$. Because of this high value, the efficiencies of $\sim 50\%$ originally predicted and recently experimentally confirmed by *Turner* et al. [3.33] are attainable only at relatively low excitation

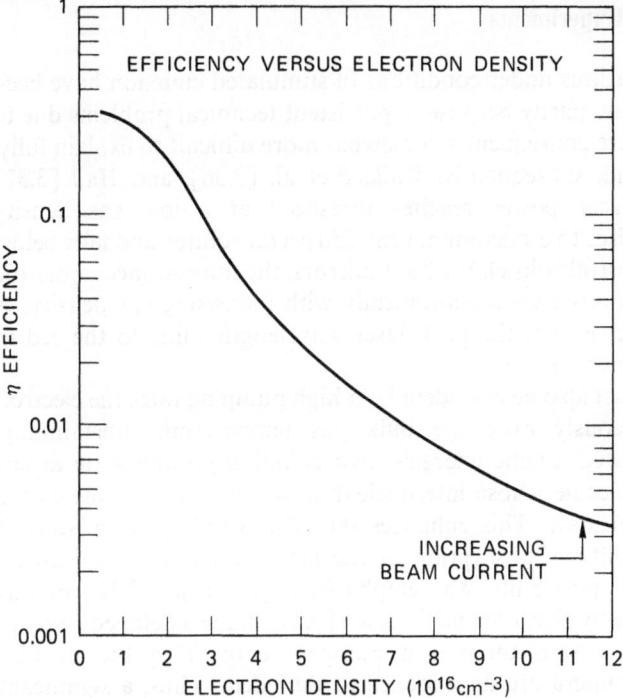

Fig. 3.2. Computed fluorescence efficiency of versus electron density in high pressure xenon gas. These results have been determined by *Werner* et al. [3.8]

rates. A dramatic demonstration of this effect was recently presented by *Powell* [3.34] who measured a distinct decrease in the Xe_2^* fluorescence signal above an electron beam current of $100 \, A \, cm^{-2}$. Laser pumping rates then are a compromise between the fast rates required to produce the high populations that are needed to offset the relatively low stimulated emission cross sections, and the low pumping rates that are needed to maintain high efficiencies.

The relationship between the fluorescence efficiency and the medium electron density, which can be related to the deposition rate, has been examined by *Werner* [3.9]. Figure 3.2 illustrates those results and clearly indicates the decrease in efficiency due to loss mechanisms involving collisions, among excited state species. The calculations suggest that at an excimer density above $\sim 10^{16} \, cm^{-3}$, the efficiency declines appreciably. This establishes maximum power P_m which can be sustained for high efficiency operation of the form

$$P_m \sim \frac{\hbar \omega N^*}{\tau} \sim 10^6 \, W \, cm^{-3} \tag{3.11}$$

where τ is the excimer radiative lifetime, N^* is the excimer density and $\hbar\omega$ is the photon energy. The limiting performance of these systems as fluorescent radiators is $\sim 1 \, MW \, cm^{-3}$.

3.1.2 Laser Kinetics Experiments

Experimental observations under conditions of stimulated emission have been more difficult to obtain, partly because of persistent technical problems due to mirror damage, and are consequently somewhat more difficult to explain fully. *Hoff* et al. [3.35] and, subsequently, *Wallace* et al. [3.36], and *Hull* [3.37] observed that the laser power reaches threshold at a low gas density ~ 110 psi (absolute), rises to a maximum near 250 psi (absolute), and falls below threshold above 380 psi (absolute). Without mirrors, the fluorescence under the same pumping conditions rises monotonically with increasing gas density. A second observed effect is that the peak laser wavelength shifts to the red in xenon [3.35] with increasing density.

Thermal effects must also be considered. At high pumping rates the electron beam can instantaneously raise the bulk gas temperature substantially. Furthermore, at elevated kinetic energies, two colliding ground state atoms may form a quasi-molecule whose internuclear separation is the same as the highly excited excimer state. This enhances the effective absorption, since it allows for the possibility of absorption of the laser radiation back into an excimer state. Such a possibility was emphasized by *Phelps* [3.38] in this context, and subsequently *Gerardo* and *Johnson* [3.39] made a detailed study of this possibility for the case of the *e*-beam pumped xenon. They showed that there was a strong temperature dependence to the effect. Thus, a significant consideration for large lasers is the relationship of the pumping rate and the thermal capacity of the medium in order to avoid significant gas heating. This, of course, establishes a bound on the excitation energy, or equivalently on the product of the excitation power P and the excitation time τ_{ex}. We note that in the discussion given above the excimer annihilation issue establishes a maximum on the excitation rate P. The Sandia group also considered the addition of diluents such as helium or argon to increase the thermal capacity and thereby avoid the extreme temperature rise.

Additional influences have also been attributed to the use of diluents. *Hoff* et al. [3.35] noted improved performance of the xenon laser when argon was added as well as a blue shift in the laser output. While such a shift is consistent with collisional absorption, they pointed out that other effects such as improved energy deposition or altered kinetic sequences could also arise under these circumstances.

An additional possibility for the laser intensity decrease at high pressure has been suggested by *Werner* et al. [3.5]. They argue that the stimulated emission for the triplets is smaller than that for the singlets and, in fact, has a value less than the computed photoionization cross section. Thus, the triplets will be a net loss for the system, while the singlets provide the net gain. If singlet-triplet relaxation via atomic collisions is large, the high pressure conditions in the triplet population will be enhanced leading to a reduction of net gain. However, as noted previously, the relevant cross sections are rather small. *Gerardo* [3.40] made measurements of the ratio of gain to photoionization loss and observed that this ratio was not strongly pressure dependent. More detailed gain and absorption studies would be particularly useful to settle this complicated dynamical issue.

Historically, the development of excimer lasers, and in particular the noble gas lasers, followed the utilization of large pulsed electron beam facilities for their excitation (see Chap. 1). Since the initial laser demonstrations there has been a coupled refinement of both the laser technology and that of the electron beam. Since the historical development of the noble gas excimer laser is examined in Chap. 1, the discussion here will emphasize the more recent experimental efforts.

At Imperial College, London, *Bradley* and his colleagues have concentrated on the design of a laboratory laser suitable to further application in laser physics such as frequency tripling into the far uv. Using a Febetron 706 diode as a pump (500 keV, 1000 A cm^{-2}, 10 J), they have demonstrated a tunable molecular xenon laser suitable for small laboratory experiments that has a bandwidth of 0.13 nm [3.41].

At Northrop Laboratories, *Ault* et al. [3.42] demonstrated oscillation on xenon with an extracted output of 10 J, at 400 MW. The pump in this case was a 100 kA, 3 MeV, 60 ns electron beam pulse. One of the most significant technical problems that they revealed in studying this system was the destruction of the dielectric mirrors by the laser pulse. They computed a laser efficiency of 6% based on energy absorbed in the gas.

Hunter et al. [3.43] reported an output of 8 J and 400 MW, with an output efficiency of 2.5%. Their source was a 5×50 cm electron beam that delivered 700–800 J in the laser volume at 850 kV, and 50 kA. Since the results of this experiment have not been widely distributed, we should consider these data in some detail. To avoid optics damage, they developed a relatively large electron beam with a pumping rate limited to a level where the optical damage threshold of the mirrors was not reached during the laser pulse. The output coupler was the surface of a MgF_2 flat window (reflectivity 0.065). The back mirror was a pure aluminum surface on a quartz substrate that was *not* coated (reflectivity ~ 0.70). Purified xenon gas was used at a fixed pressure of 100 psi (absolute).

Light emissions were observed out the side (fluorescence) and along the axis of directed emission with photodiodes. The output energy was also measured calorimetrically. Laser action was noted by a narrowing of the linewidth of the 172 nm radiation along the axis of the cell, by an increase in signal in the photodiodes along with optical axis (direction transverse to the *e*-beam), and by temporal narrowing of the transverse photodiode signal. The fluorescence photodiode signal was significantly reduced during the laser pulse (see Fig. 3.3).

In order to discuss the efficiencies that might be attainable from such a laser they analyzed the output in terms of a simple two-level model with the lower level always unpopulated (as anticipated for the repulsive state). In this analysis, they used the stimulated emission cross section of 10^{-18} cm^2 of *Gerardo* and *Johnson* [3.44] and an excimer–excimer annihilation rate of

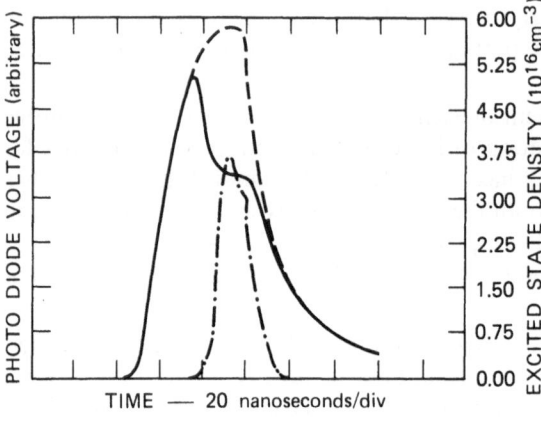

PHOTO DIODE VOLTAGE (arbitrary)

EXCITED STATE DENSITY (10^{16}cm^{-3})

6.00
5.25
4.50
3.75
3.00
2.25
1.50
0.75
0.00

TIME — 20 nanoseconds/div

Fig. 3.3. Performance of the molecular xenon laser from the work of *Hunter* et al. [3.43] at Maxwell Laboratories, Inc. The depression of the sidelight fluorescence curve during the laser pulse is particularly noticeable in this display

3.5×10^{-10} cm^3 s^{-1}. They further assumed that at the end of the optical pulse, steady-state conditions were reached with

$$n\sigma = \frac{1}{2l} \ln \left(\frac{1}{R_1 R_2} \right) \tag{3.12}$$

where l is the length of the active medium and the mirror reflectances are R_1 and R_2. This yielded a value of the steady state excimer density which could be used as a calibration for the fluorescence photodiode signal. The fluorescence curves could then be related to the total anticipated output which was 13 J, as compared with a measured value of 8 J. Perhaps most significant in this analysis was the conclusion that the dominant limits to the overall efficiency were due to optical constraints. They suggested that if high extraction efficiency can be realized, efficiencies as large as 20 % (relative to e-beam energy deposited) *might* be attainable.

3.1.3 Prognosis

Future work in this area will certainly concentrate on greater understanding of the kinetic processes. Perhaps one of the more neglected practical areas that most impedes the progress of the development of large systems is associated with the problems of optical materials (mirrors and windows) capable of surviving large energy loadings. Some efforts are being made in the areas of aerodynamic windows that sidestep these issues.

    An additional area of some promise is the use of e-beam pumped noble gases as fluorescent light sources (rather than lasers) for photolytically pumped lasers. This aspect will be continued further in Sects. 3.2.3 and 3.3.7.

3.2 The Rare Gas-Oxide Laser

At the present time one of the more vigorously pursued areas of laser application is that of laser fusion. Proposed reactor designs call for a laser with a minimum of several kilojoules of energy in pulses that are a nanosecond or less in length [3.45][1]. The optimum wavelength of such a laser is often the subject of spirited discussion, but a visible transition would certainly be valuable. To minimize risks due to self-focusing and optical damage, a gas medium is required; to maintain reasonable sizes in a gas laser, the energy should be stored at high excitation densities without loss. The stimulated emission cross section for the oscillating transition must be sufficiently low so that parasitic losses do not deplete the inverted population; however, for efficient extraction it must be large enough so that the transition can be

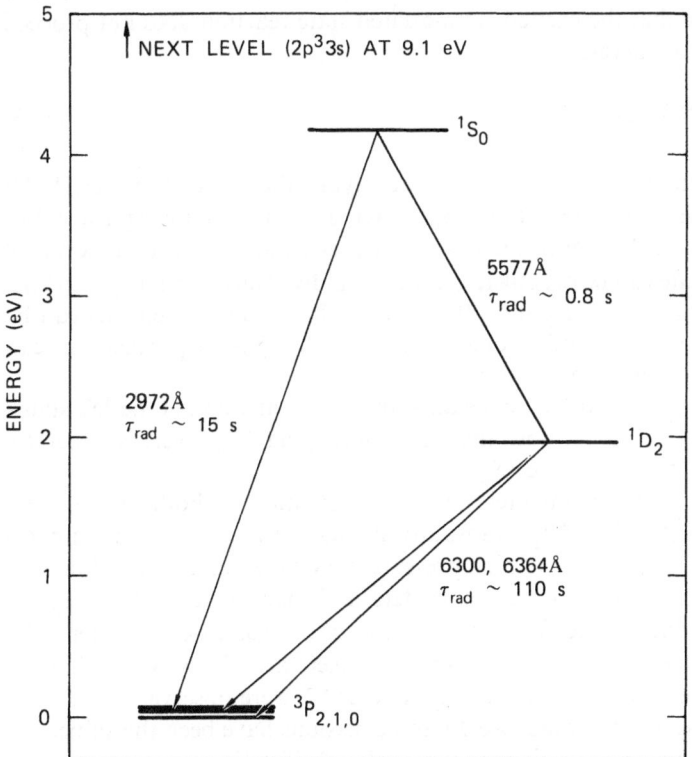

Fig. 3.4. Energy levels of oxygen atoms in the p^4 configuration. The radiative lifetime for each transition is labeled. The 5577 Å transition has been emphasized as having potential as a laser. Similar diagrams can be drawn for atomic sulphur and selenium

---

1 See also Chap. 1.

saturated as a flux level below the damage threshold for optical materials. These conditions imply that the stimulated cross section should be near $10^{-20} \, cm^2$.

3.2.1 The Atoms of Group VI-A

It was suggested some time ago [3.46, 47] that atoms with p^4 ground electron configurations could meet many of these constraints (e.g., oxygen, sulphur, selenium). Spin and angular momentum selection rules forbid electric dipole transitions among the ground state levels. Figure 3.4 shows the energy levels of ground state oxygen, and Table 3.2 specifies the energy levels of the other members of this group. A more detailed discussion of these levels and the corresponding rare gas oxide molecules is given in Chap. 2.

In all of the atoms of this configuration the highest energy level ($^1S_0$) is resistant to deactivation with many collision partners. Furthermore, there is reason to believe that the excited state-excited state reaction does not proceed rapidly at thermal energies,

$$O(^1S_0) + O(^1S_0) \rightarrow O^* + O(^2P). \tag{3.13}$$

In this expression O^* indicates excited states lying above the $O(^1S_0)$ level. All the molecular curves that lead to $O(^1S_0)$ at large separation are repulsive, but mix with $^1\Sigma_g^+$ curve. However, the lower level, $^1D_2$, and fine structure levels of the ^3P ground state can be deactivated rather rapidly. Thus, in many gas media, an inversion between the $^1S_0$ and $^1D_2$ levels of the group VI elements can be created and maintained. Under some conditions, the $^1S_0 \rightarrow ^3P_1$ transitions can have an inverted population as well.

Table 3.3 shows a list of rate constants for the deactivation of the $^1S_0$ states of oxygen at 300 K [3.61]; note that the lower lying $^1D_0$ levels is generally quenched much faster than the $^1S_0$.

Table 3.4 shows the parameters for laser oscillation on both the $^1S_0$–$^1D_2$ auroral line, and the $^1S_0 \rightarrow ^3P_1$ "transauroral" line, with an assumed temperature of 300 K. In the presence of high pressure of noble gases, the emission lines due to these atomic transitions are considerably broadened as well as slightly shifted. This is due to the formation of the noble gas-oxide excimer. The molecular potential curves for ArO, KrO and XeO are discussed in Chap. 2. The green band emissions between $2^1\Sigma^+$ and $1^1\Sigma^+$ were observed as early as 1946 [3.62]; these and the other XeO band emissions have been the subject of several spectroscopic studies since that time [3.63–65, 59].

Note that there is a relatively weak binding between $O(^1S)$ and the rare gases; this gives rise to a population equilibrium between the atoms and the rare gas oxide molecule

$$2Xe + O(^1S) \rightleftarrows XeO(^1S_0) + Xe. \tag{3.14}$$

Table 3.2. Energy levels of the P^4 lowest electron configuration of group VI-A elements

Next higher level	Wave number [cm^{-1}]			
	Oxygen	Sulfur	Selenium	Tellurium
$(p^3s\,{}^5S_0)$	73,768	52,624	48,182	44,253
${}^1S_0$	33,792	22,181	22,446	23,199
${}^1D_2$	15,858	9,239	9,576	10,599
${}^3P_0$	227	574	2,537	4,707
${}^3P_1$	159	397	1,989	4,751
${}^3P_2$	0	0	0	0

Table 3.3. Deactivation rates for atomic oxygen mestastables
Units of cm^3 molecule^{-1}s^{-1} at $\approx 300\,$K

	O (^1S)	Ref.	O (^1D)	Ref.
N_2	$<5 \times 10^{-17}$	[3.50]	3.0×10^{-11}	[3.49]
	4.2×10^{-18a}	[3.51]	Theory	[3.52]
N_2O	1.1×10^{-11}	[3.53]	1.4×10^{-10}	[3.49]
	$3.3 \times 10^{-11}e^{-840/RT}$	[3.54]		
NO	8×10^{-11}	[3.53]	2.1×10^{-10}	[3.48]
NO_2	5×10^{-10}	[3.53]	3.7×10^{-11}	[3.48]
		[3.54]		
O (^3P)	7.5×10^{-12}	[3.55]		
	Theory	[3.56]		
O_2	3.6×10^{-13}	[3.53]	4.1×10^{-11}	[3.49]
	$4 \times 10^{-12}e^{-1730/RT}$	[3.57]		
O_3	6×10^{-10}	[5.53]	2.4×10^{-10}	[3.49]
H_2	2.8×10^{-16}	[3.53]	2.1×10^{-11}	[3.49]
	3.2×10^{-18a}	[3.51]		
Xe	2.1×10^{-15a}	[3.51]	10^{-10}	[3.48]
	2.4×10^{-15}	[3.51]		
	3.7×10^{-15a}	[3.58]		
	3.8×10^{-15}	[3.58]		
	1.7×10^{-15}	[3.59]		
Kr	2.1×10^{-17a}	[3.51]		
	2.0×10^{-17a}	[3.58]		
Ar	4.7×10^{-18a}	[3.51]	10^{-12}	[3.48]
	5.2×10^{-18}	[3.60]		
	4.8×10^{-18a}	[3.58]		
	3.9×10^{-18a}	[3.59]		
Ne	3.5×10^{-19}	[3.58]		
	3.6×10^{-19}	[3.60]		
He	7×10^{-20a}	[3.51]	$<10^{-15}$	[3.48]
	1.1×10^{-19a}	[3.58]		

[a] A rate constant for collision induced emission of O(^1S)–O(^1D). The total deactivation rate is greater than this value.

See also other rates and measurements that are reviewed in [3.48, 49].

Table 3.4. Auroral and transauroral transitions of group VI-A elements[a]

Parameter	Oxygen	Sulfur	Selenium	Tellurium
$^1S_0 \rightarrow {}^1D_2$ auroral transition-electric quadrupole				
Wavelength [Å]	5577	7725	7768	7090
A coefficient [s^{-1}]	1.25	1.79	7.32	3.13
σ (10^{-20} cm^2)	8.7	45	48	60
E_s [J cm^{-2}]	4	0.6	0.5	0.4
Parameter	Oxygen	Sulfur	Selenium	Tellurium
$^1S_0 \rightarrow {}^3P_1$ transauroral transition-magnetic dipole				
Wavelength [Å]	2972	4589	4887	5419
A coefficient [s^{-1}]	0.063	0.34	7.7	37.0
σ (10^{-20} cm^2)	0.052	3	40	230
E_s [J cm^{-2}]	1300	14	1	0.16

[a] Auroral and transauroral transitions of Group VI A elements. Adapted from *J.R. Murray* and *C.K. Rhodes*: J. Appl. Phys. **47**, 5041 (1976). The peak cross section σ and saturation energy $E_s = hv/\sigma$ are calculated for 300 K Doppler-broadened lines using calculated radiative lifetimes and natural isotopic abundance.

Table 3.5. Emission parameters for rare-gas oxides[a] molecules

Rare-gas oxide	Collision-induced rate [cm^3/s-molecule]	A-Coefficient [s^{-1}]	Cross section [cm^2]	Saturation flux [J cm^{-2}]
ArO	4×10^{-18}	$95\,p$	$3 \times 10^{-21}\,p$	$120\,p^{-1}$
KrO	2×10^{-17}	$440\,p$	$1 \times 10^{-20}\,p$	$35\,p^{-1}$
XeO	3×10^{-15}	$7.3 \times 10^4\,p$	$2.9 \times 10^{-19}\,p$	$1.2\,p^{-1}$
O (^1S)	–	1.25	9×10^{-20}	4

[a] The parameters for the O (^1S) atomic transition assume a Doppler-broadened line at 300 K. The pressure p is in atmospheres.

Huestis et al. [3.66] have determined that the equilibrium constant is 5.5×10^{-22} cm^3. At room temperature and 1 atom of xenon, only about 1–2% of the $^1S_0$ atoms are bound in the molecular state. This fraction increases linearly with increasing pressure of noble gas. Because the molecular transition probability is substantially greater than that of the free atoms, the net rate of emission from the O($^1S_0$) will increase with pressure. A more detailed description of collisionally induced emission applicable to KrO or ArO where the molecular state lifetime may be limited to a collision time can be found in Chap. 2. For the molecular states, the appropriate emission parameters can be found in Table 3.5. The A coefficients will, of course, be temperature dependent. One feature illustrated here is that the presence of the high pressure buffer gas can be used to "tune" the stimulated emission cross section and thus aid laser extraction.

The green band emission of XeO between the $2^1\Sigma^+$ state and the $1^1\Sigma^+$ state is an example of a "non-traditional excimer". The lower state is bound rather than repulsive. The transitions favor terminating on high vibrational levels of the lower state. To maintain the inversion, vibrational relaxation of the lower state must be rapid, or the state must be subject to rapid quenching in subsequent collisions such as

$$XeO(^1D) + Xe \to 2Xe + O(^1D) \quad \text{or} \quad O(^3P). \tag{3.15}$$

3.2.2 The Mechanisms for Populating the Upper Levels

As was indicated previously, in a practical laser design one generally desires to exploit the high efficiency with which the pure noble gas excimers can be populated by electron beams. The fact that rare gas-oxide molecules are viable laser molecules immediately suggests that the oxygen (or other Group VI) donors be mixed in with the noble gas in the e-beam pumped cell and that the energy transfer be made by collisional transfer. This procedure has been successfully used to demonstrate laser action in the rare gas-oxides [3.67, 68]. A detailed kinetic model for this collisional mechanism with xenon and molecular oxygen has been described by the SRI (Stanford Research Institute) group and will be briefly outlined. The results for the SRI kinetic modeling program on XeO have been discussed at length in an unpublished report [3.66] as well as in condensed form elsewhere [3.54, 69]. An alternative procedure to use radiative energy transfer of the type

$$h\nu + N_2O \to O(^1S) + N_2(X) \tag{3.16}$$

where $h\nu$ is a photon emitted by the noble gas excimer, is also feasible and will be discussed in Sect. 3.2.3.

The excitation source used at SRI was a 3 ns, 600 kV, 1000 A cm^{-2} electron beam (Febetron 706). However, the conclusions appear to be applicable to other types of electron beams. The observations were made in the time following the electron pulse for which the electron temperature is relatively low (~ 0.5 eV). The kinetic modeling was based on the observations of the temporal behavior of the XeO* and Xe$_2^*$ decay as a function of Xe or O$_2$ pressure. Neither the Xe* nor the Xe$_2$ excimer produced by the electron beam has enough energy to produce O($^1S_0$) from molecular oxygen in a single step energy transfer. At most, one can produce O(^1D) + O(^3D) and, as noted before, O(^3D) will be rapidly quenched by collisions. The net effect of the first collision is to dissociate O$_2$ into two ground state atoms. In a subsequent energy transfer collision Xe$_2^*$ or Xe* produces the excited atom O(^1S). A block diagram of the whole kinetic model is shown in Fig. 3.5.

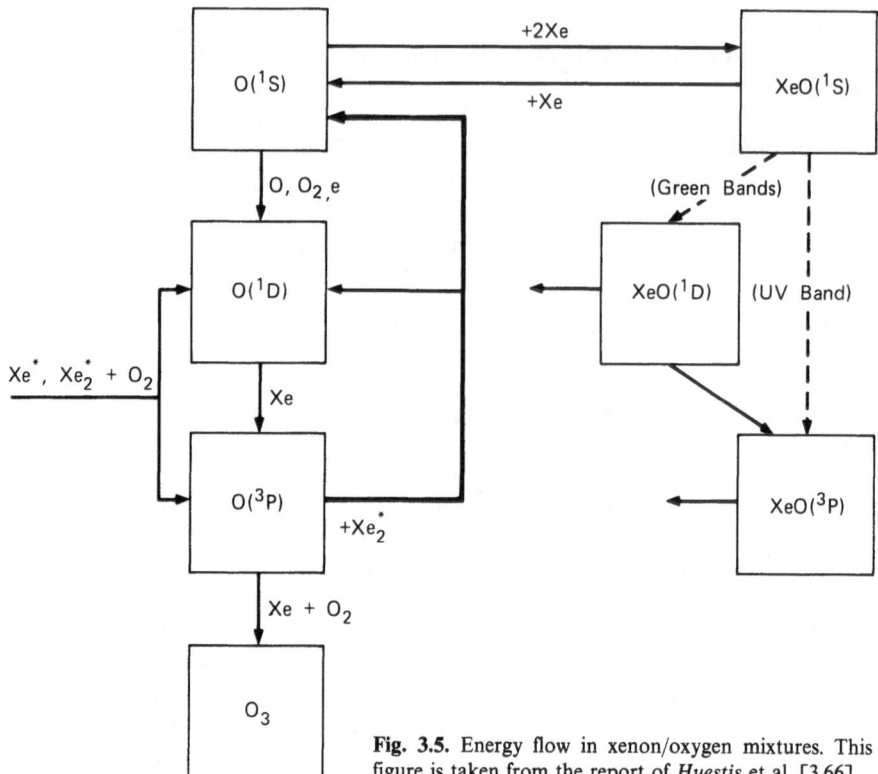

Fig. 3.5. Energy flow in xenon/oxygen mixtures. This figure is taken from the report of *Huestis* et al. [3.66]

The quenching of $O(^1S)$ and Xe_2^* by $O(^3P)$ was directly determined in this work. By measuring the total $O(^1S)$ production [by determining the fluorescence yield from $O(^1S)$] it was shown that the fraction of Xe_2^* quenched by the group state atoms produced $O(^1S)$. Thus, the optimum yield per Xe_2^* can be as large as 2/3. At the high dissociation levels where this optimum condition occurs, there will be significant quenching of $O(^1S)$ by collisions with $O(^3P)$. Table 3.6 summarizes the rate constants used in the model. Of course, the details of this model may vary somewhat if the additive molecule is altered (for example, to N_2O) or if the noble gas is changed, and continued efforts to include these effects are being pursued.

From this model, and under the experimental conditions used, it was determined that the energy efficiency for the production of $O(^1S)$ was 2%. The energy deposition determination used in that work is subject to uncertainty; however, despite its relatively low efficiency, this method of excimer production may be attractive for some applications due to its technical simplicity.

Using direct *e*-beam pumping of mixtures of xenon and oxygen and of krypton and oxygen, *Powell* et al. [3.67] successfully demonstrated laser action on the green bands of XeO and KrO. The excitation source in this case was a

Table 3.6. Reactions in xenon/oxygen mixtures

Reaction	Rate coefficient $[\text{cm}^3\,\text{s}^{-1}]$	Ref.
$Xe^* + O_2 \rightarrow$ Products	2.2×10^{-10}	[3.70]
$Xe^* + O(^3P) \rightarrow$ Products	not known	[3.66]
$Xe_2^* + O_2 \rightarrow 2O(^3P)$	$k_5 = 1.5 \times 10^{-10}$	[3.66]
$Xe_2^* + O(^3P) \rightarrow$ Products	$k_6 = 1.3 \pm 0.4 \times 10^{-9}$	[3.66]
$Xe_2^* + O(^3P) \rightarrow O(^1S)$	$k_{6a} = 6 \pm 2 \times 10^{-10}$	[3.66]
$X_2^* \rightarrow 2Xe + \gamma$	$k_7 = 6 \times 10^7\,\text{s}^{-1}$	[3.71]
$O(^1D) + Xe \rightarrow O(^3P) + Xe$	$1.0 \pm 0.4 \times 10^{-10}$	[3.48]
$Xe + O(^1S) \rightarrow \gamma$	$k_R = 3.7 \pm 5 \times 10^{-15}\,\text{s}^{-1}$	[3.66]
$XeO(2^1\Sigma^+) \rightarrow \gamma$	$k_2 = 7 \times 10^6\,\text{s}^{-1}$	[3.66]
$O(^1S) + O(^3P) \rightarrow$ Products	$k_3 = 4.5 \pm 0.5 \times 10^{-11}$	[3.66]
	8×10^{-12}	[3.72]
$O(^1S) + O_2 \rightarrow$ Products	2.1×10^{-13}	[3.73]
$O(^1S) + O_3 \rightarrow$ Products	$5.8 \pm 1 \times 10^{-10}$	[3.74]
$O + O + Xe \rightarrow O_2 + Xe$	$2.4 \pm 0.5 \times 10^{-33}$	[3.75]
$O + O_2 + Xe \rightarrow O_3 + Xe$	$5.1 \pm 1 \times 10^{-34}$	[3.75]
$O + O_3 \rightarrow 2O_2$	6.8×10^{-15}	[3.75]

2×10 cm, 1 MeV, 50 ns electron beam. High reflectivity mirrors were used on the cavity so the extraction was not optimized. Spectra from that work, both for fluorescence and for conditions of stimulated emission, are shown in Fig. 3.6. The XeO fluorescence spectrum shows structure due to the different low vibrational states. The XeO laser spectrum shows appreciable narrowing that allows resolution of substructure with laser oscillation at 557.81 nm, slightly to the red of the $O(S)$ atomic line. The laser line width was less than 0.03 nm. The LLL (Lawrence Livermore Laboratories) group has also demonstrated that ArO can oscillate [3.76, 77]; ArO has a fluorescence structure similar to KrO, the laser line for ArO was 0.4 nm. A most significant observation in that work was that the laser did not initiate until well after the excitation pulse terminates; a gain of 5% per pass was estimated. Laser oscillation was also noted with Xe/CO_2 mixtures and Xe/N_2O mixtures as well as Xe/O_2. Optimum pressure for the Kr/O_2 mixtures were at 25 atm Kr and 5 Torr O_2.

Woodworth and *Rice* [3.78] have recently reported a gain measurement of direct e-beam pumped Kr/O_2 mixtures. As suggested by the laser demonstration, the peak gain was noted to the red side of the atomic oxygen line. The technique they used was similar to that used by *Hughes* et al. [3.79] who made a gain profile measurement of photolytically pumped $O(^1S)$ in an argon buffer.

3.2.3 Photolytic Pumping

As was indicated earlier, an alternative production mechanism for the $^1S_0$ status of the group VI atoms is by photolytic pumping. It has been known for

v', v''   (0,4) (1,5) (0,5)   (0,6)   (0,7)   (0,8)

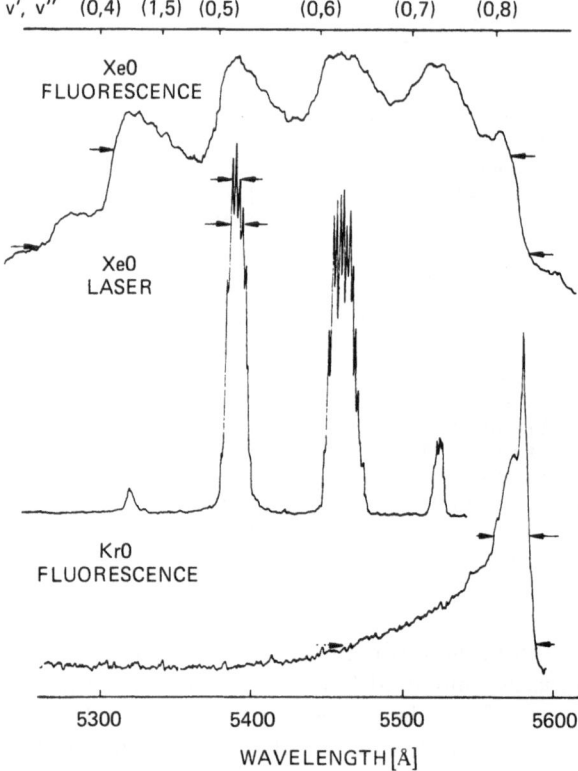

Fig. 3.6. Emission and laser spectra from XeO and KrO. These were taken from the work of *Powell* et al. [3.85] from Lawrence Livermore Laboratory

some time that there can be a very large production efficiency (for quantum yield) for $O(^1S)$ from the uv photolysis of simple oxygen bearing compounds [3.80]. These reactions

$$h\nu + RO \rightarrow O(^1S_0) + R \tag{3.17}$$

where R is a radical, are very specific and direct, and since there are a relatively small number of alternative output channels, there is hope of a substantially reduced number of excited state or radical quenchers. *Black* et al. [3.81] have studied the quantum yields of $O(^1S)$ emission as a function of wavelength for CO_2 and N_2O. More recently, molecular studies of the quantum yields for the production of $S(^1S)$ from OCS and $S(^1S_0)$ from OCSe [3.82]. Near 115 nm, the quantum yield for $O(^1S)$ from CO_2 is approximately 1.0, the quantum yield for N_2O is approximately 1 near 130 nm, the yield for $S(^1S)$ is similarly large near 145 nm and $Se(^1S_0)$ emission is produced with a quantum yield greater than 0.93 for pumping light near 172 nm. The significance here is the strong overlap with the emission bands of the pure noble gas excimers. The absorption cross section for the above mentioned molecules is relatively large in the overlap region, for example the absorption coefficient of OCS at 155 nm is approximately $8000 \, cm^{-1} \, atm^{-1}$ [3.82]. (Photolysis of several other Group IV containing molecules has been observed, but with results of less significant impact for laser development.)

This overlap has motivated a strong interest in the continued development of noble gas lasers as drivers for large photolytically pumped Group IV lasers. Particular emphasis has recently been centered around a $S(^1S)$ laser produced from carbonyl sulfide (OCS) by a molecular krypton laser. Photolytic pumping, of course, does not require a laser source; a molecular krypton fluorescence light source may offer some technical advantages.

As noted in the previous section, there remains considerable uncertainty about the long term behavior of optical materials subjected to large fluxes of uv laser radiation. An alternative is to use photolytic pumping via multiphoton processes. A particularly attractive example is to pump OCS with a CO_2 laser at 9.4 μm and with an ArF laser at 193 nm to produce the $S(^1S)$ state [3.83].

Aside from the problems associated with driver development, photolytic pumping offers some significant advantages over collisional transfer pumping. The process is, above all, simple. The production of the $O(^1S)$ or $S(^1S)$ proceeds directly through the additive, thus the choice of buffer gas is set by the need for collisional stimulated emission ("gain tuning") and not by considerations of electron stopping power. This could reduce the possible contribution of potential nonlinear optical processes that are associated with high pressure. Since there is no e-beam generated plasma, there will be very few free electrons to affect quenching rates and absorption properties of the medium. Furthermore, since the photodissociation process is very specific, there will be a limited number of species produced that may quench the $^1S_0$ atoms that are created. For example, for OCS in helium, the most significant quenchers will be OCS, CO(X), and $S(^1S_0)$ (i.e., excited state-excited state collisions). It is assumed that in a practical laser, the gas will be flowing so that $S(^3P_0)$ present after the laser radiation will be swept away between shots. Of these, the $S(^1S)$ quenching may be the most severe at the high excitation densities required for high power operation. Accurate measurements of this rate are obviously critical for laser design.

The most efficient utilization of the medium would clearly be to photolyze *every* molecule in the laser volume. A corollary to this is that one would like to exploit every (very expensive) uv photon from the uv source. To do this, the pump should be designed to have sufficient intensity in a specific direction to drive a "bleaching" or "burning wave".

In an optically thick medium of length L, we have the condition $\varrho\sigma_{abs} \gg L$; where ϱ is the absorber density and σ_{abs} is the absorption cross section. The absorption length $1/\varrho\sigma_{abs}$ characterizes the linear dimension of the transition region between the photolyzed and non-photolyzed volumes. When the pump source is on, this transition region will propagate through the medium with a velocity that is inversely proportional to the absorber density and proportional to the driver intensity. When the parent molecule quenching of the excited state atom is large, this velocity must be relatively large, i.e., the transition region must be passed in a time faster than the quenching time (denoted by $1/k_q\varrho$ where k_q is the quenching rate constant). Since the transition time is $(1\varrho\sigma_{abs})/v$ and $v = I_0/\hbar\omega\varrho$ we have the bound the driver intensity of

$$I_0 > \hbar\omega\varrho k_q/\sigma. \tag{3.18}$$

The quenching rate constants have already been discussed in Sect. 3.2.1.

Powell and *Murray* [3.84] have recently reported on a variety of experiments done at LLL to explicitly evaluate specific features of photolytically pumped systems. To study OCS photolysis, they developed a high intensity Kr_2 light source pumped by a large electron beam. They then observed the decay time of the $S(^1S)-S(^1D)$ emission while they varied the pumping light flux. Significantly, at low light flux they measured an exponential decay time for 3 Torr of OCS of 15 µs; as they increased the flux, the decay time *decreased* to 5 µs at 32 times the initial light flux. The reason for this decrease is not entirely understood. It suggests that at high pump levels, the storage time of the Group VI lasers may be limited by other considerations. Photoionization of the upper laser level by the pumping light is energetically possible, but the cross sections have not been measured. Nevertheless, the experiments did demonstrate the feasibility of using e-beam pumped noble gases as practical pumps for Group VI lasers.

3.2.4 The Absorptions

Following their initial rare gas-oxide laser demonstration experiments, *Powell* and *Murray* [3.85] initiated detailed gain measurements utilizing a cw dye laser transmitted through the laser cell. In particular, they noted that during the time that the electron beam was on, the probe laser beam was completely absorbed. This effect was independent of wavelength; after the e-beam pulse terminated, gain was measured.

Subsequently, *Zamir* et al. [3.86] investigated the nature of these absorbers and preliminary results of this study were recently reported. Probing e-beam pumped (Febetron 706) pure rare gases and rare gases with additives, three types of absorption features were observed (see Fig. 3.7). These were identified as a) atomic absorption with rapid decay (< 10 ns), b) molecular absorptions with longer decay times, and c) broad band background absorptions that decay rapidly. This broad band absorption is not yet positively identified; speculative suggestions indicate bound free transitions from highly excited molecules, or absorptions in molecular ions, may be responsible.

The rapid decay of these absorptions suggests that for high energy storage systems, with laser oscillation subsequent to the termination of the electron beam pump, there will be relatively little problem. However, for high gain quasi-cw systems operating during the pumping time, these absorptions have serious consequences. The effect of molecular additives on these absorptions must be thoroughly investigated in the future.

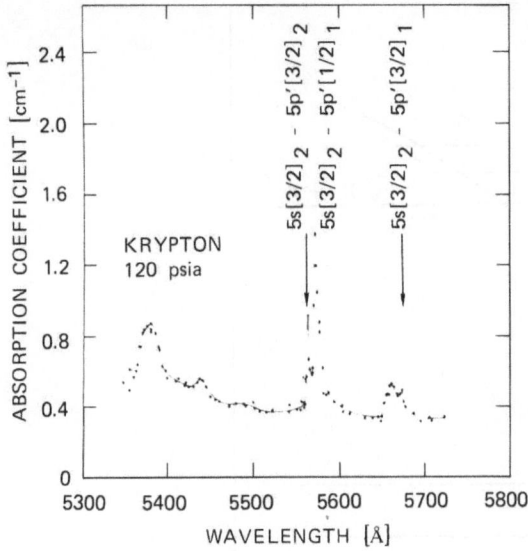

Fig. 3.7. Visible absorption spectra at 120 psi (absolute) in krypton gas. Taken from the work of *Zamir* et al. [3.86]. Note that in addition to the structured absorption features that are identified there is a broad continuum absorption that is significant during the electron beam pulse

3.3 The Homonuclear Halogens

Some excited state configurations of the homonuclear diatomic halogens behave as excimer systems. Although there is considerable similarity between the diatomic lasers and the known noble gas monohalide lasers (see Chap. 4), the two types are clearly distinct classes. As of this date, molecular iodine (I_2), bromine (Br_2) and fluorine (F_2) have exhibited stimulated emission on multiple lines at 342 nm [3.87], 293 nm [3.88] and 158 nm [3.89], respectively. There appears to be no significant reason why molecular chlorine should not behave similarly. The Sandia work on F_2^* emission at 158 nm is particularly encouraging, since that system may be a reasonable alternative to Kr_2^* as a photolytic source. These lasers can be excited either with an *e*-beam or with *e*-beam sustained discharge. The homonuclear halogens have also been shown to be efficient fluorescers; this makes them suitable candidates as relatively narrow band sources for the photolytic pumping of laser molecules that have until now been excited by broad band flashlamps.

3.3.1 Spectroscopic Features: Bound-to-Bound Transitions

As for the case of the rare gas-halide lasers, and for the mercury halides [3.90], the diatomic halogen lasers operate on transitions between an upper state that is ionic in character (i.e., I^+–I^- binding), and a lower state that is covalently bound. Thus they are similar to the rare gas-monohalide lasers, and the mercury halide systems. The rare gas-monohalides have repulsive lower states that facilitate maintenance of an inversion (with the exception of XeF).

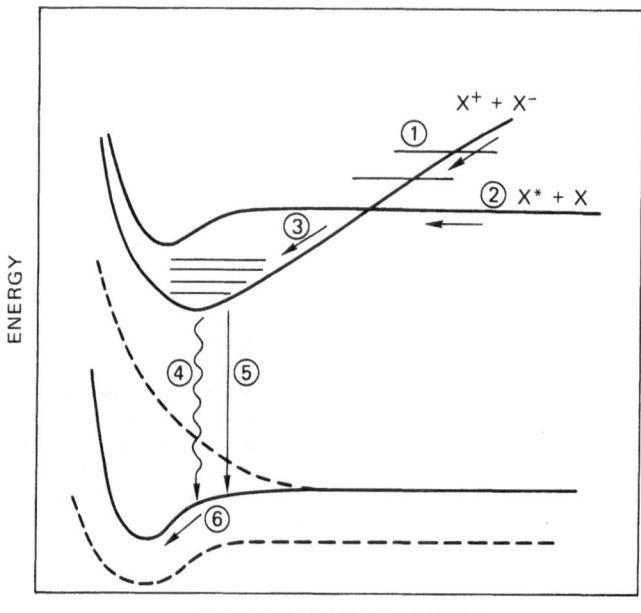

Fig. 3.8. Schematic diagram of the kinetics relevant to a diatomic halogen laser. Several different processes may populate the upper state. The lower state is not necessarily the lowest energy state of the molecule

However, the homonuclear halogens tend to terminate on high vibrational levels of the bound molecular states in the lower electronic level. The inversion, therefore, depends on rapid vibrational and electronic relaxation and possibly other processes such as predissociation to empty the lower levels.

An overview of the kinetic processes that are of most significance in halogen lasers can be obtained with the aid of Fig. 3.8. A laser pump, an electron beam, or a discharge, can rapidly and efficiently produce the precursor states in the bulk gas. We can consider two such precursors, one is the production of excited neutral argon (or other noble gas) either Ar* or Ar_2^*. Energy transfer reactions with the halogen containing additive yield excited halogens atoms X*. A possible alternative precursor production reaction would involve the creation of negative ions X^- by dissociative attachment with low energy electrons, and the simultaneous production of halogen containing positive ions (X^+ or RX^+ where R is an appropriate radical). Ion-ion neutralization reactions (process 1 in Fig. 3.8) then may yield excited states of the homonuclear halogen. The excited neutral atoms can form the molecular halogens by the well known harpooning reactions (process 2). In the high pressure bulk gas, collisions rapidly promote electronic and vibrational relaxations to the lowest lying levels of the ionic curves. For these processes to be efficient, there can be no repulsive curves that lead to ground state atoms that cross the bound upper states. The observation of high fluorescence yields for these emissions indicates that this is apparently true for the homonuclear halogens. Radiation (process 4) and quenching (process 5) are desirable and undesirable methods of depopulating the upper levels. Spectroscopic evidence indicates that radiative processes terminate on high vibrational levels of a lower curve that is *not* the ground state. Subsequent collisions with the bulk gas again promote rapid vibrational relaxation or even dissociation of the lower level, thus maintaining the inversion. The above scenario, while schematic, illustrates the essential steps that are discussed in detail in Sect. 3.2.3. There will, of course, be variations in this picture depending on the nature of the additve and the type of bulk gas used, particularly in the considerations of processes 2 and 3 of Fig. 3.8.

3.3.2 Halogen Spectra and Identification

Figure 3.9 shows the emission spectra of I_2, Br_2, and Cl_2 excited by energy transfer from electron beam excited argon done at SRI [3.91]. Except for relative intensity, these spectra appear to be essentially independent of the original source of the halogen; for example, I_2, CH_3I, HI and CF_3I have all been used as donors. With iodine, He, Ar, and Kr were used as the host gas with no important alterations in the spectrum. When small amounts of xenon are added to the Ar/I_2 mixtures, XeI* emission is observed [3.92] in addition to the I_2 343 nm band; ArCl is observed with a molecular chlorine additive as well [3.93]. These emission bands have been previously observed in emission when the halogen is excited by photoabsorption or by an electrical discharge in the presence of a foreign gas. As yet, no firm consensus has been reached on the specific spectral assignments and these bands. In the case of I_2, *Mulliken* [3.94] has summarized the various interpretations of the 343 nm bands as of 1971. He concluded that it was not possible to decide between *Verma's* [3.95] assignments as $1432\,{}^3\Pi^+ \to B\,{}^3\Pi^+$ and *Tellinghuisen's* [3.96] assignment as $1432\,{}^3\Pi_{2g} \to {}^3\Pi_{2u}$ transition[2].

The similarity of the spectra of I_2, Br_2, and Cl_2 excited under high pressure conditions clearly suggests that the same transition is being observed in all three cases. *Briggs* and *Norrish* [3.97] have observed these bands in Cl_2 and Br_2 in a transient absorption experiment using a double-flash photolysis technique. They suggested that the intermediate state formed by the first light pulse is not created by direct photoabsorption from the ground state, but involves some intermediate collision processes, which would be consistent, for example, with the suggestion that the lower state is a ${}^3\Pi_{2g}$.

On the basis of consideration of the spin orbit effect, the SRI group concluded that the lowest of the manifold ${}^3\Pi_g$ ionic states is the ${}^3\Pi_{2g}$. Since, in general, it is expected that the lowest state will be the most populated state under collision dominated conditions, the 2g–2u transition is further favored.

---

2 The four number designation for the halogen molecules refers to the number of electrons in the σ, π, π^*, σ^* orbitals. The ground state is then 2, 4, 4, 0.

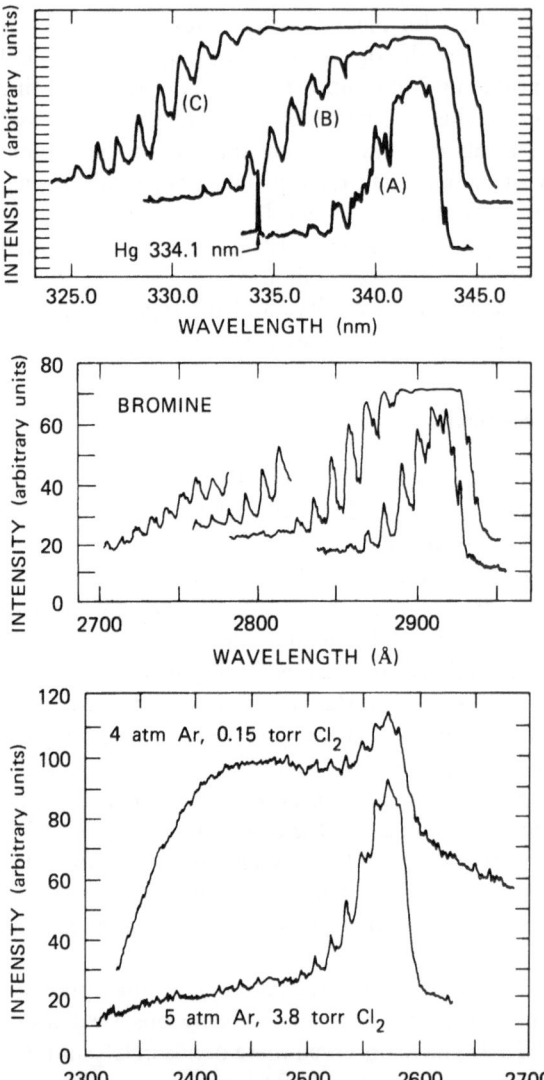

Fig. 3.9. Halogen emission spectra. These were taken from the SRI work [3.91]. Since these are densitometer traces, the vertical scale is nonlinear. The iodine curves are for one, three and five shots of the Febetron 706, the bromine curves are for one, five and ten shots. The vertical positions for each curve are adjusted to avoid confusion. Note the characteristically large number of red degraded peaks

However, this closely spaced manifold of states may also be fully mixed by collisions so that the radiation will be emitted by the transition with the largest transition probability.

In Cl_2, a rotational analysis of the spectra has been possible [3.97], and the lower state appears to be conclusively determined as the $B^3\Pi_{0u}^+$. The rotational transitions in Br_2 have not been resolved, and although the vibrational constants fit the $B^3\Pi_{0u}^+$ state, the lower state identification is not definite. Additional high resolution spectroscopy is clearly needed to resolve these questions.

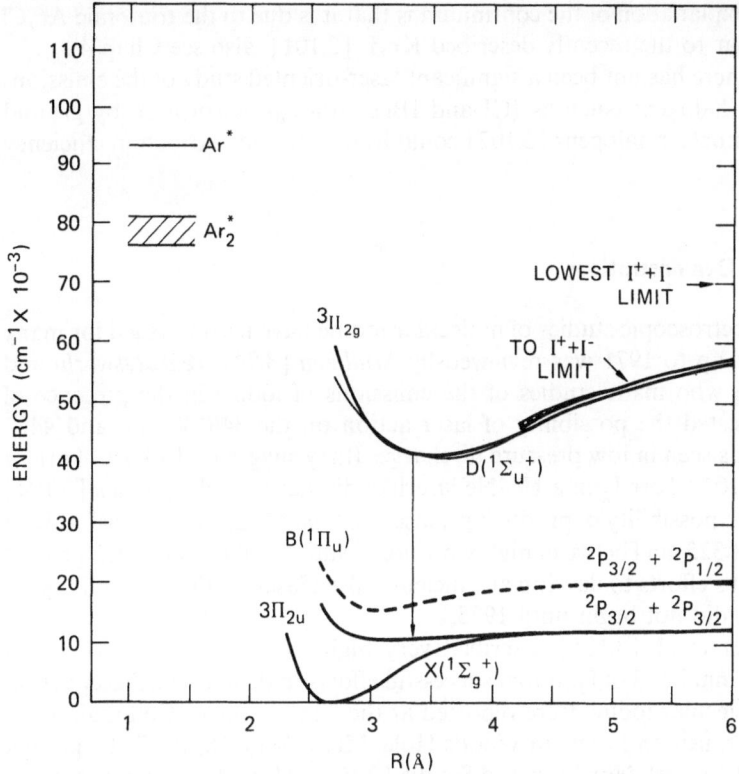

Fig. 3.10. Selected molecular iodine potentials (taken from the curves published by *Mulliken* [3.94]). The transitions believed responsible for the 342 nm laser bands are indicated

The lower state identification is important to the understanding of the ultimate performance of the halogen lasers, since it bears on the problem of lower state quenching or removal. The $B^3\Pi_{0_u}^+$ state, for example, is known to be readily predissociated by collisions and the cross sections have been measured by *Copelle* and *Broida* [3.98]. Predissociation would provide a rapid depopulation of the lower state and, therefore, avoid bottlenecking. The $^3\Pi_{2u}$, on the other hand, cannot be predissociated and may require high pressures to avoid bottlenecking. Recent studies of the Br_2^* laser at LLL have emphasized the importance of collisions in the lower laser level removal [3.99].

The Cl_2^* spectra observed in the SRI work differ markedly from those observed by *Chen* and *Payne* [3.100]. They observe a narrow unstructured peak near 260 nm at low Ar pressures, which changes to a broad continuum as the Ar density is increased to 780 Torr. They suggested that the lower states must therefore be repulsive. In the SRI work, [3.91] the observations made at 4–5 atm of Ar show the broad continuum and a narrower structured band at low Cl_2 density (0.15 Torr). At high Cl_2 density (3.8 Torr), the broad continuum is almost completely absent and the structured band is prominent (see Fig. 3.9).

A possible explanation of the continuum is that it is due to the triatomic Ar_2Cl that is similar to the recently described Kr_2F [3.101] (also see Chap. 4).

As yet, there has not been a significant laser-oriented study of the emissions from mixed halogens (such as ICl and IBr), although absorption by ground state heteronuclear halogens [3.102] could lead to serious extraction efficiency problems.

3.3.3 Laser Demonstrations

Detailed spectroscopic studies of molecular iodine have been pursued for many years; work up to 1971 was reviewed by *Mulliken* [3.94]. *Venkateswarlu* and *Rao* [3.103] who made studies of the emissions of iodine in the presence of argon, suggested the possibility of laser action on the 346–302 nm and 442–400 nm bands seen in low pressure discharges (they suggested 10 Torr of argon and $10^{-1} - 10^{-2}$ Torr I_2 in a He–Ne laserlike discharge). *Tellinghuisen* [3.104] suggested the possibility of producing a laser on "the McLennan bands" of I_2 in the region of 325 nm [between high vibrational states of the $D(O_u^+)$ $(^1\Sigma)$ ground state]. Serious efforts to develop an efficient pulsed laser on the 343 nm band of I_2, however, did not begin until 1975.

McCusker et al. [3.105] described very high ($>50\%$) fluorescence yields from the 343 nm band of I_2. Initial successful efforts to demonstrate laser action on these molecular iodine were reported at the "2nd Summer Colloquium on Electronic Transition Lasers in Woods Hole, Mass., Sept. 18, 1975", by groups from AVCO/Everett, Northrop and Sandia [3.106]. More detailed descriptions of these attempts were subsequently published [3.107–109]. All of these experiments found it easiest to produce oscillation using CF_3I or HI as an additive rather than I_2 although the AVCO group reported weak oscillation with molecular iodine additive at room temperature (vapor pressure $\simeq 0.25$ Torr [3.110]).

Ewing and *Brau* [3.107] used an electron beam of 350 kV, 150 A cm^{-2} and 100 ns pulse duration, with an area of 1 cm by 15 cm. High reflectivity mirrors ($>99\%$) were used external to a cell that was closed by uv grade quartz windows perpendicular to the cavity axis. Peak laser powers of 1 kW were measured and oscillation typically began 40 ns after the e-beam disposition command. The laser action lasted 80–100 ns, much longer than the I_2^* lifetime. Thus, they concluded that lower level bottlenecking was not a problem and that quasi-cw operation was possible. The efficiency in this attempt was only $10^{-3}\%$. There was a distinct saturation in peak laser power with argon pressures above $\simeq 35$ psi (absolute). A spectrum from that paper showing the narrowed laser line is shown in Fig. 3.11.

Ault et al. [3.108] reported oscillation on I_2 from mixtures of Ar at pressures up to 10 atm with an additive of CF_3I at partial pressures of 250:1. While they noted that the fluorescence intensity peaked at CF_3I pressures of $\simeq 15$ Torr, the laser intensity peaked at 30 Torr. They also observed that the

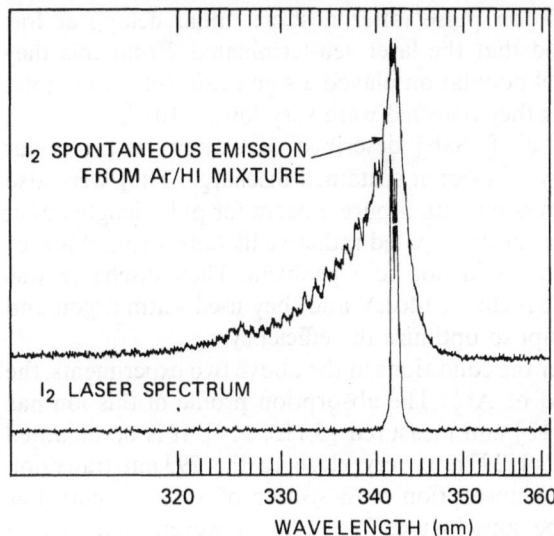

I₂ SPONTANEOUS EMISSION
FROM Ar/HI MIXTURE

I₂ LASER SPECTRUM

320   330   340   350   360
WAVELENGTH (nm)

Fig. 3.11. Molecular iodine fluorescence and laser spectrum. The laser operates on multiple lines near the peak of the fluorescence spectrum. This figure was taken from the work of *Ewing* and *Brau* [3.107]

same gas mixture could be used for many shots. They quoted an output of 3.6 MW for a 10 ns FWHM pulse length.

Hays et al. [3.109] reported peak outputs up to 1.0 J/pulse using HI as an additive. In this case, they used 1000 Torr A and 8 Torr HI; with this mixture they deduced a laser efficiency of 0.4%. This figure was based on comparing laser output with the calculated energy deposition in the gas. In general, laser action began 15 to 30 ns after the initiation of their electron beam pulse. The electron beam source produced a 140 kA, 1.2 MeV, 50 ns long pulse over an area 5 cm × 40 cm (700 A cm^{-2}). They stated that they were unable to see laser action when molecular iodine was used as an additive. They described a kinetic model based on ion-ion recombination reactions, for example

$$I^- + RI^+ \rightarrow I_2^* + R \tag{3.19}$$

to account for their observations.

The first gain measurements on the 293 nm band of molecular bromine were made by *Hunter* [3.111]. In the experiment he used an electron beam to pump a mixture of Br_2 and argon and inferred the existence of net positive gain from a measurement of the ratio of fluorescence from the long axis of his machine to the fluorescence along the short axis (side light).

Shortly after that work was announced, *Murray* et al. [3.88a] at LLL reported a direct demonstration of oscillation on molecular bromine. The excitation source for that demonstration was an electron beam of modest current density, $\simeq 5$ A cm^{-2}. Materials in the cell construction were chosen to avoid contamination or deleterious reactions prior to the test. While under optimum conditions, they were able to demonstrate laser action for the full length of their $\sim 1\,\mu s$ excitation pulse (after a short initial delay); at low bromine pressures they noted that the laser self-terminated. From this they concluded that the lower level population played a significant role in the total laser kinetics. The efficiencies they detected were very low, $< 10^{-3}$.

More recently *Ewing* et al. [3.88b] described oscillation on molecular bromine that was pumped by an *e*-beam sustained discharge. They were also able to achieve stimulated emission with a pure *e*-beam for pulse lengths of at least 200 ns, much longer than the anticipated radiative lifetime. From this they concluded that bottlenecking would not be a problem. Their discharge was sustained by an *e*-beam of 1.5 A cm^{-2}, 140 kV and they used 4 atm argon and 0.4 % Br$_2$. They did not attempt to optimize the efficiency.

We should note that under the conditions in the above two experiments, the dominant positive ion should be Ar$_2^+$. The absorption profile of this ion has recently been computed [3.112] and measured [3.113, 114]. It is dominanted by a broad feature peaking at 297 nm, very close to the 293 nm transition under discussion. With a peak absorption cross section of $6 \times 10^{-17}\,cm^2$, low extraction efficiencies will be anticipated. We note, however, that charge transfer to Br$_2^+$ should be rapid, and that the absorption features of Br$_2^+$ (Cl$_2^+$, I$_2^+$) in this region are not well known.

3.3.4 Homonuclear Halogen Kinetics

In the kinetic modeling efforts done at SRI [3.71], a short burst (3 ns) of high energy electrons from a Febetron 706 was used to excite the gas mixture. Under most conditions, the excitation pulse is short compared with subsequent chain of energy transfer reactions, and kinetic rates are obtained from the temporal behavior of the optical emissions. As in the case of the noble gas-oxides, the SRI work models the fluorescence, and the program does not include the effects of the laser radiation field. Hot electron reactions that may be present in a discharge are also neglected.

The basic block diagram of Ar/I$_2$ reaction chain that was described is shown in Fig. 3.12. The specific reactions involving the initial rare gas excitations have been discussed in Sect. 3.1.1. In the pressure range of interest, 0.5–10 atm, the excitation and the redistribution of the energy into the argon metastable states is fast compared with most of the other reactions. In effect then, the electron-beam deposition produces an almost instantaneous source of Ar*. This source may form argon excimers (Ar$_2^*$) through reaction (3.27) (see Table 3.7 for a list of all relevant reactions) or react with and transfer energy to an additive molecule as in reaction (3.20). Depending on the halogen concentration, the argon excimers may participate in subsequent energy transfer reactions, or may radiate 126 nm photons. It was observed that the I$_2^*$ emissions can persist over a time scale longer than the excitation pulse, or the decay time of the energy stored in the argon excited states. This, combined with the short I$_2^*$

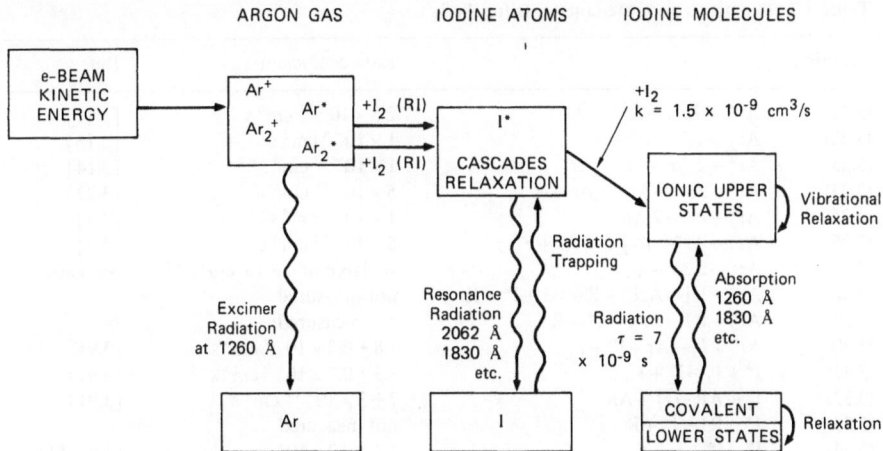

Fig. 3.12. Block diagram of argon/iodine kinetics. This kinetic sequence is appropriate to high excitation rate conditions where neutral kinetics dominate (taken from [3.91])

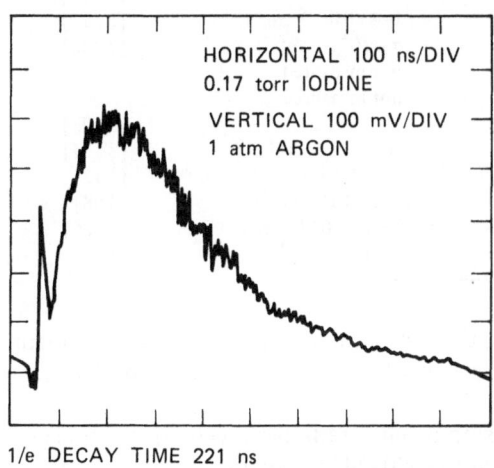

1/e DECAY TIME 221 ns

Fig. 3.13. Time evolution of the 324 nm fluorescence. The total gas pressure was atm, the iodine vapor pressure was 0.25 Torr. The exciting pulse is 3 ns long and the system resolution is approximately 3 ns. The short feature on the leading edge is due to molecular iodine to iodine excited by collisional transfer

radiative lifetime, (7 ns) required the existence of a reasonable long-lived energy transfer intermediate. This excludes the direct excitation transfer

$$Ar_2^* + I_2 \rightarrow I_2^* + 2Ar \tag{3.20}$$

as the dominant excitation pathway. Furthermore, direct excitation transfer is clearly inapplicable to mixed systems such as Ar/HI and Ar/CF$_3$I.

In homonuclear halogen systems, the excimer energy is transferred by collisions with the additive vapor to form excited neutral halogen atoms (I*). In a subsequent collision, the energy is transferred through a reactive collision to the molecular additive and excited molecular halogen (I$_2^*$) is formed. Since the

Table 3.7. Reactions and rate constants for Ar/I_2

Reaction		Rate coefficient	Ref.
(3.21)	$Ar^+ + 2\,Ar \rightarrow Ar_2^+ + Ar$	$2.5 \times 10^{-31}\ cm^6\,s^{-1}$	[3.12]
(3.22)	$Ar_2^+ + e^- \rightarrow Ar^* + Ar$	$1 \times 10^{-6}\ cm^3\,s^{-1}$	[3.13]
(3.23)	$Ar^* + 2\,Ar \rightarrow Ar_2^* + Ar$	$1 \times 10^{-32}\ cm^6\,s^{-1}$	[3.14]
(3.24)	$Ar^* + Ar^* \rightarrow Ar^+ + Ar + e^-$	$5 \times 10^{-10}\ cm^3\,s^{-1}$	[3.23]
(3.25)	$Ar_2^* + e^- \rightarrow 2\,Ar + e^-$	$1 \times 10^{-9}\ cm^3\,s^{-1}$	[3.3]
(3.26)	$Ar_2^* + Ar_2^* \rightarrow Ar_2^+ + 2\,Ar + e^-$	$5 \times 10^{-10}\ cm^3\,s^{-1}$	[3.3]
(3.27)	$Ar_2^* \rightarrow 2\,Ar + h\nu$	variable to $4 \times 10^7\ cm^3\,s^{-1}$	see Table 3.1
(3.28)	$Ar^* + RI \rightarrow ArI^* + R \rightarrow Ar + I^* + R$	not measured	
(3.29)	$Ar_2^* + RI \rightarrow 2\,Ar + I^* + R$	not measured	
(3.30)	$Ar^* + I_2 \rightarrow Ar + I^* + I$	$1.8 \pm 0.2 \times 10^{-9}\ cm^3\,s^{-1}$	[3.91]
(3.31)	$I^* + I_2 \rightarrow I_2^* + I$	$1.3 \pm 0.2 \times 10^{-9}\ cm^3\,s^{-1}$	[3.91]
(3.32)	$I_2^* + AR \rightarrow I_2^* + Ar$	$7 \pm 3 \times 10^{-11}\ cm^3\,s^{-1}$	[3.91]
(3.33)	$I^* + RI \rightarrow I_2^* + R$	not measured	
(3.34)	$I_2^* \rightarrow I_2^* + h\nu$	$1.4 \pm 0.2 \times 10^8\ cm\,s^{-1}$	[3.91, 115]
(3.35)	$I_2^* + I_2 \rightarrow$ quenching	$\lesssim 1 \times 10^{-9}\ cm^3\,s^{-1}$	[3.91]
(3.36)	$I_2^* + HI \rightarrow$ quenching	$4 \pm 2 \times 10^{-10}\ cm^3\,s^{-1}$	[3.91]
(3.37)	$I_2^* + CF_3I \rightarrow$ quenching	$\lesssim 1 \times 10^{-9}\ cm^3\,s^{-1}$	[3.91]
(3.38)	$I^+ + I^- + Ar \rightarrow I_2^* + Ar$	see Sect. 3.3.7	[3.116]
(3.39)	$e^- + I_2 \rightarrow I^- + I$	$1\text{--}4 \times 10^{-10}\ cm^3\,s^{-1}$	[3.117]
(3.40)	$h\nu(120\text{--}180\ nm) + RI \rightarrow I^* + R$	$\sigma \cong 10^{-16}\ cm^2$	[3.122]
(3.41)	$Ar^* + RI \rightarrow RI^+ + e^-$	not measured	
(3.42)	$Ar_2^* + RI \rightarrow RI^+ + e^-$	not measured	
(3.43)	$h\nu(126\ nm) + RI \rightarrow RI^+ + e^-$		[3.123]
(3.44)	$RI^+ + e^- \rightarrow I^* + R$	$10^{-6}\text{--}10^{-7}\ cm^3\,s^{-1}$	[3.124]
(3.45)	$I_2^* + h\nu(342.5\ nm) \rightarrow I_2 + 2h\nu$	$\sigma_{se} = 1.3 \pm 0.2 \times 10^{-5}\ cm^2$	[3.121]
(3.46)	$I_2 + Ar \rightarrow I_2 + Ar$	$\sim 2 \times 10^{-11}\ cm^3\,s^{-1}$	[3.98]
(3.47)	$I^- + h\nu(342\ nm) \rightarrow I + e^-$	$2 \pm 1 \times 10^{-17}\ cm^2$	[3.126]
(3.48)	$I^* + h\nu \rightarrow I^+ + e^-$	$\sigma \lesssim 10^{-17}\ cm^2$	[3.91]

excited states are ionic in character, they can radiate rapidly to lower covalent states. This simple kinetic model is consistent with all experimental observations of fluorescence.

Recently *Sauer* et al. [3.115], using a pulse radiolysis technique, observed the excitations of molecular iodine in the 320–345 nm region. They described a mechanism whereby the higher energy electrons produced excited states of I_2 which *did not* radiate in the 345 nm band. Subsequent vibrational relaxation brought the molecules to the upper levels of this band and radiation was observed. They measured a lifetime of this state of 6.7 ns and inferred an upper state vibrational deactivation rate constant of $\simeq 10^{-10}\ cm^3\,s^{-1}$.

A time history of the fluorescence from molecular iodine seen at SRI is shown in Fig. 3.13. The bulk of the emission is observed to be in the broad feature, peaking near 50–200 ns depending on the pressure, this peak is produced by collisional transfer (see Sect. 3.3.6). The small leading feature is apparently due to the direct excitation of I_2 followed by radiation as described above. The lifetime measured at SRI is 7 ± 1 ns, supporting *Sauer*'s results [3.115]. This small peak was not seen in mixtures with HI as an additive.

Black et al. [3.118] also at SRI, made a subsequent study of I_2^* produced by uv photolysis of CH_2I_2. Under these conditions (very low electron density) they determine a lifetime of 12 ns.

3.3.5 The Production of the Upper States

We can consider these mechanisms that may yield the halogen laser levels, neutral collisional transfer, ion-ion neutralization, and photolytic pumping. In collisional transfer, the electron beam produces excited noble gas atoms and molecules, which subsequently collide with the halogen, for example,

$$\left.\begin{matrix} Ar^* \\ Ar_2^* \end{matrix}\right\} + RI \rightarrow ArI^* + R \rightarrow Ar + I^* + R. \tag{3.49}$$

The suggestion of an ArI* intermediate that rapidly predissociates is consistent with observations of *Velasco* and *Setser* [3.119] who noted that only atom production in both I_2 and Br_2 interactions with metastable argon atoms. Such a process is also consistent with the observations of ArCl* that were seen when chlorine additives are used.

An alternative production mechanism that may occur is through dissociative recombination

$$Ar^+ + RI \rightarrow Ar + RI^+ \tag{3.50}$$

$$RI^+ + e^- \rightarrow I^* + R. \tag{3.51}$$

The importance of this process will be dependent on the ion density (and, therefore, the excitation). At high pressure Ar_2^* will be rapidly formed and may recombine with an electron *prior* to the charge transfer.

Whatever the mechanism, the excited iodine atoms may be produced in a variety of excited states: at SRI, iodine atom emissions were observed at 207 nm, 178.3 nm, 183 nm, and 167.3 nm. These emissions all had similar exponential decay rates. From the fluorescence data, it was determined that these I* production reactions have rate constants near $1.8 \pm 0.2 \times 10^{-9} \, cm^3 \, s^{-1}$. Since ground state $P_{\frac{1}{2}}^0 - P_{\frac{3}{2}}^0$ fine structure relaxation rates [3.120] and the recombination rate for the formation of molecular iodine from iodine atoms [3.121] is slow relative to the important kinetic time scales, radiation trapping effects will be important for the atomic iodine radiations. This extends the effective excited state lifetimes substantially and allows energy transfer re-actions from I* to proceed effectively. This trapping mechanism has been noted before in pure rare gas lasers with xenon atoms in pure xenon and in argon/xenon mixtures (see the discussion in Sect. 3.1).

At RI pressures above 1 Torr, the radiative transfer (reaction 3.40) from the argon excimers is not expected to be important, since few of the excited argon excimers or atoms will have time to radiate prior to quenching. At low I_2 or RI pressure (≤ 0.1 Torr), much of the stored energy could appear as 126 nm Ar_2^*

radiation. But the large photoabsorption cross section $(10^{-16}\,\mathrm{cm}^2)$ [3.122] of I_2 will allow absorption of the excimer radiation and a significant I^* population. Since the argon excimer can ionize I_2 (to I_2^+), this channel may proceed to I^* via dissociative recombination.

Experimentally, the dependence of the I_2^* fluorescence yield was measured as a function of argon pressure. If Ar^* were a more efficient collisional producer of the I_2^* than Ar_2^*, the fluorescence yield would be largest at low pressures. This was not observed to be the case and it was concluded that Ar_2^* is the dominant source.

There is no direct evidence with which to assess the relative importance of the ionization-recombination (reaction 3.38) pathway, as a source of I^* as opposed to dissociative energy transfer (reactions 3.28–30). Under our experimental conditions, this recombination would be sufficiently rapid to be indistinguishable from an energy transfer reaction. Assuming that transfer from Ar_2^* is the most effective production term, we note that such transfer is the least likely to yield I_2^+ by reaction (3.44). The argon excimer has enough energy to ionize I_2 (10 eV vs 9.3 eV), but the reaction mechanism should be dominated by the charge transfer channel $(Ar_2^+ + I_2^-)$. This channel should couple only weakly with ionization (it requires a two-electron interaction) and should selectively lead to the ion-pair product Ar^+I^-, which would predissociate to I^*. Energy transfer from Ar^* to I_2 is more likely to lead to ionization because of its higher excess energy, although neutral products could still be favored. There is evidence that photoabsorption of 126 nm radiation by I_2 may yield exclusively ionic products (mostly $I_2^+ + e$) [3.123].

The close similarity between the observations in I_2, HI, CF_3I, Br_2, HBr, and Cl_2 suggests a common energy flow scheme. If photo- or energy transfer ionization and subsequent recombination were dominant, greater differences caused by the different ionization potentials and presumably different recombination channels might be expected.

The kinetic scheme described above should be applicable to argon/bromine mixtures and argon/chlorine mixtures except that in the latter case, both Cl_2^* and $ArCl^*$ are observed; if triatomic Ar_2Cl is formed, its production mechanism will follow the production of $ArCl^*$. The dominant emission from argon/flourine mixtures is from ArF^* (see Chap. 4), which does not predissociate. The observed fluorescence is emitted by the homonuclear halogen.

The rate constants for the formation of chlorine and bromine and for the transfer to their respective diatomics may differ from those measured for I_2. Indeed, the fluorescence signal rise and decay times for HI and CF_3I are faster than in I_2, indicating even larger rate constants.

The reaction

$$I^* + RI \rightarrow I_2^* + R \tag{3.33}$$

will produce excited iodine in a variety of excited states. The relaxation by argon collisions suggests that the excited state population is rapidly collected in the few lowest levels. The high quantum yield indicates that quenching is unimportant at I_2 pressures ~ 1 Torr.

By observing the RI pressure dependence of the fluorescence yield, it was concluded that quenching by HI (reaction 3.36) proceeds with a rate coefficient of $4 \pm 2 \times 10^{-10}$ cm^3 s^{-1}. Such a determination is complicated in CF_3I by absorption of the I_2^* (343 nm) radiation and in I_2 by the presence of inpurities. It was concluded in both of these cases that the rate coefficient should be less than 10^{-9} cm^3 s^{-1} (reactions 3.35, 37).

Removing the lower laser level could conceivably prove difficult since it is a bound electronic state. This problem must be further assessed to assign the lower state and determine its relaxation rate. Nevertheless, relaxation in the high vibrational levels should be rapid (for example, 2×10^{-11} cm^3 s^{-1} for the $I_2(B)$ state [3.98]).

Early observations of efficient fluorescence and laser action from the ionic-covalent 343 nm transition in I_2 led many researchers to speculate that an ion–ion recombination mechanism was a likely path [3.106, 109]. For example,

$$I^+ + I^- + Ar \rightarrow I_2^* + Ar. \tag{3.38}$$

The evidence suggests that the ion recombination does not contribute in a major way to the I_2^* formation under the relatively high current excitation used in the SRI fluorescence experiments.

The production of I^- would be controlled by competition for free electrons between the two reactions

$$e^- + Ar_2^+ \rightarrow Ar^* + Ar \tag{3.22}$$

$$e^- + I_2 \rightarrow I^- + I \tag{3.39}$$

$$(e^- + RI \rightarrow I^- + R). \tag{3.52}$$

The rate constant for reaction (3.22) varies from less than 10^{-7} cm^3 s^{-1} at 1 eV to 10^{-6} cm^3 s^{-1} [3.124] at thermal energies. The rate constant for reaction (3.43) is 1.8×10^{-10} cm^3 s^{-1} [3.117] at thermal energies and increases to 2.8×10^{-10} cm^3 s^{-1} at 0.27 eV electron energy, and also increases with higher gas temperature. In Ar at pressures above 1 atm, a Febetron 706 produces electron densities $n_e \geq 10^{15}$ cm^{-3} [3.125], giving a production rate for reaction (3.22) of $\geq 10^8$ s^{-1}. The attachment rate at 1 Torr I_2 is $\leq (4 \times 10^{-10} \times 3 \times 10^{16} \simeq 10^7$ s$^{-1})$. The electrons then recombine with Ar_2^+ to yield Ar* far more rapidly than they attach to I_2 to yield I^-. Thus, the kinetic rates support the conclusion that I_2^* is produced mainly by way of neutral reactions. With lower excitation density conditions where the electron density is lower and electron-ion recombination slowed, electron attachment will become more important. The attachment rate for different additive compounds also varies.

The observed I_2^* decay observed and its pressure dependence provides evidence for an I* intermediate. If the ion-ion recombination processes were dominant, the decay rate would be proportional to $1/t$ and nonexponential while the observed decays were exponential. In addition, each of the decays showed a linear dependence on the $I_2(RI)$ pressure (and very little dependence on argon pressure) as would be expected from the neutral transfer reactions. Experimental determinations [3.116] of the three-body Thompson ion-ion recombination rates in oxygen and air indicate that the three-body rate coefficient is $1-2 \times 10^{-6} \, cm^3 \, s^{-1}$. At higher pressures, the recombination is further slowed by the reduction of ion mobility [3.127]; this effect is presumably present in noble gas-halogen mixtures. Several measurements of the $I_2^+ + I^-$ neutralization rate have been made and indicate a reaction rate near $1.4 \times 10^{-7} \, cm^3 \, s^{-1}$ at 300 °C [3.128–130]. This behavior should be further explored at lower pumping rates than provided by a Febetron 706.

The possibility of ion clustering

$$I_2^+ + I_2 + M \rightarrow I_4^- + M \tag{3.53}$$

and

$$I^- + I_2 + M \rightarrow I_3^- + M \tag{3.54}$$

further complicates the question. The positive ion clustering reaction rate for Br_2^+ in xenon is large, $3.9 \times 10^{-29} \, cm^6 \, s^{-1}$ [3.131]. The products from a reaction of the type

$$I_4^+ + I_3^- \rightarrow Products \tag{3.55}$$

are not well known.

3.3.6 Extraction Efficiency

In an analysis of a simple model of a rare gas/halide laser system, *Hunter* et al. [3.133] derived the following expression for this laser extraction efficiency in the presence of a nonsaturating absorber,

$$\eta_e = \frac{1}{(\varphi_s/\varphi + 1)} \cdot \frac{\varphi}{\varphi_s} \frac{1}{(g_0/g_a + 1)} \tag{3.56}$$

where φ is the light flux in the medium, φ_s is the saturation flux, g_0 is the small signal gain and g_a is the absorption coefficient. In this derivation the efficiency term is defined as a *local* extraction efficiency, and is essentially the ratio of the laser light energy to the pump energy necessary to produce it. Also the absorber in this formula must be a nonsaturating absorber, not depleted by the laser light. For example, the photoabsorption by the upper laser level in the Xe_2^* exciter must be described by an alternative procedure.

A given laser medium will have a specific ratio of g_0/g_a. In iodine for example there will be I^- photodetachment, absorption by ground state I_2 (possibly vibrationally excited), absorption by I* and others. For a given ratio, the equation then indicates that there is an optimum value of the flux ratio φ/φ_s for optimum extraction (note that this establishes the length scaling of the laser). Even when g_0/g_a is as large as 10, the optimum extraction efficiency, formed at $\varphi/\varphi_s \simeq 2$, will be less than 50 %.

In a subsequent experiment *Hunter* et al. [3.133] measured the absorption coefficients of Ar_2^+ and Kr_2^+ in e-beam pumped noble gas mixtures at a total pressure of 1500 Torr. The peak absorption in Ar_2^+ is near 293 nm and is 30 nm wide, while in Kr_2^+ it is near 325 nm.

This has particularly serious consequences for the diatomic halogen lasers. The argon absorption completely overlaps the strongest Br_2^* emission lines. The argon absorption is also reasonably strong near 340 nm, the peak of the I_2^* emission. I^- and Br^- also have strong photodetachment cross sections in this wavelength region. In any e-beam pumping scheme the Ar_2^+ density is expected to be reasonably large. If charge transfer to I_2^* occurs, one expects similar bound-free absorption processes to occur.

3.3.7 Photolytic Pumping

By analogy with the Group VI elements we might consider the possibility of photolytic pumping of the molecular halogen. However, depending on the halogen compound this may take a variety of pathways, only a few of which are thoroughly understood. *Okabe* et al. [3.134] noted that the photodissociations of $COCl_2$ below 140 nm led directly to the formation of Cl_2^* which could subsequently radiate in the 258 nm band. *Black* [3.118] repeated some of these experiments and noted that this state of Cl_2^* had a lifetime of 10 ± 1 ns, but had a quenching rate by argon of $10^{-14} cm^3 s^{-1}$. Helium, however, did not quench this state. The quantum yield for this mechanism has not been measured. *Black* also considered photodissociation of CH_2I_2 into I_2^* but found peak quantum yields that were disappointingly low (<1 %) even at wavelengths rather far into the ultraviolet (<120 nm).

The most promising route to a photolytically driven halogen laser appears to be via direct photo-excitation of the homonuclear halogen. i.e.

$$h\nu + I_2 \rightarrow I_2^* \tag{3.57}$$

or

$$h\nu + Cl_2 \rightarrow Cl_2^* . \tag{3.58}$$

More study in this area is necessary to establish the magnitude of the quantum yields.

3.3.8 The Fluorescence Yields

One of the most significant features of all these molecular halogen systems is their high fluorescence yields. This makes them suitable candidates as efficient light sources for applications such as laser pumping. For example, the 290 nm band of Br_2 overlaps the absorption band of CF_3I, the parent molecule in the efficient photolytically pumped I* laser at 1.3 µm. *Rice* [3.89] noted that the F_2^* 154 nm emission overlaps the OCS photoabsorption band. By making a comparison with the well calibrated second positive $(C^3\Pi \rightarrow B^3\Pi)$ emission from Ar/N_2 mixtures [3.135], the SRI group determined the fluorescence yield (defined as the ratio of the number of observed photons to the number of excited species produced in the lost gas by the electron beam) of Ar/I_2 to be as high as 67%, of Ar/Br_2 to be $\simeq 50\%$ and of Ar/Cl_2 (including the broad peak that is speculated to be Ar_2Cl) to be $100 \pm 25\%$. Energy efficiencies for these systems are then approximately 12, 11, and 26%, respectively [3.91]. This assumes Ar_2^* is produced with 50% efficiency. When other additives such as HI were used, the fluorescence yields were generally lower by 50% or more, from the case for the homonuclear diatomic additive.

These large values clearly indicate that there is a very efficient mechanism for populating the upper laser levels. These systems may indeed be useful fluorescence sources for applications such as photo-pumping other lasers.

3.3.9 Prognosis

The homonuclear halogens have been demonstrated to be efficient fluorescers, but rather inefficient lasers. The reasons for the relatively poor laser performance are not well known, but the presence of strong optical absorptions in the media studied so far is one possibility. Future work should include a study of the optical absorptions. Additional studies should be made of the kinetic processes; the two main routes to populating the upper laser level, ion-ion recombination or neutral transfer will vary in importance depending on the pumping rate and upon the additive chosen. Finally, considerably greater understanding of the mechanisms of lower level removal are necessary.

Acknowledgements. In preparing this manuscript I have particularly benefited from conversations with David Huestis of SRI International and with Robert Hunter of Maxwell Laboratories Inc. The editorial advice of Charles Rhodes and Paul Hoff is also gratefully acknowledged. The work on molecular iodine that was done at SRI and which was described here was the result of a team effort and I would like to thank my colleagues R. M. Hill, D. L. Huestis, D. C. Lorents, R. Gutcheck, and H. Nakano for their efforts and cooperation.

References

3.1   D.C.Lorents: Physica **82**C, 19 (1976)
3.2   D.C.Lorents, D.J.Eckstrom, D.L.Huestis: "Excimer Formation and Decay Processes in Rare Gases", Final Report MP 73-2. Contract N00014-72-C-0457, SRI Proj. 2018. Stanford Res. Inst., Menlo Park, Calif (1973)
3.3   D.C.Lorents, R.E.Olson: "Excimer Formation and Decay Processes in Rare Gases", Semiannual Tech. Rpt. No. 1, Contract N00014-72-C-0457, SRI Proj. 2018. Stanford Res. Inst., Menlo Park, Calif. (1972)
3.4   G.R.Fournier: Opt. Commun. **13**, 385 (1975)
3.5   C.W.Werner, E.V.George, P.W.Hoff, C.K.Rhodes: Appl. Phys. Lett. **25**, 235 (1974)
3.6   E.V.George, C.K.Rhodes: Appl. Phys. Letts. **23**, 139 (1973)
3.7   C.K.Rhodes: IEEE J. QE-**10**, 153 (1974)
3.8   M.H.R.Hutchinson: Inst. Phys. Conf. Ser. **29**, 109 (1976)
3.9   C.W.Werner, E.V.George, P.W.Hoff, C.K.Rhodes: IEEE J. QE-**13**, 769 (1977)
3.10 R.S.Mulliken: J. Chem. Phys. **52**, 5170 (1970)
3.11 D.C.Lorents, R.E.Olson, G.M.Conklin: Chem. Phys. Lett. **20**, 589 (1973)
3.12 K.T.Gillen, R.P.Saxon, D.C.Lorents, G.E.Ice, R.E.Olson: J. Chem. Phys. **64**, 1925 (1976)
3.13 W.F.Liu, D.C.Conway: J. Chem. Phys. **62**, 3070 (1975) (Argon)
      D.Smith, P.R.Cromey: J. Phys. B**1**, 638 (1968) (Argon)
      A.P.Vitols, H.J.Oskam: Phys. Rev. A**8**, 1860 (1973) (Xenon)
3.14 H.J.Oskam B.R.Mittlestadt: Phys. Rev. **132**, 1445 (1963)
      J.N.Bardsley, M.A.Biondi: Adv. At. Mol. Phys. **6**, 2 (1970)
      Y.J.Shiu, M.A.Biondi, D.P.Sipler: Phys. Rev. A**15**, 494 (1977)
      Y.J.Shiu, M.A.Biondi: Phys. Rev. A**16**, 1817 (1977)
3.15 E.Ellis, N.D.Twiddy: J. Phys. B**2**, 1366 (1969)
      A.H.Futch, F.A.Grant: Phys. Rev. **104**, 356 (1956)
      N.Thonnard, G.S.Hurst Phys. Rev. A**5**, 1110 (1972)
3.16 P.K.Leichner: Phys. Rev. A**8**, 815 (1973)
3.17 B.Schneider, J.S.Cohen: J. Chem. Phys. **61**, 3240 (1974)
3.18 B.Schneider, J.S.Cohen: J. Chem. Phys. **61**, 3240 (1974) [Ref. 3.3]
3.19 T.Oka, K.Rama Rao, J.Redgath, R.Firestone: J. Chem. Phys. **61**, 4740 (1974)
3.20 N.Thonnard, G.S.Hurst: Phys. Rev. A**5**, 1110 (1972)
3.21 R.Bocique, P.Mortier: J. Phys. D**3**, 1905 (1970)
3.21 J.W.Keto, R.E.Gleason, G.K.Walters: Phys. Rev. Lett. **33**, 1375 (1974)
3.23 P.K.Leichner, R.J.Ericson: Phys. Rev. A**9**, 251 (1974)
3.24 H.A.Koehler, L.S.Ferderber, D.L.Redhead, P.J.Ebert: Appl. Phys. Lett. **21**, 198 (1972)
3.25 J.B.Gerardo, A.W.Johnson: J. Chem. Phys. **59**, 1738 (1973)
3.26 S.C.Wallace, R.T.Hodgson, R.W.Dreyfus: Appl. Phys. Lett. **23**, 22 (1973)
3.27 D.J.Bradley, M.H.R.Hutchinson, H.Koetser: Opt. Commun. **7**, 187 (1973)
3.28 J.W.Keto, R.E.Gleason, Jr., T.D.Bonifield, G.K.Walters: Chem. Phys. Lett. **42**, 125 (1976)
3.29 J.W.Keto, R.E.Gleason, Jr., G.K.Walters: Phys. Rev. Lett. **23**, 1365 (1974); see also R.E.Gleason, T.D.Bonifeld, J.W.Keto, G.K.Walters: J. Chem. Phys. **66**, 1589 (1977)
3.30 F.B.Dunning, R.F.Stebbings: Phys. Rev. A**9**, 2378 (1974)
3.31 H.A.Hyman: Appl. Phys. Lett **31**, 14 (1977)
3.32 E.Zamir, C.W.Werner, J.P.Lapatovich, E.V.George: Appl. Phys. Lett. **27**, 56 (1975)
3.33 C.W.Turner,Jr., P.W.Hoff, J.Taska: Proc. Int. Topical Conf. on Electron Beam Res. Techn., Albuquerque, NM, 1975 (SAND-76-5122); see also Lawrence Livermore Lab. Rpt. CRL-50021-75-(1975) Chap. 9
3.34 H.Powell,Jr.: 8th Winter Colloquuim on High Power Visible Lasers, February 16–18, 1977, Park City, Utah
3.35 P.W.Hoff, V.C.Swingle, C.K.Rhodes: Opt. Commun. **8**, 128 (1973)
3.36 S.C.Wallace, R.T.Hodgson, R.W.Dreyfus: Appl. Phys. Lett **23**, 672 (1973)

3.37   D.R.Hull: "A Tunable Vacuum Ultraviolet Xenon Laser", Ph.D. Thesis, Department of Physics, Imperial College of Science and Technology, University of London (1975) (unpublished)

3.38   A.V.Phelps: "Tunable Gas Lasers Utilizing Ground State Dissociation", JILA Rpt. No. 110 (1972)

3.39   J.B.Gerardo, W.W.Johnson: Phys. Rev. A**10**, 1204 (1974)

3.40   J.B.Gerardo, W.W.Johnson: Appl. Phys. Lett. **26**, 582 (1974); IEEE J. QE-**9**, 748 (1973)

3.41   D.J.Bradley, D.R.Hull, M.H.R.Hutchinson, M.W.McGeoch: Opt. Commun. **14**, 1 (1975)

3.42   E.R.Ault, R.S.Bardford,Jr., M.L.Bhaumik: "uv Gas Laser Investigations Semiannual Technical Report", ARPA Contract N00014-72-C-0456 (1975)

3.43   R.O.Hunter, J.Shannon, W.Hughes: "Scaling and Efficient Operation of the 1720 Å Xenon Laser", Maxwell Lab. Rpt. MLR-413 (1975)

3.44   J.B.Gerardo, A.W.Johnson: J. Appl. Phys. **44**, 4120 (1973)

3.45   J.Nuckolls, J.Emmett, L.Wood: Phys. Today **26**, 46 (1973)

3.46   J.R.Murray, C.K.Rhodes: J. Appl. Phys. **47**, 5041 (1976)

3.47   S.D.Rockwood: Los Alamos Scientific Lab. Res. Rpt. LA-UR-73-1031 (1973)

3.48   R.J.Donovan, D.Husain: Chem. Rev. **70**, 489 (1970)

3.49   J.A.Davidson,   C.M.Sadowski,   H.I.Schiff,   G.E.Streit,   C.J.Howard,   D.A.Jennings, A.L.Schmeltekopf: J. Chem. Phys. **64**, 57 (1976)

3.50   R.Atkinson, K.H.Welge: J. Chem. Phys. **57**, 3689 (1972

3.51   G.Black, R.L.Sharpless, T.G.Slanger: J. Chem. Phys. **63**, 4546 (1975)

3.52   G.E.Zahr, R.K.Preston, W.H.Miller: J. Chem. Phys. **62**, 1127 (1975)

3.53   S.V.Filseth, F.Stuhl, K.H.Welge: J. Chem. Phys. **52**, 239 (1970)

3.54   T.G.Slanger, G.Black: J. Chem. Phys. **65**, 2025 (1976)

3.55   W.Felder, R.A.Young: J. Chem. Phys. **56**, 6028 (1972)

3.56   R.E.Olsen: Chem. Phys. Lett. **19**, 137 (1973)
       M.Krauss, D.Neumann: Chem. Phys. Lett. **36**, 372 (1975)

3.57   T.G.Slanger, B.J.Wood, G.Black: Chem. Phys. Lett. **17**, 401 (1972); see also [Ref. 3.12]

3.58   K.H.Welge, R.Atkinson: J. Chem. Phys. **64**, 531 (1976)

3.59   D.L.Cunningham, K.C.Clark: J. Chem. Phys. **61**, 1118 (1974)

3.60   A.Corney, O.M.Williams: Phys. Rev. B**5**, 686 (1972)

3.61   J.R.Murray, C.K.Rhodes: J. Appl. Phys. **47**, 5091 (1976)

3.62   C.Kenty, J.O.Aicker, E.B.Noel, A.Poritsky, V.Paolius: Phys. Rev. **69**, 36 (1946)

3.63   R.Herman, L.Herman: J. Phys. Radium **11**, 69 (1950)

3.64   C.D.Cooper, G.C.Cobb, E.L.Tolnas: J. Mol. Spectrosc. **7**, 223 (1961)

3.65   J.R.Wilt: Thesis, Department of Chemistry, University of California at Los Anageles, Calif. (1965) (unpublished)

3.66   D.L.Huestis, R.A.Gutcheck, R.M.Hill, M.V.McCusker, D.C.Lorents: "Studies of E-Beam Pumped Molecular Lasers", SRI Rpt. No. MP75-18, ARPA Contract N00014-72-0478, Stanford Res. Inst., Menlo Park, Calif. (1975)

3.67   H.T.Powell, J.R.Murray, C.K.Rhodes: Appl. Phys. Lett. **25**, 730 (1974); see also IEEE J. QE-**11**, 27D (1975)

3.68   A.W.Johnson, F.W.Bingham, J.R.Rise: IEEE J. QE-**11**(9), 560 (1975)

3.69   D.C.Lorents, D.L.Huestis: In *Laser Spectroscopy*, ed. by S.Haroche, Lecture Notes in Physics, Vol. 43 (Springer, Berlin, Heidelberg, New York 1975) p. 100

3.70   J.E.Velazco, D.W.Setser: Chem. Phys. Lett. **25**, 197 (1974)

3.71   D.C.Lorents, D.J.Eckstrom, D.L.Huestis: "Excimer Formation and Decay Processes in Rare Gases", Final Rpt. MP 73-2, ARPA Contract N00014-72-C-0457, Stanford Res. Inst., Menlo Park, Calif. (1973)

3.72   T.G.Slanger: Stanford Res. Inst., private communication

3.73   T.G.Slanger, B.J.Wood, G.Black: Chem. Phys. Lett. **17**, 401 (1972)

3.74   G.London, R.Gilpin, H.I.Schiff, K.H.Welge: J. Chem. Phys. **54**, 4512 (1971)

3.75   H.S.Johnson: "Gas Phase Reaction Kinetics of Neutral oxygen Species", Nat. Std. Ref. Data Series-NBS-20 (1968)

3.76   H.Powell: Paper presented at the 8th Winter Colloquium on High Power Visible Lasers, Park City, Utah (1977)

3.77   H.Powell: Lawrence Livermore Lab., private communication

3.78   J.R.Woodworth, J.K.Rice: Bull. Am. Phys. Soc. **22**, 189 (1977)

3.79   W.M.Hughes, N.T.Olson, R.O.Hunter: Appl. Phys. Lett. **28**, 81 (1976)

3.80   G.M.Lawrence: J. Chem. Phys. **57**, 5617 (1972)

3.81   G.Black, R.L.Sharpless, T.G.Slanger, D.C.Lorents: J. Chem. Phys. **62**, 4266 (1975)

3.82   G.Black, R.L.Sharpless, T.G.Slanger, D.C.Lorents: J. Chem. Phys. **62**, 4274 (1975)
       G.Black, R.L.Sharpless, T.G.Slanger: J. Chem. Phys. **64**, 3993 (1976)

3.83   C.K.Rhodes: Stanford Res. Inst., private communication

3.84   H.T.Powell, J.R.Murray: Paper presented at the 8th Winter Colloquium on High Power Visible Lasers, Park City, Utah (1977)

3.85   H.T.Powell, J.R.Murray: Lawrence Livermore Lab. Laser Program Annual Rpt. 1974 (March 1975, unpublished)

3.86   E.Zamir, D.L.Huestis, D.C.Lorents, H.H.Nakano: "Visible Absorptions by Electron-Beam Pumped Rare Gases", Summer Colloquium on Electronic Transition Lasers, Sept. 7–10, Snowmass, Colo. (1976)

3.87   J.J.Ewing, C.A.Brau: Appl. Phys. Lett. **27**, 557 (1975)
       R.S.Bradford,Jr., E.R.Ault, M.L.Bhaumick: Appl. Phys. Lett. **27**, 546 (1975)
       A.K.Hays, J.M.Hoffman, G.C.Tisone: Chem. Phys. Lett. **39**, 353 (1976)

3.88a J.R.Murray, J.C.Swingle, C.E.Turner,Jr.: Appl. Phys. Lett. **28**, 530 (1976)

3.88b J.J.Ewing, J.H.Jacob, J.A.Mangano, H.A.Brown: Appl. Phys. Lett. **28**, 656 (1976)

3.89   J.Rice: "Power Conditioning for Production of Group VI $A(^1S)$ Atoms", 7th Winter Colloquium, High Power Visible Lasers, February 16–18, Park City, Utah (1977)

3.90   H.H.Parks: Appl. Phys. Lett. **31**, 192 (1977)

3.91   M.V.McCusker, D.C.Lorents, D.L.Huestis, R.M.Hill, H.H.Nakano, J.A.Margevicius: "New Electronic Transition Laser Systems", Tech. Rpt. No. 4A, Contract DAAH 01-74-C-0624, SRI MP76-46, Stanford Res. Inst. (1976)

3.92   J.J.Ewing, C.A.Brau: Phys. Rev. A **12**, 129 (1975)

3.93   M.F.Golde, B.A.Thrush: Chem. Phys. Lett. **29**, 486 (1974)

3.94   R.S.Mulliken: J. Chem. Phys. **55**, 288 (1971)

3.95   R.D.Verma: Proc. Indian Acad. Sci. A **48**, 197 (1958); see also
       P.Venkatewarlu, R.D.Verma: Proc. Indian Acad. Sci. **46**, 251 (1977) for a discussion of Br_2^*

3.96   K.Wieland, J.B.Tellinghuisen, A.Nobs: J. Mol. Spectrosc. **41**, 69 (1972)

3.97   A.G.Briggs, R.G.W.Norrish: Proc. R. Soc. A **276**, 51 (1963)

3.98   G.A.Copelle, H.P.Broida: J. Chem. Phys. **58**, 4212 (1973)

3.99   J.R.Murray, J.C.Swingle, C.E.Turner: Appl. Phys. Lett. **28**, 530 (1976)

3.100 C.H.Chen, M.G.Payne: Appl. Phys. Lett. **28**, 219 (1976)

3.101 H.H.Nakano, R.M.Hill, D.C.Lorents, D.L.Huestis, M.V.McCusker: "New Electronic Transition Laser Systems", Final Tech. Rpt., Contract No. DAAH 01-74-C-0524, Stanford Res. Inst. Rpt., MP 76–99 (1976)

3.102 D.J.Seery, D.Britton: J. Phys. Chem. **68**, 2263 (1964)

3.103 P.Venkateswarlu, D.R.Rao: Proc. Indian Acad. Sci. **64**, 9 (1966)

3.104 J.Tellinghuisen: Chem. Phys. Lett. **29**, 359 (1974)

3.105 M.V.McCusker, R.M.Hill, D.L.Huestis, D.C.Lorents, R.A.Gutcheck, H.H.Nakano: Appl. Phys. Lett. **27**, 363 (1975)

3.106 The abstracts from this conference are published in *Electronic Transition Lasers*, ed. by J.I.Steinfeld (MIT Press, Cambridge, Mass. 1976)

3.107 J.J.Ewing, C.A.Brau: Appl. Phys. Lett. **27**, 557 (1975)

3.108 E.R.Ault, R.S.Bradford,Jr., N.L.Bhaunik: Appl. Phys. Lett. **27**, 413 (1975)

3.109 A.K.Hays, J.M.Hoffman, G.C.Tisone: Chem. Phys. Lett. **39**, 353 (1976)

3.110 A.N.Nesmeyanov: *Vapor Pressure of the Elements*, (Academic Press, New York 1963)

3.111 R.O.Hunter,Jr.: Rpt. at "1975 ARPA Contractor's Meeting", Stanford Res. Inst., Menlo Park, Calif. (1975) (unpublished)

3.112 W.Stephens, M.Gardner, A.Karo: Submitted to J. Chem. Phys. (1977)
3.113 R.O.Hunter, J.Oldnettel, C.Howton, M.V.McCusker: Submitted to Chem. Phys. Lett. (1977)
3.114 J.T.Moseley, R.P.Saxon, B.A.Huber, P.C.Cosby, R.Abouaf, M.Tadjeddine: J. Chem. Phys. **67**, 1969 (1977)
3.115 M.C.Sauer, W.A.Mulac, R.Cooper, F.Grieser: J. Chem. Phys. **64**, 4587 (1976)
3.116 H.Jatting, R.Koepp, J.Booz, H.G.Ebert: Z. Naturforsch. **204**, 213 (1965)
3.117 F.K.Truby: Phys. Rev. **188**, 508 (1969)
3.118 G.Black: "Research on High Energy Storage for Laser Amplifiers", Final Rpt., Contract AT(04-3)-115, USERDA, Stanford Res. Inst. Rpt., MP 76–107 (1976)
3.119 J.E.Velasco, D.W.Setser: J. Chem. Phys. **62**, 1990 (1975)
3.120 R.J.Donovan, D.Hussain: Trans. Faraday Soc. **62**, 11 (1966)
3.121 J.K.Ip, G.Burns: J. Chem. Phys. **56**, 3155 (1972)
3.122 J.A.Meyer, J.A.R.Samson: J. Chem. Phys. **52**, 716 (1970)
3.123 J.D.Morrison, H.Heirzeler, M.G.Ingraham, H.E.Stanton: J. Chem. Phys. **33**, 821 (1960)
3.124 J.N.Bardsley, M.A.Biondi: *Advances in Atomic and Molecular Physics* (Academic Press, New York 1970) Chap. 1
3.125 C.W.Werner, E.Zamir, E.V.George: Appl. Phys. Lett. **29**, 236 (1976)
3.126 H.Jitting, R.Koepp, J.Booz, H.G.Ebert: Z. Naturforsch. **20**A, 213 (1965)
3.127 D.R.Bates: J. Phys. B: Atom Molec. Phys. **8**, 2722 (1975)
3.128 T.H.Y.Yeung: Proc. Phys. Soc. (London) **71**, 341 (1958)
3.129 G.Greaves: J. Electron. Control **17**, 171 (1964)
3.130 T.S.Carlton, B.H.Mahan: J. Chem. Phys. **40**, 3683 (1964)
3.131 B.Huber, T.Miller: J. Appl. Phys. **48**, 1798 (1977)
3.132 R.O.Hunter, C.Howton, J.Oldenettel: AIAA, 15th Aerosp. Sci. Meeting, Paper no. 77-26, Los Angeles, Calif. (1977)
3.133 R.O.Hunter, J.Oldenettel, C.Howton, M.V.McCusker: "UV absorption of rare gas ions in E-beam pumped rared gases", submitted to Chem. Phys. Lett. (1977)
3.134 H.Okabe, A.H.Laufer, J.J.Ball: J. Chem. Phys. **55**, 373 (1971)
3.135 R.M.Hill, R.A.Gutcheck, D.L.Huestis, D.Mukherjee, D.C.Lorents: "Studies of E-beam pumped Molecular Lasers", Tech. Rpt. No. 3., Contract N00014-72-C-0478, Stanford Res. Inst. Rpt., MP 74-39 (1974)

4. Rare Gas Halogen Excimers

Ch. A. Brau

With 8 Figures

Although the rare gas halide lasers were discovered only about three years ago [4.1–6], they have received a great deal of attention since then and are presently the most highly developed excimer lasers. Moreover, the rare gas monohalides themselves represent an intriguing new class of molecules which (except for XeF [4.7]) have only recently been observed [4.8–12]. In spite of their recent discovery, a great deal is now understood about these molecules and lasers, and it is worth reviewing what is known.

As summarized in Table 4.1, ten rare gas monohalide species have now been observed, of which six have exhibited stimulated emission. Of these, ArF, KrF, XeCl and XeF have been observed to oscillate rather efficiently, which is the principal reason for the great interest in these molecules. At the same time, the fundamental structure and chemistry of these molecules is both interesting and important for understanding the behavior of the lasers. For these reasons, both the fundamental and applied aspects of the rare-gas halides are discussed in this chapter. As their headings suggest, Sects. 4.1 (Spectroscopy) and 4.2 (Reaction Kinetics) are devoted to the fundamental physical structure and chemical behavior of these molecules. Sections 4.3 and 4.4 discuss the principles of operation of the lasers themselves, some of the results which have been achieved, and the limitations which must be faced in scaling these lasers to higher power and efficiency. Finally, in Sect. 4.5 the properties of these lasers are summarized in such a way as to provide a framework within which to view future applications.

Table 4.1. Observed emission from rare gas monohalides

	He	Ne	Ar	Kr	Xe
F	(Ionic	Fluorescence	Lasing	Lasing	Lasing
Cl	state		Fluorescence	Lasing	Lasing
Br	not lowest		(predissociated?)	Fluorescence	Lasing
I	excited state)			(predissociated?)	Fluorescence

4.1 Spectroscopy

The general structure of the rare gas halide molecules is shown in Fig. 4.1. The ground state is covalently bonded and correlates to ground state ^1S rare gas and ^2P halogen atoms at infinite internuclear separation. Because of the P character of the halogen atom (net orbital angular momentum of one), the ground state manifold consists of two states. Of these, the $^2\Sigma$, has the lowest energy since in this configuration the singly occupied halogen orbital is directed toward the rare gas atom (see Chap. 2). Since it forms the ground state, this state is referred to as the X state. This state is generally nearly flat or at most weakly bound (255 cm^{-1} in XeCl [4.13]). The exception is XeF which has a binding energy of approximately 1065 cm^{-1} [4.14–16]. The strong bonding in this molecule has proved to be a puzzle for theorists [4.17]. The $^2\Pi$ state of the ground state manifold is always repulsive as indicated. Since it forms the first state above the ground state, it is referred to as the A state[1].

The upper laser level is an ionically bound, charge-transfer state correlating to the ^2P rare gas positive ion and ^1S halogen negative ion at infinite internuclear separation. Beginning at an energy equal to the ionization potential of the rare gas less the electron affinity of the halogen, the potential of this state follows a Coulomb curve which crosses, in a diabatic sense, the covalent curves correlating to excited states of the rare gas and halogen atoms. These crossings generally occur at large internuclear separations [4.10], where the overlap between orbitals centered on the rare gas and the halogen is small. Since any interaction between the covalent states, which correlate to excited atoms, and the ionic states will involve a matrix element between an orbital

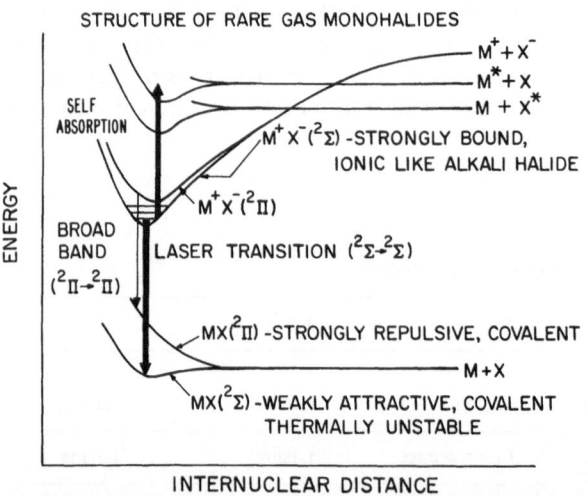

Fig. 4.1. Schematic potential energy diagram illustrating the electronic structure of the rare gas monohalides

---

[1] Actually, due to spin-orbit coupling, the A state is split into $^2\Pi_{3/2}$ and $^2\Pi_{1/2}$ branches. However, by convention these are lumped together and referred to as the A state

centered on the halogen and one centered on the rare gas atom, such interactions will be small at the crossing point. Thus, even curves of the same symmetry, which cannot cross in the adiabatic limit, will nearly intersect [4.18]. At close internuclear separation, the potential energy curve splits into $^2\Sigma$ and $^2\Pi$ states as indicated. The ionic species which form the upper laser level have ^1S (halogen ion) and ^2P (rare gas ion) character, like the ground state species, but with the rare gas and halogen atom roles reversed. Thus, the structure of the ionic upper level manifold is similar to that of the ground state manifold, with the $^2\Sigma$ state lying lowest. By convention this is referred to as the B state. In the interesting cases, the Coulomb curve crosses *all* the excited covalent curves, and the ionic state forms the lowest excited state (aside from the A state, which correlates to ground state atoms). However, if the rare gas ionization potential is too high, states correlating to the lowest excited states of the halogen may lie lower, as in HeI (see Table 4.1). In the cases of ArBr and KrI, the ionic state barely crosses the lowest covalent state correlating to an excited halogen atom [4.10, 12] and the ionic state may predissociate. This would account for the fact that no emission has been observed from these species.

Since the Coulomb potential is a rather slowly varying function in the neighborhood of the equilibrium internuclear separation, the excitation energy E^* of the B state may be approximated by the formula

$$E^* \approx IP(\text{rare gas}) - EA(\text{halogen atom}) - \frac{2\,IP(\text{hydrogen})}{r_e(\text{Bohr radii})}, \tag{4.1}$$

where IP stands for ionization potential and EA stands for electron affinity. The equilibrium internuclear separation r_e may be approximated by comparing the ionic rare gas halide molecule with the alkali halide molecule adjacent to it in the periodic table, viz. by comparing KrF* (asterisks are used to indicate electronic excitation) with RbF. In their lowest excited states, the rare gases bear considerable resemblance to the alkali metals in that they have an ionization potential of about 4 eV, and a single s-electron in a valence orbital about a nearly closed shell ionic core. Thus, it is possible to improve on (4.1) by equating the binding energy of the B state (relative to the separate positive and negative ions) to that of the closest alkali halide [4.10, 12]. In this way, we are able to predict the energy of the upper laser level to within a few percent. Even the vibrational frequencies of the corresponding molecules agree within about 20 %, the vibrational frequencies of the rare gas halides being slightly lower (see Table 4.2). This analogy is also useful for understanding the kinetics of the rare gas halides, since many of the reactions which they undergo are similar to reactions observed among the alkali halides.

The emission spectrum of the rare gas halides consists of several bands, as shown in Fig. 4.2 [4.19]. The strongest band is assigned to the $B(^2\Sigma) \to X(^2\Sigma)$ transition, and gives rise to the laser transitions observed to date (see Table 4.1). At high pressures this emission band is structured as shown in Fig. 4.2, with successive peaks due to higher vibrational levels of the upper electronic state

Table 4.2. Spectroscopic features of the charge transfer states of the rare gas halides. Only those species are included for which the charge transfer state is the lowest excited state (predictions of the alkali-halide model [a] are shown in parentheses)

Species	r [Å]	E [eV]	ω_e [cm^{-1}]	$\lambda(\Sigma \to \Sigma)$ [nm]	$\sigma\tau(\Sigma \to \Sigma)$ [Å^2-ns]	$\tau(\Sigma \to \Sigma)$ [ns]	$\lambda(Rg_2X)$ [nm]	$\tau(Rg_2X)$ [ns]
XeI	(3.3)	(4.85)	(119) 112[b]	(256) 253[b]	17[b]	12[l]		
XeBr	(3.1)	(4.25)	(150) 120[b]	(292) 282[b]	26[b]	12[l] 17.5[m]		
XeCl	(2.9)	(3.83)	(214) 194[c]	(324) 308[c]	50[a]	11[l]	450[u]	
XeF	(2.4)	(3.12)	(365) 309[d, e]	(397) 351[d]	64[t]	12[l] 18.8[n] 16.5[o] 13.5[p] 8[q] 15[r]		
KrI	(3.2)	(6.69)	(138)	(185)				
KrBr	(2.9)	(6.12)	(170) 166[f]	(203) 206[f]			325[u]	
KrCl	(2.8)	(5.65)	(233) 210[f]	(219) 222[g]				
KrF	(2.3)	(4.86)	(370) 310[b]	(256) 249[b]	17[b]	6.7[l] 9[r,z] 6.8[aa]	400[v, w]	313[s] 181[z]
ArBr	(2.8)	(7.71)	(220)	(161)				
ArCl	(2.7)	(7.20)	(280) 230[f]	(172) 175[h]			450[u]	
ArF	(2.2)	(6.41)	(430)	(193) 193[i]	12[k]	4.2[l]	290[v, x]	128[y] 185[s]
NeF	(1.9)	(11.56)	(536)	(107) 108[j]		2.6[l]		

[a] C.A. Brau, J.J. Ewing: J. Chem. Phys. **63**, 4640 (1975)

[b] J. Tellinghuisen et al.: J. Chem. Phys. **65**, 4473 (1976)

[c] J. Tellinghuisen et al.: J. Chem. Phys. **64**, 2484 (1976)

[d] J. Tellinghuisen et al.: J. Chem. Phys. **64**, 4796 (1976)

[e] A.L. Smith, P.C. Kobrinsky: J. Mol. Spectrosc. **69**, 1 (1978)

[f] M.F. Golde: J. Mol. Spectrosc. **58**, 261 (1975)

[g] J.R. Murray, H.T. Powell: Appl. Phys. Lett. **29**, 252 (1976)

[h] R.W. Waynant: Appl. Phys. Lett. **30**, 234 (1977)

[i] J.M. Hoffman et al.: Appl. Phys. Lett. **28**, 538 (1976)

[j] J.Rice: 7th Winter Colloquium on High Power Visible Lasers, Park City, Utah (1977)

[k] Yu.A. Kudryavtsev, N.P. Kuzmina: Appl. Phys. **13**, 107 (1977)

[l] P.J. Hay, T.H. Dunning,Jr.: Submitted to J. Chem. Phys.

[m] G.A. Hart, S.K. Searles: J. Appl. Phys. **47**, 2033 (1976)

[n] R.Burnham, N.W. Harris: J. Chem. Phys. **66**, 2742 (1977)

[o] J.G. Eden, S.K. Searles: Appl. Phys. Lett. **30**, 287 (1977)

[p] J.J. Ewing et al.: 2nd Winter Colloquium on Laser Induced Chemistry, Park City, Utah (1977)

[q] J. Goodman, L.E. Bruce: J. Chem. Phys. **65**, 3808 (1976)

[r] R. Burnham, S.K. Searles: J. Chem. Phys. (in press)

[s] C.-H. Chen, M.G. Payne, J.P. Judish: Private communication

[t] M. Rokni et al.: Appl. Phys. Lett. **30**, 458 (1977)

[u] D.C. Lorents et al.: Submitted to J. Chem. Phys.

[v] H.H. Nakano et al.: SRI Rpt. MP 76–99, Stanford Res. Inst., Menlo Park, Calif. (1976) (unpublished)

[w] J.A. Mangano et al.: Appl. Phys. Lett. **31**, 26 (1977)

[x] M. Rokni et al.: Appl. Phys. Lett. **31**, 79 (1977)

[y] W.R. Wadt, P.J. Hay: Appl. Phys. Lett. **30**, 573 (1977)

[z] G.P. Quigley, W.M. Hughes: Submitted Appl. Phys. Lett, **32**, 649 (1978)

[aa] J.G. Eden et al.: Appl. Phys. Lett. **32**, 733 (1978)

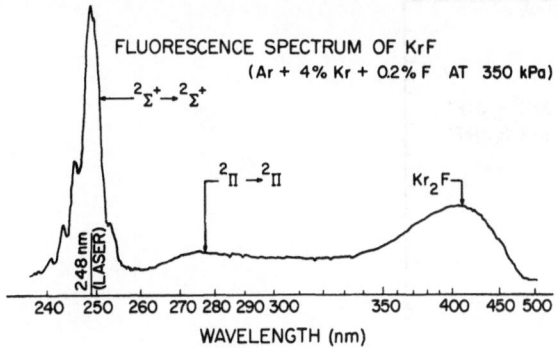

Fig. 4.2. Fluorescence spectrum of KrF showing the sharp $(^2\Sigma \to {}^2\Sigma)$ and broad $(^2\Pi \to {}^2\Pi)$ bands

[4.10, 12]. At low pressures the vibrational structure becomes smeared out and the band shifts toward the blue [4.9, 11]. This is due to emission from higher vibrational levels which are quenched (relaxed to lower levels) at high pressures. A broad continuum lies to the red of the sharp band, as shown in Fig. 4.2, and corresponds to transitions to the repulsive $A(^2\Pi)$ lower electronic state. All the emission bands from the charge-transfer states correspond to electron-jump transitions in which the radiating (valence) electron jumps from the negative halogen ion (upper state) to the rare gas atom (lower state). The $B(^2\Sigma) \to X(^2\Sigma)$ transition is strongest because the initial and final $p\sigma$ orbitals which the electron occupies have the largest overlap of any of the valence orbitals [4.10, 20, 21].

   In those rare gas halide molecules, such as XeI, in which the halogen has a high atomic number, the A state is split by spin-orbit effects [4.10]. In those molecules, such as KrF, in which the rare gas atom has a high atomic number, the excited state manifold is complicated by spin-orbit effects since the ^2P rare gas ion state is split by spin-orbit coupling. The effects of spin-orbit coupling in KrF are shown in Fig. 4.3 [4.20, 21] (see also the discussion in Sect. 2.2.2). The result of spin-orbit coupling is to split the $^2\Pi$ state into states of Ω (total axial angular momentum) = 1/2 and 3/2. The $\Omega = 1/2$ component lies higher, correlating with the higher lying $^2P_{1/2}$ rare ion state, and is referred to as the D state. The lower lying $\Omega = 3/2$ component is referred to as the C state. At close internuclear separations the D state becomes mixed with the B state (also $\Omega = 1/2$), allowing it to radiate to the ground X state [4.21]. Since this excited state lies above the lowest excited (ionic) state by approximately a constant amount, independent of internuclear separation, the $D \to X$ band is blue-shifted from the $B \to X$ band by roughly the spin-orbit splitting of the rare gas ion. Although it is not shown in Fig. 4.2, the emission from KrF in this band appears at 220 nm [4.22, 23]. The separation of the $D \to X$ band from the $B \to X$ band at 248 nm corresponds to 0.64 eV, which compares well with the $Kr^+\,{}^2P_{1/2}$–$^2P_{3/2}$ spin-orbit splitting of 0.67 eV [4.24]. Similar bands are observed from XeF at 260 nm [4.12, 23], KrCl at 199 nm [4.22, 23], XeBr at

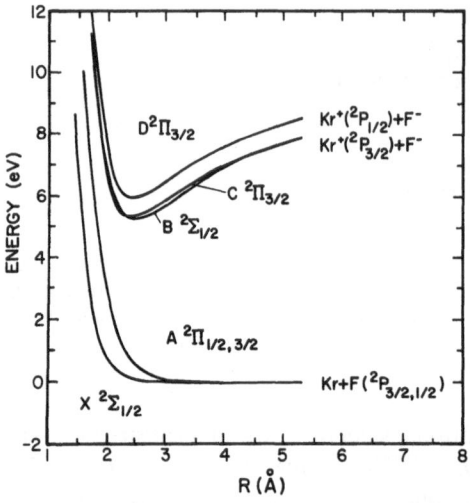

Fig. 4.3. Potential energy diagram showing the molecular structure of KrF

221 nm [4.23], XeI at 203 nm [4.23] and XeCl at 235 nm [4.23, 25]. In general, the emission in these bands is structured, like that in the $B \rightarrow X$ band. However, its intensity is much less, indicating that the upper level is quenched by collisions to lower states in the ionic manifold. In XeF, 1 % Xe in Ar is observed to quench this emission [4.12]. Thus, the $D \rightarrow X$ bands are probably not interesting as high efficiency laser candidates.

Transitions to the $^2\Pi$ state of the ground state manifold are substantially weaker [4.20, 21, 26]. These bands form relatively broad continua, due to the repulsive nature of the lower level. Until recently they have been somewhat difficult to understand due to confusion with other broad bands which are now attributed to emission from triatomic species such as Kr_2F. The emission from Kr_2F is seen in Fig. 4.2 as a broad continuum centered near 400 nm. The experimental evidence supporting the identification of this as Kr_2F emission is the observation that this band becomes relatively strong, compared with the $B \rightarrow X$ band, as the pressure is increased [4.27, 28]. The kinetics of formation are discussed in the next section. Similar bands are observed from Ar_2F at 290 nm [4.27, 29], Ar_2Cl at 450 nm, Kr_2Cl at 325 nm and Xe_2Cl at 450 nm [4.30]. Relatively little is known about the structure of the upper levels of these triatomic species. Recent ab initio molecular orbital calculations for Ar_2F indicate that the triangular configuration is most stable [4.31] (also see the discussion in Sect. 2.10). The predicted wavelength, about 270 nm, supports the identification of the 290 nm band as Ar_2F.

An apparently similar band at around 450 nm has been observed in mixtures of Ar, Xe and F_2 [4.12, 27]. However, this band becomes relatively stronger at *low* Xe pressure [4.19], suggesting that it may be due to a higher electronic state of XeF which is quenched at high Xe pressures. Quenching of the upper $\Omega = 1/2$ state by Xe (but apparently not Ar) has been mentioned above [4.12]. Alternatively, it may be a state of ArXeF [4.27].

The radiative lifetimes of the upper laser levels of the rare gas halides are difficult to measure since they are short (of the order of nanoseconds, see Table 4.2) and it is difficult to prepare the initial state in such a short time [4.32]. Exceptions are provided by XeF* and KrF*, which may be prepared by electron impact or photodissociation of XeF_2 [4.33–35] or KrF_2 [4.36]. The lifetimes measured in this way are found to be about 15 ns for XeF and 9 ns for KrF. These results are in excellent agreement with the theoretical value of 12 ns for XeF [4.26] and 6.7 ns for KrF [4.20, 21]. For the other species, particularly the shorter lived ones, it may be necessary to rely on the theoretical calculations [4.20, 21, 26]. Alternatively, it may be possible to measure the lifetime of the other rare gas halides by preparing them in frozen rare gas matrices [4.37, 38]. The transition moment is a strongly decreasing function of the internuclear separation, so that higher vibrational levels, which have larger average internuclear separations, have longer radiative lifetimes [4.17]. This explains why observations of the relaxation of low pressure spectra from high vibrational levels indicate a strong variation of the radiative lifetime [4.23].

A parameter of central significance for lasers is the stimulated emission cross section. In addition to the lifetime, this cross section depends on the shape and width of the emission band. Although rotationally resolved spectra of most of the rare gas halides have not yet been obtained, the individual rotation-vibration lines probably overlap sufficiently strongly for even the $B \to X$ bands to be regarded as continua with no structure beyond that indicated in Fig. 4.2. XeF, which has a relatively strongly bound ground state, is the exception [4.16]. Careful analysis of the $B \to X$ bands of XeBr, XeI, and KrF has provided values of the product of the gain cross section σ_v and the lifetime τ_v for these molecules [4.39]. For the others, the cross section-lifetime product may be estimated by assuming that near the peak the band shape is approximately Gaussian and using the formula [4.40]

$$\sigma_v \tau_v = \frac{1}{4\pi} \left(\frac{\ln 2}{\pi}\right)^{1/2} \frac{\lambda^4}{c \Delta \lambda}, \tag{4.2}$$

where λ is the wavelength, c is the speed of light and $\Delta \lambda$ is the width (FWHM) of the band. For example, the effective width (FWHM) of the $B \to X$ spectrum in KrF is about 2 nm [4.12]. From (4.2) we find that $\sigma_v \tau_v \approx 25$ ns $\cdot$ Å^2, in good agreement with the more accurate value of 17 ns $\cdot$ Å^2. Taken together with the measured lifetime of 9 ns [4.36], the latter value corresponds to a simulated emission cross section of 2×10^{-16} cm^2. The corresponding cross sections in the other species are expected to be of the same order of magnitude.

The gain on the broad bands is likely to be much smaller since these bands are both weaker, in general, and much broader. Theoretical calculations indicate that the dipole matrix elements for transitions to the A state are perhaps an order of magnitude smaller than those for the $B \to X$ transition [4.20, 21, 26]. This suggests that the gain cross sections are more than an order of magnitude smaller, less than 10^{-17} cm^2. These transitions will be inverted due to the repulsive nature of the lower laser level, and may have sufficient gain to oscillate in large systems. However, the gain may be offset by parasitic absorption processes (see below). In addition, the laser efficiency may be low if the upper laser level is too far above the $B(^2\Sigma)$ state, making it difficult to pump. In any event, it will presumably be necessary to suppress oscillation on the high gain $B{\to}X$ transitions.

Likewise, the gain on the broad bands of the triatomic species such as Kr_2F and Ar_2F is likely to be smaller than that on the relatively narrow $B{\to}X$ bands of the diatomic species. This is particularly true since the lifetimes of the triatomic species are predicted to be rather long (of the order of 100 ns or longer [4.31]), like the $^2\Pi{\to}^2\Pi$ transitions in the diatomic species. The long lifetimes are due to the symmetry of the triatomic species. Due to the low gain cross sections, parasitic absorption processes in the gas will make net gain difficult to achieve. In fact, recent attempts to observe gain on Kr_2F and the broad band of XeF (or ArXeF) near 450 nm showed net (weak) absorption due to some unidentified transient species rather than gain [4.41].

Several photoabsorption processes are believed to be at least potentially important in rare gas halide laser plasmas. These include photodissociation of the halogen precursor from which the rare gas halide is formed, such as

$$F_2 + hv \to 2F, \qquad (4.3)$$

photodetachment of negative ions formed in the plasma, such as

$$F^- + hv \to F + e^-, \qquad (4.4)$$

photoionization of excited rare gas atoms and molecules, such as

$$Ar^* + hv \to Ar^+ + e^-, \qquad (4.5)$$

$$Ar_2^* + hv \to Ar_2^+ + e^-, \qquad (4.6)$$

photodissociation of rare gas dimer ions, such as

$$Ar_2^+ + hv \to Ar + Ar^+, \qquad (4.7)$$

and even photoabsorption by the excited diatomic and triatomic rare gas halide molecules themselves. Photoabsorption by the halogens as in reaction (4.3) has been measured in the visible and near uv down to about 220 nm [4.42, 43]. The available results are summarized in Table 4.3, where we see that this problem is most important in XeCl and KrF lasers. The problem can be avoid in XeCl by using HCl or CH_2Cl_2 in place of Cl_2, and some improvement in performance has been achieved this way [4.13, 19]. In KrF lasers NF_3, SF_6 or N_2F_4 can be used in place of F_2 [4.44–46], but the performance is reduced by unfavorable kinetics, as described in the next section. Photodetachment, through reaction (4.4), has been investigated

Table 4.3. Photoabsorption processes in rare gas halide plasmas (Rg = rare gas, X = halogen)

RgX	λ [nm]	$\sigma(X_2)$ [cm²]	$\sigma(X^-)$ [cm²]	$\sigma(Rg^*)$ [cm²]	$\sigma(Rg^{**})$ [cm²]	$\sigma(Rg_2^*)$ [cm²]	$\sigma(Rg_2^+)$ [cm²]
XeI	253		4.5×10^{-17c}			8×10^{-18g}	
XeBr	282	$< 10^{-21a}$	2.3×10^{-17c}			1.1×10^{-17g}	1.7×10^{-18h}
XeCl	308	1.7×10^{-19a}	2.2×10^{-17d}			1.4×10^{-17g}	1.6×10^{-17h}
XeF	351	6.6×10^{-21b}	2×10^{-17e}			(?)	4.8×10^{-17h}
							3.8×10^{-17i}
KrI	185		8.7×10^{-17c}	1×10^{-21f}	2×10^{-18f}	3×10^{-18g}	
KrBr	206		4.1×10^{-17c}	1×10^{-20f}	3×10^{-18f}	4×10^{-18g}	
KrCl	222	$< 10^{-21a}$	1.9×10^{-17c}	4×10^{-20f}	3×10^{-18f}	5×10^{-18g}	
KrF	249	1.5×10^{-20b}	5.5×10^{-18e}	6×10^{-20f}	4×10^{-18f}	3×10^{-18g}	3.1×10^{-19h}
ArBr	161		5.5×10^{-17c}	3×10^{-20f}	2×10^{-18f}	2×10^{-18g}	
ArCl	175	$< 10^{-21a}$	3.2×10^{-17c}	2×10^{-20f}	2×10^{-18f}	3×10^{-18g}	
ArF	193		9.3×10^{-18c}	1×10^{-20f}	3×10^{-18f}	4×10^{-18g}	
						7×10^{-19j}	
NeF	108		1.3×10^{-17c}			6×10^{-19g}	

[a] D.J. Seery, D. Britton: J. Phys. Chem. **68**, 2263 (1964)
[b] R.K. Steunenberg, R.C. Vogel: J. Am. Chem. Soc. **78**, 901 (1956)
[c] E.J. Robinson, S. Geltman: Phys. Rev. **153**, 4 (1967)
[d] D.E. Rothe: Phys. Rev. **177**, 93 (1969)
[e] A. Mandl: Phys. Rev. A**3**, 251 (1971)
[f] H.A. Hyman: Appl. Phys. Lett. **31**, 14 (1977)
[g] D.C. Lorents et al.: SRI Rpt. MP 73–2, Stanford Res. Inst., Menlo Park, Calif., (1973) (unpublished)
[h] W. Wadt et al.: Appl. Phys. Lett. **31**, 672 (1977)
[i] J.A. Vanderhoff: J. Chem. Phys. **68**, 3311 (1978)
[j] T.N. Rescigno et al.: J. Chem. Phys. **68**, 5283 (1978)

theoretically [4.47] and experimentally [4.48, 49] with good agreement. The results for rare gas halide lasers are summarized in Table 4.3. Photoionization of excited rare gas atoms, (4.5), and excimers, (4.6), has received less attention. Theoretical calculations indicate that atoms excited to the first few excited states [those arising from the $...np^5(n+1)s$ configuration] have rather small photoionization cross sections, as summarized in Table 4.3 [4.50]. However, the results are sensitive to the model used since they depend upon cancellations in the integrals for the dipole matrix elements. On the other hand, these same calculations indicate that the next few excited states [those arising from the $...np^5(n+1)p$ configuration] have much larger cross sections, as summarized in Table 4.3. Continuing the analogy between the excited rare gases and the alkalis, we would expect large cross sections for electron-impact excitation from the $...np^5(n+1)s$ levels to the $np^5(n+1)p$ levels [4.51]. Thus, these levels are likely to be highly populated in laser plasmas. The cross sections for photo-ionization of the rare gas excimers have been estimated using the quantum defect method but ignoring the phase shift of the outgoing electron [4.52] (this approximation can introduce large errors near threshold, as in the alkali atoms). Using this formula, we obtain the results in Table 4.3. In the case of XeF (and possible XeCl), the laser photon does not have sufficient energy to photoionize Xe_2^* from its lowest levels. Higher lying excimer levels probably predissociate or are collisionally quenched at high pressure [4.53]. Theoretical calculations indicate that photodissociation of the rare gas dimer ions is confined to two bands, one in the visible and one the near ultraviolet centered around 250 nm (He_2^+) to 370 nm (Xe_2^*) [4.54, 55]. These results, summarized in Table 4.3, indicate that dimer ion photodissociation may be an important absorption mechanism in XeF, XeCl, and XeBr lasers [4.56].

Finally, there is the possibility of absorption by the excited rare gas halide molecules themselves. The transitions responsible for self-absorption from the upper laser level are indicated schematically in Fig. 4.1. Very little is known quantitatively about this absorption. Qualitatively, transitions from the ionic states to states correlating to excited halogen atoms should be very weak since they correspond to 2–electron transitions (one electron from the halogen ion transfers to the rare gas atom, while the second undergoes a transition to an excited halogen level), and can be ignored. However, photons from the $B \rightarrow X$ bands of all the molecules (except for XeF and, perhaps, XeCl) have enough energy to reach covalent states correlating to excited rare gas atoms, although the Franck-Condon factors may not be favorable. The rare gas trimers may also absorb at the laser wavelength. Since these molecules are essentially ionic complexes of the type ($Kr_2^+ \ldots F^-$), the absorption will be similar to that of the rare gas dimer ions and halogen negative ions [4.57].

4.2 Reaction Kinetics

The reaction kinetics of the rare gas halide lasers are rather complex, involving several ground state atomic and molecular species, several ionic species, and a large number of excited atomic and molecular species. As a result, the kinetics of these lasers are not yet fully understood. However, much interesting chemistry has been learned, enough, in fact, to adequately describe the most important aspects of laser performance. This chemistry is described in detail in the present section. We begin with a discussion of the rare-gas halide reactions (since they are the most important) and follow this with a discussion of the kinetics of the rare gases.

4.2.1 Formation and Quenching of Rare Gas Halides

The result of electrical excitation of predominantly rare gas mixtures is to produce both excited rare gas atoms and rare gas ions. Both these products lead very rapidly to excited rare gas halides. The excited species do this by direct reaction, for example,

$$Kr^* + F_2 \rightarrow KrF^* + F. \tag{4.8}$$

Reactions of this type occur with large cross sections, and frequently have near unit efficiency for producing excited products [4.23]. Extending the analogy developed in the previous section between excited rare gas atoms and alkali atoms, we may compare reactions of this type to the "harpooning" reactions observed between alkali atoms and halogen molecules [4.58]. In such reactions the readily ionized alkali atom can transfer its outermost electron to the halogen at a rather large distance (of the order of 5–10 Å), "harpooning" it and forming an ionic bond. Reactions of the type (4.8), like the alkali halide reactions, have rate coefficients generally of the order of 10^{-9} cm^3 s^{-1}. Thus, in a typical KrF laser mixture consisting of 2 atm of Ar containing 5 % Kr and 0.2 % F_2, the lifetime of an excited Kr^* atom against reaction (4.8) is of the order of 10 ns. Some measured cross sections for reactions of this type are summarized in Table 4.4.

As shown in Table 4.4, not all reactions of the type (4.8) have unit branching ratios for producing excited rare gas halide molecules. For example, in the reaction of Ar^* with Cl_2, excited states of Cl are also formed [4.23, 59, 60]. This may explain, at least in part, why it has not yet been possible to make $ArCl^*$ oscillate. Such effects are expected whenever the rare gas excitation energy is sufficient to form the excited products, that is, for the lighter rare gases and the heavier halogens.

In addition to the halogen molecules themselves, other halogen-bearing molecules may be used to produce excited rare gas halides, as shown in Table 4.4. This is useful when it is necessary to avoid corrosive chemicals, such as F_2, or photoabsorption of the laser radiation by the halogen species, as discussed above. Good branching ratios for forming excited rare gas halides can frequently be obtained from simple molecules such as NF_3 and OF_2 [4.23]. However, the branching ratios for forming excited rare gas halides from complex molecules are generally small, especially for more complex oxidizers and more energetic (lighter) excited rare gas species. For example, NF_3 has a near unity branching ratio for forming excited XeF^* in reactions with Xe^* but a smaller branching ratio for forming excited KrF^* in reactions with Kr^*.

The mechanism for forming excited rare gas halides from ions and electrons begins with dissociative attachment of the electrons to the halogen to form negative ions. This is followed by three-body recombination of the negative halogen ions with the positive rare gas ions, for example

$$e^- + F_2 \rightarrow F^- + F, \tag{4.9}$$

$$F^- + Kr^+ + Ar \rightarrow KrF^* + Ar. \tag{4.10}$$

Exothermic dissociative attachment reactions such as (4.9) are frequently very rapid, with rate coefficients as large as 10^{-7} cm^3 s^{-1} at room temperature, but

Table 4.4. Reactions forming excited rare gas halides [J.E. Velazco et al.: J. Chem. Phys. **65**, 3468 (1976)]

Reaction	Rate $[cm^3 s^{-1}]$	Branching ratio for forming excited rare gas halide
$Xe^* + F_2 \rightarrow XeF^* + F$	7.5×10^{-10}	1.0
$Xe^* + OF_2 \rightarrow XeF^* + (?)$	5.7×10^{-10}	0.92
$Xe^* + NF_3 \rightarrow XeF^* + (?)$	9×10^{-11}	1.06
$Xe^* + NOF \rightarrow XeF^* + (?)$	3.9×10^{-10}	0.77
$Xe^* + Cl_2 \rightarrow XeCl^* + Cl$	7.2×10^{-10}	1.0
$Xe^* + HCl \rightarrow XeCl^* + H$	5.6×10^{-10}	
$Xe^* + Br_2 \rightarrow XeBr^* + Br$	6.0×10^{-10}	1[a]
$Kr^* + F_2 \rightarrow KrF^* + F$	7.2×10^{-10}	1.0
	5.2×10^{-10e}	
$Kr^* + OF_2 \rightarrow KrF^* + (?)$	5.3×10^{-10}	1.06
$Kr^* + NF_3 \rightarrow KrF^* + (?)$	8.9×10^{-11}	0.57
	1.1×10^{-10f}	
$Kr^* + NOF \rightarrow KrF^* + (?)$	4.7×10^{-10}	< 0.09
$Kr^* + Cl_2 \rightarrow KrCl^* + Cl$	7.3×10^{-10}	0.90
$Ar^* + F_2 \rightarrow ArF^* + F$	7.5×10^{-10}	
	9×10^{-10}	$\leqq 0.6^e$
$Ar^* + Cl_2 \rightarrow ArCl^* + Cl$	7.1×10^{-10}	$\sim 0.5^b$
$Ar_2^* + F_2 \rightarrow ArF^* + Ar + F$	3×10^{-10c}	
	5.2×10^{-10}	$< 0.3^e$
$Kr_2^* + F_2 \rightarrow KrF^* + Kr + F$	2.1×10^{-10d}	
	4.0×10^{-10e}	

[a] J. Velasco, D.W. Setser: J. Chem. Phys. **62**, 1990 (1975)
[b] L.A. Gundel et al.: J. Chem. Phys. **64**, 4390 (1976)
[c] H.H. Nakano et al.: SRI Rpt. MP 76–99, Stanford Res. Inst. Menlo Park, Calif. (1976) (unpublished)
[d] G.P. Quigley, W.M. Hughes: Appl. Phys. Lett. **32**, 649 (1978)
[e] C.-H. Chen, M.G. Payne, J.P. Judish: Private communication
[f] M.J. Shaw, J.D.C. Jones: Appl. Phys. **14**, 393 (1977)

decreasing somewhat at elevated electron temperatures [4.61, 62]. Unfortunately, few data are available for the halogen compounds of interest for rare gas halide lasers. Even for F_2 the rate of dissociative attachment at elevated electron temperatures is not firmly established, but the best evidence indicates that the rate coefficient is of the order of $0.5-1 \times 10^{-9} cm^3 s^{-1}$ at an electron temperature of a few electron volts, which is typical of rare gas halide lasers [4.63, 64]. Thus, in a typical KrF laser mixture consisting of 2 atm of Ar with 5% Kr and 0.2% F_2, the lifetime of an electron against attachment to F_2 is of the order of 10 ns.

The three-body recombination of ions (4.10) is likewise extremely rapid. Although the rate of recombination of rare gas and halogen ions has not been directly measured, experimental data are available for a variety of other systems, including thermal dissociation of alkali halides [4.65], and the data are generally in good agreement with theoretical models describing these reactions [4.66]. The pressure dependence of these reactions is rather unusual. At low pressures (below about 1 atm), the pressure dependence is typical of three-body reactions with a rate coefficient typically of the order of $10^{-25}\,\mathrm{cm^6\,s^{-1}}$. The recombination rate in this regime was first described by *Thompson* [4.67]. The reason for the large value of this rate coefficient is that the ions attract one another with a long-range Coulomb interaction. For recombination to occur, the ion pair must be stabilized by momentum transfer to the third body at a depth in the Coulomb potential well of the order of $(3/2)kT$, where T is the gas temperature and k is Boltzmann's constant. At room temperature this corresponds to an internuclear separation of the order of $400\,\text{Å}$. The three-body recombination rate coefficient is roughly the product of the corresponding volume (of the order of $10^{-16}\,\mathrm{cm^3}$) with the momentum transfer collision rate coefficient between the ions and the third body (typically of the order of $10^{-9}\,\mathrm{cm^3\,s^{-1}}$). As derived by *Thompson* [4.67], the effective two-body recombination rate at low pressures is given by the formula

$$\alpha = 2\pi\, b^3 \bar{v}_\pm (\lambda_+^{-1} + \lambda_2^{-1}), \tag{4.11}$$

where $b = 2/3(q_e^2/kT)$ is the radius of the stable region of the potential well, q_e is the electron charge, $\bar{v}_\pm$ is the rms relative thermal velocity of the ions and λ_+ and λ_- are the mean free paths of the positive and negative ions in the background gas. Corrections to this formula are discussed by *Flannery* [4.68]. At high pressures (above about 1 atm) the rate coefficient begins to fall off with increasing pressure. The reason for this is that at high density a large number of third-body particles will be found in the volume within which the ions attract one another, and the ions must migrate toward one another through this background medium. This process was first described by *Langevin* [4.69], who gave the formula

$$\alpha = 4\pi\, q_e^2 (\mu_+ + \mu_-), \tag{4.12}$$

where μ_+ and μ_- are the positive and negative ion mobilities, respectively. At around one atmosphere the effective two-body recombination rate reaches a maximum, with a rate coefficient of the order of $10^{-6}\,\mathrm{cm^3\,s^{-1}}$. A more general theory, valid in this regime as well as at low and high pressures, is described by *Flannery* [4.68], who has calculated some useful rates [4.70].

The branching ratio for forming excited rare gas halide molecules from three-body ion–ion recombination is believed to be very high, based on the high efficiencies observed from rare gas halide lasers. This is also likely on theoretical grounds since the Coulomb potential energy curve along which the ions approach one another crosses most of the covalent curves at rather large internuclear separations. At such large separations it is difficult for an electron to jump from the negative ion to the positive ion, which suppresses transitions to the covalent curves. This same effect is seen in thermal dissociation of the

Table 4.5. Rare gas halide quenching reactions

Reaction	Lifetime × Rate [cm³]	Reference
$ArF^* + F_2 \rightarrow$ products	7.6×10^{-18}	M. Rokni et al.: Appl. Phys. Lett. **31**, 79 (1977)
	8.0×10^{-18}	C.-H. Chen, M.G. Payne, J.P. Judish: Private communication. Lifetime is 4.2 ns
$KrF^* + F_2 \rightarrow$ products	5×10^{-18}	M. Rokni et al.: Phys. Rev. A **16**, 2216 (1977)
	3.2×10^{-18}	J.G. Eden et al.: Appl. Phys. Lett. **32**, 733 (1978) Lifetime is 6.8 ns
	5.1×10^{-18}	G.P. Quigley, W.M. Hughes: Appl. Phys. Lett. **32**, 649 (1978). Lifetime is 9.0 ns
$XeF^* + F_2 \rightarrow$ products	5.3×10^{-18}	H.C. Brashears et al.: Chem. Phys. Lett. **48**, 84 (1977)
$XeF^* + NF_3 \rightarrow$ products	2.4×10^{-19}	H.C. Brashears et al.: Chem. Phys. Lett. **48**, 84 (1977)
$XeF^* + Xe \rightarrow$ products	4.6×10^{-19}	H.C. Brashears et al.: Chem. Phys. Lett. **48**, 84 (1977)
$XeF^* + Ar \rightarrow$ products	1.3×10^{-20}	M. Rokni et al.: Appl. Phys. Lett. **30**, 458 (1977)
	2.6×10^{-20}	H.C. Brashears et al.: Chem. Phys. Lett. **48**, 84 (1977)
$KrF^* + Ar \rightarrow$ products	1.2×10^{-20}	J.G. Eden et al.: Appl. Phys. Lett. **32**, 733 (1978). Lifetime is 6.8 ns
$ArF^* + Ar \rightarrow$ products	3.6×10^{-20}	M. Rokni et al.: Appl. Phys. Lett. **31**, 79 (1977)
$KrF^* + Kr \rightarrow$ products	$\leqq 1 \times 10^{-20}$	M. Rokni et al.: Phys. Rev. A **16**, 2216 (1977)
	6×10^{-20}	J.G. Eden et al.: Appl. Phys. Lett. **32**, 733 (1978). Lifetime is 6.8 ns
$Ar_2F^* + F_2 \rightarrow$ products	3.9×10^{-17}	C.-H. Chen, M.G. Payne, J.P. Judish: Private communication. Lifetime is 185 ns
$Kr_2F^* + F_2 \rightarrow$ products	7.8×10^{-17}	C.-H. Chen, M.G. Payne, J.P. Judish: Private communication. Lifetime is 313 ns

alkali halide molecules, which tend to dissociate to ionic, rather than neutral, products [4.65].

Just as the rare gas halides are formed by very rapid processes, they are quenched by very rapid processes as well. The most important quenching process, at least at low pressures, is direct quenching by the halogen-bearing species, typified by the reaction

$$KrF^* + F_2 \rightarrow Kr + 3F . \qquad (4.13)$$

This evidently proceeds with a rate coefficient of the order of $10^{-9}\,\mathrm{cm^3\,s^{-1}}$, that is, at every collision [4.28]. Thus, in a typical laser mixture (as above) the quenching lifetime is of the order of 10 ns. Like the radiative lifetimes, the quenching times are generally too short to measure directly. However, by observing the quenching of fluorescence as a function of quenchant density, the relative rates of quenching and fluorescence (specifically, the product of the quenching rate coefficient and the radiative lifetime) can be determined. Some results are summarized in Table 4.5. Such rapid quenching of electronic excitation by molecules is frequently observed, and presumably occurs by energy transfer to the quenching molecule. The products of reaction (4.13) have

Table 4.6. Rare gas halide recombination reactions

Reaction	Lifetime × rate	Reference
$ArF^* + 2\,Ar \rightarrow Ar_2F^* + Ar$	1.6×10^{-39}	M. Rokni et al.: Appl. Phys. Lett. **31**, 79 (1977)
	2.2×10^{-39}	C.-H. Chen, M.G. Payne, J.P. Judish: Private communication. Lifetime is 4.2 ns
$KrF^* + 2\,Kr \rightarrow Kr_2F^* + Kr$	4.4×10^{-39}	M. Rokni et al.: Phys. Rev. A **16**, 2216 (1977)
	4.5×10^{-39}	V.H. Shui: Appl. Phys. Lett. **31**, 50 (1976). Lifetime assumed to be 9 ns
	6.6×10^{-39}	J.G. Eden et al.: Appl. Phys. Lett. **32**, 733 (1978). Lifetime is 6.8 ns
	2.6×10^{-39}	G.P. Quigley, W.M. Hughes: Appl. Phys. Lett. **32**, 649 (1978). Lifetime is 9.0 ns
$KrF^* + 2\,Ar \rightarrow ArKrF^* + Ar$	5.2×10^{-40}	J.A. Mangano et al.: Appl. Phys. Lett. **31**, 28 (1977)
	7.5×10^{-40}	J.G. Eden et al.: Appl. Phys. Lett. **32**, 733 (1978). Lifetime is 6.8 ns
	8.1×10^{-40}	V.H. Shui: Appl. Phys. Lett. **31**, 50 (1976). Lifetime assumed to be 9 ns
$KrF^* + Kr + Ar \rightarrow Kr_2F^* + Ar$	4.2×10^{-39}	J.A. Mangano et al.: Appl. Phys. Lett. **31**, 28 (1977)
$XeF^* + 2\,Ar \rightarrow$ products	2.4×10^{-40}	M. Rokni et al.: Appl. Phys. Lett. **30**, 458 (1977)
$XeF^* + Xe + Ar \rightarrow$ products	4.8×10^{-39}	M. Rokni et al.: Appl. Phys. Lett. **30**, 458 (1977)
$XeF^* + Xe + Ne \rightarrow$ products	1.25×10^{-38}	M. Rokni et al.: Appl. Phys. Lett. **32**, 223 (1978)
$XeF^* + 2\,Ne \rightarrow$ products	4.0×10^{-41}	M. Rokni et al.: Appl. Phys. Lett. **32**, 223 (1978)

never actually been identified, but are expected from the repulsive nature of the F_2 states to which the excited KrF can transfer its energy.

Direct quenching of the excited rare gas halide molecules by rare gas atoms is much slower, as indicated in Table 4.5 [4.71]. However, at high pressures three-body reactions of the rare-gas atoms effectively quench the excited rare gas halides by forming triatomic species in reactions of the type [4.28, 29]

$$KrF^* + Kr + Ar \rightarrow Kr_2F^* + Ar, \qquad (4.14)$$

$$KrF^* + 2Ar \rightarrow ArKrF^* + Ar. \qquad (4.15)$$

The species Kr_2F^* has been observed in fluorescence, as discussed in the previous section, and its formation reduces the intensity of the KrF* emission. However, the branching ratio for forming excited triatomic species is not known. The species ArKrF* has not been observed directly. Its existence is postulated as an intermediate to explain the quenching of KrF* emission and appearance of Kr_2F^* emission at high Ar pressures [4.28]. In this case, reaction (4.15) is believed to be followed by the very rapid reaction

$$ArKrF^* + Kr \rightarrow Kr_2F^* + Ar. \qquad (4.16)$$

Some data on the rates of reactions like (4.14, 15) are summarized in Table 4.6. Direct formation of the pure rare gas triatomics (e.g., Kr_2F^*) is evidently quite fast, while formation of the mixed species (e.g., ArKrF*) is perhaps an order of magnitude slower.

4.2.2 Kinetics of Pure Rare Gases

The kinetics of the precursor species (rare-gas ions and excited rare-gas atoms) have been extensively studied, partly out of interest in the rare gas excimer lasers (see also Chap. 2). Thus, at the present time, the kinetics of the pure rare gases are fairly well understood, although some questions remain [4.52, 53]. Although the excited atoms are initially formed in many excited states, at high pressures most of these states are rapidly quenched to lower states, so that the bulk of the excitation resides in the lowest excited levels of the various atomic and molecular species [4.72]. Thus, in the following the various rare gas species are assumed to have only one excited state. The ions and excited atoms formed by electrical excitation rapidly dimerize by three-body recombination reactions of the type

$$Ar^* + 2Ar \rightarrow Ar_2^* + Ar, \tag{4.17}$$

$$Ar^+ + 2Ar \rightarrow Ar_2^+ + Ar. \tag{4.18}$$

Rate coefficients for some reactions of this type are summarized in Tables 4.7 and 4.8. The dimer ions can recombine with electrons to form excited atoms in dissociative recombination reactions of the type [4.73]

$$Ar_2^+ + e^- \rightarrow Ar^* + Ar. \tag{4.19}$$

Table 4.7. Rare gas excimer recombination reactions

Reaction	Rate $[cm^6 \, s^{-1}]$	Reference
$Ar^* + 2 Ar \rightarrow Ar_2^* + Ar$	6.6×10^{-33}	R. Boucique, P. Mortier: J. Phys. D3, 1905 (1970)
	9×10^{-33}	L. Colli: Phys. Rev. 95, 892 (1954)
	$1.14 - 1.70 \times 10^{-32}$	E. Ellis, N.D. Twiddy: J. Phys. B2, 1366 (1969)
	1.3×10^{-32}	A.H. Futch, F.A. Grant: Phys. Rev. 104, 356 (1956)
	$1 - 2 \times 10^{-32}$	N. Thonnard, G.S. Hurst: Phys. Rev. A5, 1110 (1972)
	$1 - 1.2 \times 10^{-32}$	P.K. Leichner et al.: Submitted to Phys. Rev. A
	1.2×10^{-32}	G.S. Hurst et al.: J. Chem. Phys. 61, 3680 (1974)
	6×10^{-33}	J. LeCalve, M. Bourene: J. Chem. Phys. 58, 1446 (1973)
	8.7×10^{-33}	A.V. Phelps, J.P. Molnar: Phys. Rev. 89, 1202 (1953)
	$2.6 - 3.3 \times 10^{-32}$	J.H. Kolts, D.W. Setser: 7th Winter Colloquium on High Power Visible Lasers, Park City, Utah (1977)
	2.8×10^{-33}	J.H. Keto et al.: Phys. Rev. Lett. 33, 1365 (1974)
	1.0×10^{-32}	C.-H. Chen, M.G. Payne, J.P. Judish: Private communication
$Kr^* + 2 Kr \rightarrow Kr_2^* + Kr$	2.59×10^{-32}	C.J. Tracy, H.J. Oskam: J. Chem. Phys. 65, 1666 (1976)
	4.4×10^{-32}	R. Boucique, P. Mortier: J. Phys. D3, 1905 (1970)
$Kr^* + 2Ar \rightarrow (ArKr)^* + Ar$	$\sim 1 \times 10^{-32}$	H.H. Nakano et al.: SRI Rpt. MP 76-99, Stanford Res. Inst., Menlo Park, Calif. (1976) (unpublished)
	1.2×10^{-32}	J.H. Kolts, D.W. Setser: 7th Winter Colloquium on High Power Visible Lasers, Park City, Utah (1977)

Table 4.7 (continued)

Reaction	Rate $[cm^6 s^{-1}]$		Reference
$He^* + 2\,He \rightarrow He_2^* + He$	2	$\times 10^{-34}$	A. V. Phelps, J. P. Molnar: Phys. Rev. **82**, 1202 (1953)
$Xe^* + 2\,Ar \rightarrow (ArXe)^* + Ar$	1	$\times 10^{-33}$	J. H. Kolts, D. W. Setser: 7th Winter Colloquium on High Power Visible Lasers, Park City, Utah (1977)
	7	$\times 10^{-34}$	P. K. Leichner et al.: Submitted to Phys. Rev. A
$Xe^* + 2\,Xe \rightarrow Xe_2^* + Xe$	2.5	$\times 10^{-32}$	R. Boucique, P. Mortier: J. Phys. D **3**, 1905 (1970)
	5.0	$\times 10^{-32}$	J. K. Rice, A. W. Johnson: J. Chem. Phys. **63**, 5235 (1975)
	8.5	$\times 10^{-32}$	P. R. Timpson, J. M. Anderson: Can J. Phys. **48**, 1817 (1970)
	4.0	$\times 10^{-32}$	P. K. Leichner et al.: Phys. Rev. A **13**, 1787 (1976)
$Xe^* + Xe + Ar \rightarrow Xe_2^* + Ar$	2.15	$\times 10^{-32}$	R. E. Gleason et al.: J. Chem. Phys. **66**, 1589 (1977)
	2.3	$\times 10^{-32}$	J. K. Rice, A. W. Johnson: J. Chem. Phys. **63**, 5235 (1975)
	3.4	$\times 10^{-32}$	P. K. Leichner et al.: Submitted to Phys. Rev. A
$Xe^* + Xe + He \rightarrow Xe_2^* + He$	1.4	$\times 10^{-32}$	J. K. Rice, A. W. Johnson: J. Chem. Phys. **63**, 5238 (1975)
$Xe^* + 2\,Ar \rightarrow (ArXe)^* + Ar$	1	$\times 10^{-33}$	J. H. Kolts, D. W. Setser: 7th Winter Colloquium on High Power Visible Lasers, Park City, Utah (1977)
	7	$\times 10^{-34}$	P. K. Leichner et al.: Submitted to Phys. Rev. A
$Xe^* + 2\,Xe \rightarrow Xe_2^* + Xe$	2.5	$\times 10^{-32}$	R. Boucique, P. Mortier: J. Phys. D **3**, 1905 (1970)
	5.0	$\times 10^{-32}$	J. K. Rice, A. W. Johnson: J. Chem. Phys. **63**, 5235 (1975)
	8.5	$\times 10^{-32}$	P. R. Timpson, J. M. Anderson: Can. J. Phys. **48**, 1817 (1970)
	4.0	$\times 10^{-32}$	P. K. Leichner et al.: Phys. Rev. A **13**, 1787 (1976)
$Xe^* + Xe + Ar \rightarrow Xe_2^* + Ar$	2.15	$\times 10^{-32}$	R. E. Gleason et al.: J. Chem. Phys. **66**, 1589 (1977)
	2.3	$\times 10^{-32}$	J. K. Rice, A. W. Johnson: J. Chem. Phys. **63**, 5235 (1975)
	3.4	$\times 10^{-32}$	P. K. Leichner et al.: Submitted to Phys. Rev. A
$Xe^* + Xe + He \rightarrow Xe_2^* + He$	1.4	$\times 10^{-32}$	J. K. Rice, A. W. Johnson: J. Chem. Phys. **63**, 5238 (1975)

Rate coefficients for some reactions of this type are summarized in Table 4.9. The excimers formed by reaction (4.17) can fluoresce with lifetimes which depend on the electron density due to collisional mixing of the $^1\Sigma$ and $^3\Sigma$ states [4.53]. Atomic radiation is trapped at high pressures. Some data on excimer lifetimes are presented in Table 4.10. Both excited atoms and excimers can be quenched by electrons [4.53, 74], as well as by other excited species in reactions

Table 4.8. Rare gas ion recombination reactions

Reaction	Rate $[\mathrm{cm}^6\,\mathrm{s}^{-1}]$	Reference
$He^+ + 2\,He \rightarrow He_2^+ + He$	$6.3\ \times 10^{-32}$	A. V. Phelps, S. C. Brown: Phys. Rev. **86**, 102 (1952)
	$10.2\ \times 10^{-32}$	E. C. Beaty, P. L. Patterson: Phys. Rev. **137**, A 364 (1965)
	$9.7\ \times 10^{-32}$	R. Hackam, J. J. Lennon: Proc. Phys. Soc. London **84**, 133 (1964)
	$10.7\ \times 10^{-32}$	F. E. Niles, W. W. Robertson: J. Chem. Phys. **42**, 3277 (1965)
	6.78×10^{-32}	D. Smith, M. J. Copsey: J. Phys. B **1**, 650 (1968)
	$9.8\ \times 10^{-32}$	H. J. Oskam, V. R. Mittelstadt: Phys. Rev. **132**, 1435 (1963)
$Ne^+ + 2\,Ne \rightarrow Ne_2^+ + Ne$	$7\ \ \times 10^{-32}$	E. C. Beaty, P. L. Patterson: Phys. Rev. **170**, 116 (1968)
	7.9×10^{-32}	D. Smith, P. R. Cromey: J. Phys. B **1**, 638 (1968)
	$5.7\text{--}8.9 \times 10^{-32}$	J. P. Gaur, L. M. Chanin: Phys. Rev. **182**, 167 (1969)
$Ne^+ + 2\,He \rightarrow (HeNe)^+ + He$	$2.1\ \times 10^{-32}$	G. E. Veach, H. J. Oskam: Phys. Rev. A **2**, 1422 (1970)
$Ne^+ + Ne + He \rightarrow Ne_2^+ + He$	$3\ \ \times 10^{-31}$	G. E. Veach, H. J. Oskam: Phys. Rev. A **2**, 1422 (1970)
$Ar^+ + 2\,Ar \rightarrow Ar_2^+ + Ar$	$4.4\ \times 10^{-31}$	P. F. Knewstubb: In *Advances in Mass Spectrometry*, ed. by E. Kendrich, Vol. 4 (Elsevier Publishing, New York 1968) p. 391
	$2.5\ \times 10^{-31}$	D. Smith, P. R. Cromey: J. Phys. B **1**, 638 (1968)
	1.46×10^{-31}	C. B. Kritschmer, H. L. Peterson: J. Appl. Phys. **34**, 3209 (1963)
	$5\ \ \times 10^{-32}$	P. Keberle et al.: J. Chem. Phys. **47**, 1684 (1966)
	$1.9\ \times 10^{-31}$	A. K. Bhattacharya: J. Appl. Phys. **41**, 1707 (1970)
	2.07×10^{-31}	W.-C. F. Liu, D. C. Conway: J. Chem. Phys. **60**, 784 (1974)
	$2.3\ \times 10^{-31}$	W.-C. F. Liu, D. C. Conway: J. Chem. Phys. **62**, 3070 (1975)
	$3.0\ \times 10^{-31}$	J. C. Cronin, M. C. Sexton: J. Phys. D **1**, 889 (1968)
$Ar^+ + Ar + He \rightarrow Ar_2^+ + He$	$1\ \ \times 10^{-31}$	D. K. Bohme et al.: J. Chem. Phys. **51**, 863 (1969)
$Kr^+ + Kr + He \rightarrow Kr_2^+ + He$	$6.1\ \times 10^{-32}$	C. L. Chen: Phys. Rev. **131**, 2550 (1963)
$Xe^+ + 2\,Xe \rightarrow Xe_2^+ + Xe$	$1.8\ \times 10^{-31}$	A. K. Bhattacharya: Appl. Phys. Lett. **17**, 521 (1970)
	3.57×10^{-31}	D. Smith et al.: J. Phys. B **5**, 2134 (1972)
	$2.0\ \times 10^{-31}$	A. P. Vitols, H. J. Oskam: Phys. Rev. A **8**, 1860 (1973)
$Xe^+ + Xe + He \rightarrow Xe_2^+ + He$	$1.1\ \times 10^{-31}$	C. L. Chen: Phys. Rev. **131**, 2250 (1963)

Table 4.9. Rare gas dissociative recombination reactions

Reaction	Rate $[cm^3\ s^{-1}]$	Reference
$e^- + He_2^+ \rightarrow He^* + He$	$< 10^{-8}$	J.N. Bardlsey, M.A. Biondi: Adv. At. Mol. Phys. **6**, 2 (1970)
$e^- + Ne_2^+ \rightarrow Ne^* + Ne$	$1.8 \times 10^{-7} \left(\dfrac{T_e}{300\,K}\right)^{-0.43}$	W.H. Kasner: Phys. Rev. **164**, 194 (1967) J. Philbrick et al.: Phys. Rev. **181**, 271 (1969) M.A. Biondi, S.C. Brown: Phys. Rev. **75**, 1700 (1949) J.N. Bardsley, M.A. Biondi: Adv. At. Mol. Phys. **6**, 2 (1970)
$e^- + Ar_2^+ \rightarrow Ar^* + Ar$	$7.5 \times 10^{-7} \left(\dfrac{T_e}{300\,K}\right)^{-0.67}$	H.J. Oskam, V.R. Mittelstaedt: Phys. Rev. **132**, 1445 (1963) F.J. Mehr, M.A. Biondi: Phys. Rev. **176**, 322 (1968) J.N. Bardsley, M.A. Biondi: Adv. At. Mol. Phys. **6**, 2 (1970)
$e^- + Kr_2^+ \rightarrow Kr^* + Kr$	$1.2 \times 10^{-6} \left(\dfrac{T_e}{300\,K}\right)^{-0.5}$	H.J. Oskam, V.R. Mittelstad: Phys. Rev. **132**, 1445 (1963) J.N. Bardsley, M.A. Biondi: Adv. At. Mol. Phys. **6**, 2 (1970)
$e^- + Xe_2^+ \rightarrow Xe^* + Xe$	$1.4 \times 10^{-6} \left(\dfrac{T_e}{300\,K}\right)^{-0.5}$	H.J. Oskam, V.R. Mittelstadt: Phys. Rev. **132**, 1445 (1963) J.N. Bardsley, M.A. Biondi: Adv. At. Mol. Phys. **6**, 2 (1970)

Table 4.10. Radiative lifetimes of rare gas excimers

Excimer	Lifetime [s]	Reference
Ar_2^*	3.7×10^{-6}	R. Boucique, P. Mortier: J. Phys. D **3**, 1905 (1970)
	3.4×10^{-6}	L. Colli: Phys. Rev. **95**, 892 (1954)
(1_u)	3.2×10^{-6}	J.W. Keto et al.: Phys. Rev. Lett. **33**, 1365 (1974)
(0^+)	4.2×10^{-9}	J.W. Keto et al.: Phys. Rev. Lett. **33**, 1365 (1974)
	2.8×10^{-6}	N. Thonnard, G.S. Hurst: Phys. Rev. A **5**, 110 (1972)
Kr_2^*	1.7×10^{-6}	R. Boucique, P. Mortier: J. Phys. D. **3**, 1905 (1970)
Xe_2^*	5×10^{-7}	R. Boucique, P. Mortier: J. Phys. D. **3**, 1905 (1970)
(1_u)	9.6×10^{-8}	J.W. Keto et al.: Phys. Rev. Lett. **33**, 1365 (1974)
	9.9×10^{-8}	P.K. Leichner et al.: Phys. Rev. A. **13**, 1787 (1976)
(0_u^+)	5.5×10^{-9}	J.W. Keto et al.: Phys. Rev. Lett. **33**, 1365 (1974)

of the type

$$Ar_2^* + Ar^* \rightarrow Ar_2^+ + Ar + e^- , \tag{4.20}$$

$$Ar_2^* + Ar_2^* \rightarrow Ar_2^+ + 2Ar + e^- , \tag{4.21}$$

$$Ar^* + Ar^* \rightarrow Ar^+ + Ar + e^- , \tag{4.22}$$

which generally occur at every collision.

When halogen-containing species are added to the rare gas, the excited atoms and atomic ions react to form rare gas halides in reactions like (4.8, 10). Likewise, the rare gas excimers and dimer ions can react to form rare gas halide species in analogous reactions of the type

$$Ar_2^* + F_2 \rightarrow ArF^* + Ar + F,$$ (4.23)

$$Ar_2^+ + F^- + M \rightarrow ArF^* + Ar + M.$$ (4.24)

Some data on the excimer reactions are included in Table 4.4. The ion reaction (4.24) has never been measured, but should occur with a rate coefficient given by (4.11,12). Although both reactions (4.23, 24) could, in principle, form triatomic species such as Ar_2F^* [4.27], the evidence indicates that diatomic products predominate [4.29, 71, 75].

4.2.3 Kinetics of Rare Gas Mixtures

The kinetics of rare gas mixtures are much more complex, and are not completely understood at the present time. In general, rare gas halide laser mixtures consist of a large amount of a lighter rare gas (He, Ne, or Ar), with a small amount ($\leq 10\%$) of a heavier rare gas (Ar, Kr or Xe) added. Much (or most) of the effect of the electrical excitation is to form excited atoms and atomic ions of the lighter (dominant) rare gas species. These can dimerize, recombine with electrons, and react to form diatomic and triatomic rare gas halides as in the case of pure rare gases. In addition, if the excited species are sufficiently energetic they may react with the heavier rare gas atoms to form ions in Penning or associative ionization reactions of the type

$$He^* + Ar \nearrow \text{He} + Ar^+ + e^-,$$ (4.25a)

$$\searrow \text{HeAr}^+ + e^-.$$ (4.25b)

Generally, reactions of this type are very fast, proceeding at every collision if there is sufficient energy.

If the energy of excitation is not sufficient to produce Penning or associative ionization, energy transfer to the heavier species may occur in reactions of the type

$$Ar^* + Xe \rightarrow Ar + Xe^*,$$ (4.26)

$$Ar_2^* + Xe \rightarrow 2Ar + Xe^*.$$ (4.27)

Energy transfer between rare gas atoms, reaction (4.26), is generally slow unless there is a near resonance between the initial and final states, and the rate depends sensitively on the closeness of the resonance [4.59]. Some data are summarized in Table 4.11. Reactions of the type (4.27) are generally rapid since

Table 4.11. Energy transfer reactions in rare gas mixtures

Reaction	Rate [$cm^3 s^{-1}$]	Reference
$Ar^* + Kr \rightarrow Kr^* + Ar$	1.4×10^{-11}	M. Bourene, J. LeCalve: J. Chem. Phys. **58**, 1452 (1973)
	5×10^{-12}	A.V. Phelps, J.P. Molnar: Phys. Rev. **89**, 1202 (1953)
	$2.3–6.2 \times 10^{-12}$	L.G. Piper et al.: J. Chem. Phys. **59**, 3323 (1973)
$Ar^* + Xe \rightarrow Xe^* + Ar$	$1.9–3.4 \times 10^{-10}$	P.K. Leichner et al.: Submitted to Phys. Rev. A
	$1.8–3.0 \times 10^{-10}$	L.G. Piper et al.: J. Chem. Phys. **59**, 3323 (1973)
	2.1×10^{-10}	M. Bourene, J. LeCalve: J. Chem. Phys. **58**, 1452 (1973)
	$2.2–3.3 \times 10^{-10}$	C.H. Chen, J.P. Judish, M.G. Payne: Private communication
$Kr^* + Xe \rightarrow Xe^* + Kr$	1.6×10^{-10}	J.H. Kolts, D.W. Setser: 7th Winter Colloquium on High Power Visible Lasers, Park City, Utah (1977)
$Ar_2^* + Xe \rightarrow Xe^* + 2 Ar$	4.39×10^{-10}	R.E. Gleason et al.: J. Chem. Phys. **66**, 1589 (1977)
	2.4×10^{-10}	C.H. Chen, J.P. Judish, M.G. Payne: Private communication
$Ar_2^* + Kr \rightarrow Kr^* + 2 Ar$	1.5×10^{-9}	A. Gedanken et al.: J. Chem. Phys. **58**, 3456 (1972)
$Kr_2^* + Xe \rightarrow Xe^* + 2 Kr$	4.4×10^{-10}	W. Hughes: 7th Winter Colloquium on High Power Visible Lasers, Park City, Utah (1977)

a close resonance is not needed [4.76]. The excess energy can be disposed of in the dissociation of the molecular species. Relatively few firm data are available regarding processes like (4.27), but they are evidently very fast. Some data are included in Table 4.11.

Asymmetric charge transfer between atomic ions is generally very slow at thermal energies and may be ignored. On the other hand, charge transfer from dimer ions to atoms and molecules is generally very fast, as typified by the reactions

$$He_2^+ + Ar \rightarrow 2He + Ar^+ , \qquad (4.28)$$

$$Ar_2^+ + Xe \rightarrow 2Ar + Xe^+ . \qquad (4.29)$$

Some data are presented in Table 4.12. Most such reactions occur at nearly the ion-molecule collision frequency predicted by *Langevin* [4.77], and *Gioumousis* and *Stevenson* [4.78], with a rate given by the formula

$$k = 2\pi q(\alpha/\mu)^{1/2} , \qquad (4.30)$$

where q is the ionic charge, α is the polarizability of the neutral body, and μ is the reduced mass of the reaction. Charge transfer from Ne_2^+ to Ar and Kr are apparently exceptions [4.79].

The ions and excited atoms of the heavier rare gas species can dimerize in reactions like

$$Ar^+ + Ar + He \rightarrow Ar_2^+ + He , \qquad (4.31)$$

$$Xe^* + Xe + Ar \rightarrow Xe_2^* + Ar . \qquad (4.32)$$

Table 4.12. Charge transfer reactions in rare gas mixtures

Reaction	Rate $[\text{cm}^3 \text{s}^{-1}]$	Reference
$He_2^+ + Ne \rightarrow Ne^+ + 2He$	6.0×10^{-10}(200 K)	D.K. Bohme et al.: J. Chem. Phys. **52**, 5094 (1970)
	1.4×10^{-10}	F.C. Fehsenfeld et al.: J. Chem. Phys. **44**, 4087 (1966)
	1.7×10^{-10}	H.J. Oskam, V.R. Mittelstadt: Phys. Rev. **132**, 1435 (1963)
$He_2^+ + Ar \rightarrow Ar^+ + 2He$	2.0×10^{-10}(200 K)	D.K. Bohme et al.: J. Chem. Phys. **52**, 5094 (1970)
$He_2^+ + Kr \rightarrow Kr^+ + 2He$	1.85×10^{-11}(200 K)	D.K. Bohme et al.: J. Chem. Phys. **52**, 5094 (1970)
$Ne_2^+ + Ar \rightarrow Ar^+ + 2Ne$	$\leqq 5 \times 10^{-14}$(200 K)	D.K. Bohme et al.: J. Chem. Phys. **52**, 5094 (1970)
$Ne_2^+ + Kr \rightarrow Kr^+ + 2Ne$	$\leqq 5 \times 10^{-13}$(200 K)	D.K. Bohme et al.: J. Chem. Phys. **52**, 5094 (1970)
$Ar_2^+ + Kr \rightarrow Kr^+ + 2Ar$	7.5×10^{-10}(200 K)	D.K. Bohme et al.: J. Chem. Phys. **52**, 5094 (1970)
$HeNe^+ + Ne \rightarrow Ne_2^+ + He$	3×10^{-11}	G.E. Veach, H.J. Oskam: Phys. Rev. A **2**, 1422 (1970)
$ArKr^+ + Kr \rightarrow Kr_2^+ + Ar$	3.2×10^{-10}(200 K)	D.K. Bohme et al.: J. Chem. Phys. **52**, 5094 (1970)
$Kr_2^+ + Xe \rightarrow Xe^+ + 2Kr$	2×10^{-10}	P. Kebarle et al.: J. Chem. Phys. **47**, 1684 (1967)
$KrXe^+ + Xe \rightarrow Xe_2^+ + Kr$	2×10^{-10}	P. Kebarle et al.: J. Chem. Phys. **47**, 1684 (1967)
$Ar_2^+ + Xe \rightarrow$ products	7.2×10^{-11}	C.H. Chen, J.P. Judish, M.G. Payne: Private communication

The recombination rates with other third bodies are comparable to those in the pure gases, as indicated in Table 4.8. However, at low mole fractions of the heavier rare gas, two-step processes such as

$$Ne^+ + 2He \rightarrow (HeNe)^+ + He, \tag{4.33}$$

$$(HeNe)^+ + Ne \rightarrow Ne_2^+ + He, \tag{4.34}$$

and

$$Xe^* + 2Ar \rightarrow (ArXe)^* + Ar, \tag{4.35}$$

$$(ArXe)^* + Xe \rightarrow Xe_2^* + Ar, \tag{4.36}$$

are much faster. The heteronuclear ions are stable enough to be observed in a mass spectrometer, and the two-step mechanism [reactions (4.33, 34)] serves to explain the formation of $(HeNe)^+$ and Ne_2^+ or $(ArKr)^+$ and Kr_2^+ with the addition of Ne to He afterglows or Kr to Ar afterglows [4.80]. However, the rate of recombination of heteronuclear dimer ions [reaction (4.33)] has been measured in only one case and is evidently slightly slower than that of homonuclear ions (see Table 4.8). The displacement reactions, like (4.34), are very fast (nearly gaskinetic), with rate coefficients of the order of $10^{-10} \text{cm}^3 \text{s}^{-1}$ or larger. Some data are included in Table 4.12. The heteronuclear rare gas excimer recombination reaction (4.35) has been only recently measured, and is evidently fast for Kr* in Ar but about an order of magnitude slower for Xe* in Ar [4.27, 81, 82] (see Table 4.7). No direct data are available regarding the

Table 4.13. Rare gas halide displacement reactions
[M. Rokni et al.: Appl. Phys. Lett. **31**, 79 (1977)]

Reaction	Lifetime × Rate [cm³]
$ArF^* + Kr \rightarrow KrF^* + Ar$	6.1×10^{-18}
$ArF^* + Xe \rightarrow XeF^* + Ar$	1.8×10^{-17}

displacement reactions like (4.36), but considering the weak bonding of the heteronuclear rare gas excimers and the large rate of the corresponding ion reactions [like reaction (4.34)], these reactions are expected to be fast, with coefficients of the order of $10^{-10}\,cm^3\,s^{-1}$. This is consistent with indirect evidence from KrF fluorescence [4.27].

When halogens are added to a rare gas mixture, the kinetics become even more complex. The rare gas positive ions can react with the halogen negative ions in reactions like (4.10, 24), as in pure rare gases, and the excited rare gas atoms and excimers can react with the halogen in reactions like (4.8, 23). In addition, displacement reactions of the type

$$ArF^* + Xe \rightarrow XeF^* + Ar, \tag{4.37}$$

may occur, shifting the fluorescence from the lighter rare gas halide to the heavier species as the concentration of the heavier rare gas is increased. The rate of reactions like (4.37) is evidently very fast, of the order of gaskinetic [4.28, 83], (see Table 4.13). Thus far, there has been no direct, unambiguous evidence for mixed rare gas halide trimers such as ArKrF*. Nevertheless, they are postulated to explain the quenching of KrF* fluorescence and appearance of Kr_2F^* fluorescence by reactions (4.15, 16) [4.28, 29, 71]. Other exchange reactions are also possible involving the triatomic rare gas halides such as Ar_2F. For example, reactions of the type

$$Ar_2F^* + Xe \rightarrow XeF^* + 2Ar \tag{4.38}$$

could occur.

An interesting conclusion to draw from the kinetic described above is that all mechanisms for forming the excited rare gas halides begin with molecular halogen species such as F_2 or NF_3, rather than atomic halogens. This suggests that oscillation should cease when the molecular halogens in the gas are consumed, since three-body recombination of halogen molecules in rare gases is rather slow, with rate coefficients of the order of $10^{-32}\,cm^6\,s^{-1}$. For a typical halogen concentration of $10^{17}\,cm^{-3}$ in two atmospheres of a rare gas this corresponds to a recombination time of the order of 20 µs, which is long compared to laser pulse lengths achieved to date. This sets a limit on the available total laser pulse energy per unit volume corresponding to no more than one photon per initial halogen molecule. For a typical KrF laser mixture containing 0.2% F_2, this limit is about 40 J/l-atm. One can think of reactions which would form excited rare gas halides from halogen atoms and thereby recycle them to gain additional energy. For example, the reaction

$$Kr_2^* + F \rightarrow KrF^* + Kr, \tag{4.39}$$

is exothermic, and the alkali metals are observed to undergo analogous reactions with very large cross sections [4.84]. Likewise, three-body (neutral stabilized) electron attachment to halogen atoms is known to occur at low electron temperatures [4.61]. However, recent evidence suggests that such reactions may not be important in rare gas halide lasers [4.85]. In experiments with electron beam pumped $Ne/Kr/F_2$ mixtures, the pump power was increased by increasing the Ne density at constant Kr and F_2 density. The laser power was observed to increase in proportion, but the laser pulse energy was observed to saturate at a value which was proportional to the F_2 concentration – up to 20 torr, although not beyond. As the pump power increased above the saturation point, the laser pulse terminated earlier, evidently corresponding to consumption of the initial F_2. However, in these experiments, the laser output corresponded to only one photon for every eight halogen molecules initially present. This is probably explained in part by inefficient laser extraction and other losses (the intrinsic efficiency was only 7%). When these same experiments were carried out in $Ar/Kr/F_2$ mixtures the total energy was observed to *decrease* as the Ar pressure was increased beyond the saturation value. This could be due to formation of Ar_2F^*, which may have too little available energy to form KrF^* in displacement reactions like (4.38). Alternatively it could be due to absorption by Ar_2F^*, Ar_2^* or Ar_2^+ formed at high Ar pressures [4.57], as is also observed in XeF lasers [4.56].

In summary, it is fair to say that the reaction kinetics of rare gas halide lasers are rather complex and not completely understood at the present time. There remain qualitative uncertainties regarding some important reactions, such as (4.37, 38, 39), and quantitative uncertainties in some important reaction rates such as the rate of attachment of electrons to F_2. Nevertheless, the basic mechanisms by which the excited rare gas halides are formed are fast and evidently quite efficient. We turn our attention, therefore, to methods of pumping rare gas halide lasers.

4.3 Electron Beam and Nuclear Pumped Lasers

The first rare gas halide lasers were pumped by high intensity, relativistic electron beams [4.1–4]. Such apparatuses offer a relatively straightforward, brute force method for pumping high pressure gases with the very high power densities needed to achieve threshold. Compared with discharges, electron beams eliminate problems of discharge stability and minimize problems of

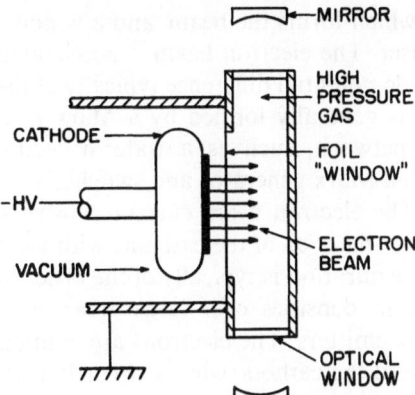

Fig. 4.4. Schematic diagram of a laser pumped by a high intensity electron beam

circuit inductance since they operate at higher impedances than transverse discharges. As discussed in the next section, discharges are potentially more efficient and powerful, and are simple to construct in small sizes. Consequently, they are currently receiving considerable attention. Nevertheless, up to the present time the best efficiency reported with electron beam pumping [4.44] is comparable to that achieved with discharges (of the order of 1 % overall [4.86]), and the largest pulse energy achieved with an electron beam (about 350 J [4.87]) is still almost an order of magnitude greater than the best result reported from a discharge (about 50 J [4.87]). Moreover, while the Marx generator type electron beams used in the early experiments are awkward to pulse repetitively, microwave linear accelerators (electron LINAC's) show considerable promise as efficient, reliable, high average power electron beam sources for repetitively pulsed lasers [4.88].

In the future, a third source of excitation may become important, that is, direct nuclear excitation [4.89, 90]. Although the rare gas halides have not yet been made to operate in this way, CO, Xe, N, and Ar have [4.91–94]. For direct nuclear pumping, a pulsed nuclear reactor is used to generate a high intensity pulse of neutrons lasting between about 0.1 and 10 ms. The neutrons react with nuclear species in the laser gas [4.93, 94] or on the walls [4.91, 92] to produce high energy charged particles in reactions of the type

$$^{3}\text{He} + n \rightarrow {}^{2}\text{H} + {}^{2}\text{H} + 0.8\,\text{MeV}, \tag{4.40}$$

$$^{235}\text{U} + n \rightarrow Z_1 + Z_2 + 175\,\text{MeV}, \tag{4.41}$$

where Z_1 and Z_2 are fission fragment nuclei. The high energy charged particles produced in these reactions ionize and excite the laser gas in much the same way as do electrons. Thus, for the remainder of this section we consider only pumping by relativistic electrons, keeping in mind that much of the discussion applies equally to heavy charged particles formed from nuclear reactions.

A laser pumped by a high intensity, relativistic electron beam is shown schematically in Fig. 4.4. The electron-beam gun consists of a pulsed high voltage power supply, a vacuum diode, which forms the beam, and a window through which the electrons enter the laser. The electron beam is accelerated from the cathode to the anode through a dc potential difference typically of the order of 0.2 to 2 MV. The high voltage is generally formed by a Marx L–C generator. Frequently a pulse-forming network, such as a water-dielectric transmission line, is impulse charged with a Marx generator and switched into the diode to give a better pulse shape. The electron beam current density is typically of the order of 5–500 A cm^{-2} over the area of the cathode, with total currents of the order of 5–50 kA. The pulse duration is typically of the order of 50 ns to 1 μs. To obtain such high current densities over large areas, cold cathodes are used, rather than thermionic emitters. The electrons are emitted by a plasma which forms over the surface of the cathode when extremely high voltage gradients are applied [4.95]. The current density is space-charge limited, in accordance with Child's law [4.96]. The electrons enter the laser cavity through a thin metallic foil of the order of 25–50 μm thick (sometimes aluminized mylar or kapton is used) with only a small loss of energy [4.97, 98]. Some sort of structure is generally required to support the foil, consisting, typically, of a thick plate covered with closely spaced holes. Most of the energy lost by the beam in passing from the diode into the laser is absorbed by the foil support structure. The transverse pumping geometry shown in Fig. 4.4 is most commonly used [4.1, 2, 4, 6, 22, 85, 87, 99, 100], but coaxial geometries (in which the electron beam is accelerated radially inward, entering the laser cell through a cylindrical foil) [4.3, 44] and axial geometries (in which the electron beam is injected near one end of the laser and guided along the axis by a magnetic field) [4.88, 101] have also been used. The transverse geometry is simplest, but the target gas in the laser cavity is usually so thin (in mass per unit area facing the electron beam) that most of the electrons pass through the laser cell and deposit their energy in the far wall. Typically, of the order of 5–50 % of the electron-beam energy entering the laser cell is deposited in the gas in the laser cavity.

As they pass through the laser gas the incident electrons scatter off the nuclei and electrons, transferring energy to the electrons in the gas and creating excited atoms and secondary electrons. The secondary electrons further ionize the gas as they slow down and create some electronically excited atoms as well. This complex process has been investigated both theoretically and experimentally, with fairly good agreement where comparison is possible [4.53, 62, 102, 103]. A measure of the overall efficiency of this process is provided by the average energy W required to form one electron–ion pair (W: energy of one incident electron $\div$ number of electron–ion pairs produced). For electrons incident on Ar at energies above about 1 keV, the result is $W \approx 26$ eV/ion-electron pair [4.53, 62]. Since both electron-ion pairs and excited rare gas atoms can form excited rare gas halides, as discussed in the previous section, a more useful number is W^*, the average energy required to form either an ion-electron pair or an excited atom (W^*: energy of one incident electron $\div$ number of ions and excited atoms produced). This number, which has been calculated theoretically, is of the order of 20 eV/precursor for electrons above about 1 keV in Ar [4.53, 102]. At the electron densities characteristic of laser plasmas the secondary electrons rapidly relax by electron-electron and electron-atom collisions to a non-Maxwellian distribution having a mean energy of a few electron volts [4.74, 104].

To understand the efficiency of this process as a method of pumping a rare gas halide laser, we observe that the most efficient (lowest energy) precursor is a rare gas atom in its lowest excited state, with an energy E^*. We therefore define the pumping efficiency η_P (the efficiency for producing precursors) by the expression

$$\eta_P = E^*/W^*. \tag{4.42}$$

For example, in a KrF laser [$E^*(Kr) = 9.9$ eV] operating in an Ar diluent [$W^*(Ar) = 20$ eV], the pumping efficiency is $\eta_P = 50\%$. The maximum possible intrinsic efficiency η_{IN} (defined as the laser energy out $\div$ pump energy deposited in the gas in the laser cavity), corresponding to perfect kinetics and complete extraction of the laser energy (one laser photon out for each precursor) is then

$$\eta_{IN}(\text{max}) = \eta_P \eta_Q. \tag{4.43}$$

The quantum efficiency η_Q is defined by the expression

$$\eta_Q = h\nu/E^*, \tag{4.44}$$

in which $h\nu$ is the energy of a laser photon. For a KrF laser ($h\nu = 5.0$ eV), for example, the quantum efficiency is $\eta_Q = 50\%$, and the maximum possible intrinsic efficiency for an electron beam pumped mixture in Ar is $\eta_{IN}(\text{max}) = 25\%$ [4.4, 105]. Of course, the achievable overall efficiency will be much lower, as discussed below.

Experimentally, the largest pulse energy reported to date is 350 J obtained from KrF [4.87]. This was achieved in a 600 ns pulse using a large area (2 m by 20 cm), relatively low current density (10 A cm^{-2}) electron beam to pump a 60 l volume in a transverse geometry. Of the 7 kJ entering the laser cell, approximately 50% was deposited in the laser gas (1 atm Ar with 7% Kr and 0.1% F_2). This was determined by theoretical calculations checked against the measured pressure rise in the cell. Thus, the intrinsic efficiency was $\eta_{IN} \sim 10\%$. The low efficiency is attributed to low laser power extraction efficiency, possibly due to photoabsorption in the gas, as discussed in Sect. 4.2.

The highest reported intrinsic efficiency is 15%, obtained from KrF [4.44]. This was achieved in a 1.5 J, 125 ns pulse from a coaxial electron beam. The energy deposition in the gas (Ar with 10% Kr and 0.1% NF_3 at a total pressure of two atmospheres) was estimated theoretically using a one dimensional electron transport code. Comparable efficiency (13%) has been obtained using a transverse electron beam [4.105]. In these experiments a 0.6 J pulse was obtained, and the energy deposited in the gas was measured by observing the pressure rise following the electron beam pulse with a fast transducer. In all these experiments the *overall* efficiency was much lower than the intrinsic efficiency, due principally to inefficient deposition of the electron beam energy in the gas. For example, in the coaxial electron beam discussed above the energy stored in the capacitors was 200 J, of which 10 J (5%) was deposited in the gas, giving an efficiency (laser energy ÷ energy stored in the capacitors) of 0.8% [4.44]. It should be possible to do much better than this in carefully designed experiments, but losses to the foil and foil support and losses in the formation of the electron beam will probably approach 50% in any case.

The ultimate pulse energy and pulse repetition frequency to which electron beam pumped devices can be scaled are limited in several ways. As mentioned in Sect. 4.3, the maximum pulse energy *per unit volume* is limited by the halogen density (no more than one photon out per initial halogen molecule). However, there are also physical restrictions which limit the *total* pulse energy. To begin with, the pulse duration achievable with cold cathodes is limited by a phenomenon called "closure" of the diode, that is, collapse of the diode impedance as the cathode plasma expands across the vacuum gap. The closure velocity v_c is typically of the order of a few centimeters per microsecond [4.95]. Since the current density J [A cm^{-2}] depends on the vacuum gap d [cm] and the diode voltage V [MV] in accordance with the Child-Langmuir law [4.96],

$$J = 2.3 \times 10^3 \, V^{3/2}/d^2 \,, \tag{4.45}$$

we see that the closure time $\tau_c = d/v_c$ is a function of the voltage and current density. For a current density of 100 A cm^{-2}, and a voltage of 400 kV, the closure time is of the order of 0.5–1 µs, but the diode impedance will drop 50% in one-fourth this time, or about 200 ns.

The total electron beam current is limited by self-pinching of the electron beam by its own magnetic field. Self-pinching sets in when the orbit radius of the electrons in the magnetic field of the beam itself is of the order of the diode gap. For relativistic electrons the orbit radius r [cm] is given by the formula [4.106].

$$r = [T(T + 2T_0)]^{1/2}/3B \,, \tag{4.46}$$

where T [MeV] is the electron energy, $T_0 = 0.511$ MeV is the rest mass of an electron, and B [T] is the magnetic field strength. For a long, narrow transverse geometry, as in Fig. 4.4, the magnetic field at the edge of the beam is given by Ampere's law,

$$B = \mu_0 I/2L \,, \tag{4.47}$$

where L [m] is the length of the laser and $\mu_0 = 4\pi \times 10^{-7}$ H m^{-1} is the permeability of free space. To get an order-of-magnitude estimate of the pinching limitation, we may set $r \sim 2d$ and $V \sim 2T$. Then, for a pulse duration

$\Delta t = \tau_c/4$, and a closure velocity $v_c \sim 3\,\mathrm{cm\,\mu s^{-1}}$ we see that the maximum electron beam energy E_{beam} [kJ] per pulse is

$$E_{\mathrm{beam}} = IV\Delta t \approx \frac{LV^{3/2}(V+4T_0)^{1/2}}{24\mu_0 v_c} \sim 10V^{3/2}(V+4T_0)^{1/2}L \; . \tag{4.48}$$

If 10 % of this can be passed through the foil support, deposited in the gas in the laser cavity, and converted to laser energy, then laser pulse energies of the order of a kilojoule per meter of laser length are possible, at least in principle, with megavolt electron beams. Higher energies may be possible if externally generated magnetic fields can be used to suppress pinching. In coaxial geometries, self-pinching limits the energy per unit length around the circumference of the laser.

The length of the laser along the optical axis is limited by the onset of parasitic oscillations, under high gain conditions, and by losses due to photoabsorption, as discussed in Sect. 4.2. The gain may be reduced to suppress parasitics by pumping at reduced current densities for longer periods, but must remain large compared with the absorption for efficient laser energy extraction (see Sect. 4.5). Overall laser lengths of a few meters appear to be possible, corresponding to pulse energies as large as several kilojoules.

Finally, there is the problem of foil heating. As the high energy electrons pass through the foil they lose energy in collisions with the electrons in the foil material. Collisions with the nuclei in the foil scatter the electron beam, but absorb relatively little energy due to the mass of the nuclei. For 250 keV electrons in Al, the rate of loss of energy (called the "stopping power" of the foil material) is about $2\,\mathrm{MeV\text{-}cm^2\,g^{-1}}$ [4.97]. The stopping power is a slowly decreasing function of the electron energy (between 100 keV and 1 MeV) and the atomic number of the absorber [4.98]. Taking the specific heat of Al to be $0.2\,\mathrm{cal\,g^{-1}\cdot{}^\circ C^{-1}}$ we see that the temperature rise in an Al foil is $2 \times 10^6\,\mathrm{cm^2\cdot{}^\circ C\,C^{-1}}$. For a temperature rise of 200 °C, this corresponds to a charge density of $10^{-4}\,\mathrm{C\,cm^{-2}}$, or $100\,\mathrm{J\,cm^{-2}}$ per pulse at 1 MeV. The temperature rise affects the performance of the laser in two ways. At high repetition rates or large single pulse energies the heating may be enough to destroy the foil. Although complete destruction of the foil in a single shot requires a large temperature rise and a large electron beam pulse, damage sufficient to destroy a foil after hundreds or thousands of shots is observed to set in much sooner [4.107]. This damage may be due to the fact that the voltage is low at the beginning and end of the pulse, and the low energy electrons deposit all their energy near surface of the foil, causing nonuniform heating. At lower repetition rates the warm foil heats the laser gas, causing index of refraction gradients and degrading the laser beam quality. Cooling large area, high transmission foil support structures to reduce these effects is a difficult engineering problem.

Clearly, scaling electron beam pumped rare gas halide lasers to large pulse energy and high average power is difficult, requiring careful optimization of many factors, not all of which are well understood. The problems of beam pinching, diode closure and foil heating are avoided by direct nuclear pumping. However, for the near term, especially for smaller lasers, a more promising method of pumping is by means of electrical discharges.

4.4 Discharge Pumped Lasers

Compared with electron beams, discharges offer the potential for higher pumping efficiency and higher average power. The reason for the higher pumping efficiency is that it is possible, using a carefully controlled discharge, to pump the metastable levels of the rare gases directly by electron impact with very high efficiency. The higher average power is achieved by avoiding or (in electron beam stabilized discharges) reducing the problems of foil heating and beam pinching discussed above.

4.4.1 Discharge Physics

Basically, in rare gas discharges the low energy electrons suffer only elastic encounters with the rare gas atoms, in which they lose very little energy. Thus, as they drift along the electric field they increase in energy until they reach the threshold for excitation of the first excited state of the rare gas atoms. Above this energy they experience inelastic collisions in which the rare gas atom is left in an excited state while the electron returns to a low energy. Thus, the electron distribution function, which is fairly flat up to the first excitation threshold even at rather low electric field strengths (a few $kV\,cm^{-1} \cdot atm^{-1}$), drops off rapidly at higher energies [4.108]. Due to its small mass, He is exceptional in this regard, and elastic losses are important [4.109].

In addition to losses due to the elastic encounters mentioned above, the low energy electrons lose energy due to inelastic collisions with the halogen (or halogen-bearing) molecules, and with excited species present in the plasma [4.51]. These latter processes, which become important at large excitation densities, are discussed further below. Also, because the electron distribution function extends beyond the threshold for exciting the lowest state of the rare gas, higher levels are directly excited by the electrons. In particular, in rare gas mixtures such as Kr in Ar the lowest excited levels of the Ar, which lie above those of the Kr, are strongly excited. Although the excitation is efficiently transferred to the Kr, this process is inefficient in the sense that more electron energy is consumed than is necessary to excite the lowest Kr levels directly. Fortunately, theoretical and experimental evidence indicates that the sum of all these losses can be minimized in a carefully controlled discharge, and that in Ar–Kr mixtures, for example, discharge efficiencies η_D (defined as the energy in Kr excitation ÷ discharge energy) in excess of 80 % can be achieved [4.51, 108].

The maximum possible intrinsic efficiency η_{IN}, corresponding to perfect kinetics and complete extraction of the laser energy (one photon out for each excited rare gas atom produced), is the product of the discharge efficiency and the quantum efficiency [see (4.44)],

$$\eta_{IN}(\text{max}) = \eta_D \eta_Q. \tag{4.49}$$

For KrF lasers with an Ar diluent this product can exceed 40 %. Of course, the achievable overall efficiency will be lower due to mirror and window losses, photoabsorption in the gas, power supply losses and imperfect kinetics.

The gain in discharge pumped lasers may be estimated in the following way [4.110]. The total discharge power per unit volume is given by the expression

$$P_D = J\mathcal{E} = n_e q_e \mu_e \mathcal{E}^2, \tag{4.50}$$

where J is the current density and $\mathcal{E}$ the electric field, n_e is the electron density and q_e the electron charge. The quantity $\mu_e p$ depends on the gas composition and $\mathcal{E}/p$, where p is the total pressure [4.111]. However, for typical discharge laser mixtures, which contain mostly Ar or He, the mobility is only weakly dependent on $\mathcal{E}/p$. Thus, in mixtures dominated by Ar, $\mu_e p \approx 4 \times 10^5$ (cm$^2 \cdot$ atm kV$^{-1} \cdot$ s^{-1}) [4.111]. For the case of ideal kinetics (one excited rare gas halide molecule formed from each excited rare gas atom produced by the discharge) the density of excited rare gas halide molecules is given by the expression

$$n^* = P_D \eta_D \eta_Q \tau / h\nu, \tag{4.51}$$

where τ is the effective lifetime of the upper laser level and $h\nu$ is the laser photon energy. In the absence of stimulated emission the lifetime is given by the expression

$$\frac{1}{\tau_0} = \frac{1}{\tau_\nu} + \frac{1}{\tau_Q}, \tag{4.52}$$

where τ_ν is the spontaneous radiative decay time and τ_Q is the quenching lifetime. The small signal gain is then given by the formula

$$g_0 = \sigma_\nu n_0^* = \eta_D \eta_Q \frac{\sigma_\nu \tau_0}{h\nu} n_e q_e \mu_e \mathcal{E}^2. \tag{4.53}$$

In a well designed laser (η_D near unity), we see that $g_0 \propto n_e \mathcal{E}^2$.

Figure 4.5 shows a calculation of laser gain as a function of the electron density n_e and electric field $\mathcal{E}$, for a typical laser mixture consisting of 1 atm of Ar containing 4 % Kr and 0.2 % F$_2$, calculated using (4.53). The discharge efficiency η_D and electron mobility μ_e are calculated including the effects of excited rare gas atoms on the discharge [4.51]. The cross sections for exciting

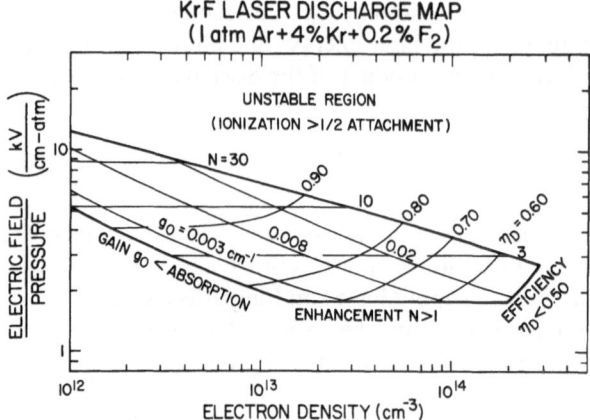

Fig. 4.5. Map showing the region in parameter space (electric field vs. electron density) where KrF lasers can be pumped by an electron beam stabilized discharge. These calculations are for a mixture of Ar with 4% Kr and 0.2% F_2 at a total pressure of 1 atm

and ionizing the excited rare atoms are estimated by extending the analogy developed earlier between excited rare gases and alkali metals, and adopting the appropriate alkali atom cross sections [4.51]. To estimate the number of excited atoms present in the discharge, a greatly simplified kinetic model has been assumed, in which the discharge forms excited Kr* atoms directly with an efficiency η_D. These react with the F_2 to form KrF* by reaction (4.8). We see that under conditions of low gain (low excitation) and high electric field, where the discharge efficiency is high, the lines of constant small signal gain g_0 obey the approximate proportional relationship $g_0 \propto n_e \mathscr{E}^2$. Under conditions of high gain and low electric field the gain is reduced somewhat by losses from inelastic collisions of low energy electrons with excited species in the gas. Because of the large number of low energy electrons in a discharge, especially when the electric field is small, the effect of excited species on the discharge efficiency can be appreciable at excitation densities as low as 1 part in 10^5, corresponding to a gain of only $2\% \, cm^{-1}$ under the conditions of Fig. 4.5 [4.51].

4.4.2 Electron Beam Stabilized Discharges

From Fig. 4.5 it is clear that to obtain the highest efficiency it is desirable to operate at the highest electric field. However, the extent to which this can be done is limited by discharge stability. The best way to achieve large pulse energy and high discharge efficiency is to use discharges which are stabilized by electron beam ionization, as shown schematically in Fig. 4.6. In devices of this type, the discharge electrons are formed by electron beam irradiation of the laser gas, as discussed above, rather than by the discharge itself. In fact, without the ionizing radiation the discharge is self-extinguishing. The purpose of the

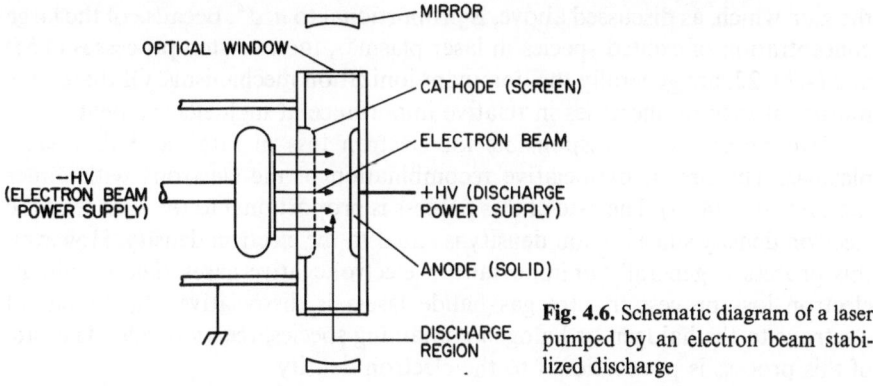

Fig. 4.6. Schematic diagram of a laser pumped by an electron beam stabilized discharge

ionizing radiation is to make up the difference between the rate of ionization due to the discharge and the rate of electron loss due to dissociative attachment and recombination. In this way, it is possible to achieve absolute discharge stability [4.112] and the electron density and electric field may be separately controlled to optimize the performance.

Four processes are responsible for ionization of the gas by the discharge. The first of these is direct ionization of ground state atoms and molecules in the gas by the discharge electrons, for example,

$$Kr + e^- \rightarrow Kr^+ + 2e^- . \tag{4.54}$$

The second is Penning or associative ionization, as in (4.25). The rate of these processes is proportional to the electron density. Although they are relatively more important in Ne dominated mixtures [4.113], these processes are unimportant in most electron-beam stabilized discharges because there are too few electrons with enough energy. The third ionization process is electron impact ionization of excited species, for example,

$$Kr^* + e^- \rightarrow Kr^+ + 2e^- . \tag{4.55}$$

These processes can be very fast since highly excited species are easily ionized by low energy electrons. In fact, even the lowest excited states of the rare gases are within 4 eV of ionization. Since the concentration of excited species in the gas is proportional to the gain, the rate of this process is proportional to the product of the gain and the electron density. However, as discussed above, the gain is itself proportional to the product of the electron density and the square of the electric field. Thus, the rate of this ionization process is proportional to $n_e^2 \mathscr{E}^2$. The fourth ionization process is mutual quenching of two excited species, reactions (4.20–22), in which two excited species pool their energy to form an ion or dimer ion plus an electron. Since the concentration of excited species is proportional to the gain, the rate of this process is proportional to the square of the gain which, as discussed above, is proportional to $n_e^2 \mathscr{E}^4$. Because of the large concentration of excited species in laser plasmas, the two-step processes (4.55) and (4.20–22) are generally the dominant ionization mechanisms. Of these two, mutual quenching increases in relative importance at high electric fields.

Two processes are responsible for electron loss in rare gas halide laser plasmas. The first is dissociative recombination of the electrons with dimer ions, reaction (4.19). The rate of this process is proportional to the square of the electron density since the ion density is equal to the electron density. However, this process is generally unimportant in electronegative gases. The dominant electron loss process in rare gas halide lasers is dissociative attachment of electrons to the halogen or halogen-containing species, reaction (4.9). The rate of this process is proportional to the electron density.

Since the dominant ionization processes are quadratic in the electron density, whereas the dominant electron loss process is linear in the electron density, the discharge is stable only so long as the discharge ionization rate is less than one-half the electron loss rate [4.112]. This limit is shown in Fig. 4.5. Clearly, the discharge is most stable, that is, the highest electric field strengths are possible, under conditions of low electron density and low gain. These are also the conditions for the highest discharge efficiency. However, the laser energy extraction efficiency under low gain conditions is low, as we shall see.

Another limit circumscribing the "window" in Fig. 4.5 within which electron beam stabilized rare gas halide lasers can operate is that of "discharge enhancement" [4.112]. Because of the limitations of electron-beam pumping, it is desirable to put most of the energy into the laser by means of the discharge rather than the electron beam used to produce ionization. The ratio of the discharge power to the electron beam power is called the enhancement ratio. The power which must be provided by the electron beam is given by the product of the energy W required to produce one electron–ion pair and the rate of dissociative attachment,

$$P_{EB} = n_e n_a k_a W, \tag{4.56}$$

where k_a is the dissociative attachment rate and n_a is the density of attaching species. The enhancement ratio, N, is therefore given by the expression

$$N = \frac{P_D}{P_{EB}} = \frac{q_e \mu_e \mathscr{E}^2}{n_a k_a W} \propto \mathscr{E}^2. \tag{4.57}$$

The point at which the discharge power is equal to the electron beam power ($N = 1$) is shown as the lower limit of electron-beam stabilized discharge operation in Fig. 4.5. As indicated in Fig. 4.5, large enhancement ratios are possible under the same conditions where the discharge efficiency is high but the gain is relatively low.

These results show that to achieve high discharge efficiency and a large enhancement ratio, it is necessary to operate at conditions of low gain and low specific power. However, under conditions of low gain the laser power extraction efficiency is reduced by parasitic losses, which include window and mirror losses and photoabsorption in the laser gas. Of these, photoabsorption is a serious problem even for large lasers. To show this, we consider a laser medium having a gain coefficient g (in the absence of absorption) and a parasitic absorption coefficient α. The extraction efficiency η_{EX} (defined as the photons extracted by the laser flux ÷ the upper laser level molecules produced by the discharge), is given by the expression

$$\eta_{EX} = \frac{\Phi(g-\alpha)}{n_0^*/\tau_0}, \tag{4.58}$$

where Φ is the laser flux, n_0^* is the density of upper laser level molecules in the absence of stimulated emission, (4.51), and τ_0 is the effective lifetime of an upper laser level molecule in the absence of stimulated emission, (4.52). Generalizing from (4.51–53), we see that the gain in the presence of the laser flux (the saturated gain) is given by the formula

$$g = n^*\sigma_v = n_0^*\sigma_v \frac{1/\tau_0}{1/\tau_0 + \Phi\sigma^*} = \frac{g_0}{1+\Phi/\Phi_s}, \tag{4.59}$$

where g_0 is the small signal gain (4.52) and $\Phi_s = 1/\sigma_v\tau_0$ is the saturation flux, that is, the flux at which the stimulated emission time is equal to the lifetime τ_0. Combining this with (4.58), we obtain the expression [4.87]

$$\eta_{EX} = \frac{\Phi}{\Phi_s}\left(\frac{1}{1+\Phi/\Phi_s} - \frac{\alpha}{g_0}\right). \tag{4.60}$$

This shows that the extraction efficiency is maximized by adjusting the laser flux to a value which depends on the absorption coefficient α and its bleaching behavior. If we assume that α is not bleachable (i.e., not dependent on Φ), then we can easily optimize Φ/Φ_s to obtain the maximum extraction efficiency, given by the expression

$$\eta_{EX}(\max) = 1 - 2\left(\frac{\alpha}{g_0}\right)^{1/2} + \frac{\alpha}{g_0}. \tag{4.61}$$

From this formula we see that to achieve an extraction efficiency greater than 50% it is necessary that g_0/α be greater than about 12. For the conditions of Fig. 4.5, where the F_2 density is $10^{17}\,\mathrm{cm}^{-3}$, corresponding to an absorption coefficient of about $1.5 \times 10^{-3}\,\mathrm{cm}^{-1}$ (see Table 4.2), we see that to achieve an extraction efficiency greater than 50% it is necessary to operate with a small signal gain in excess of about $2\%\,\mathrm{cm}^{-1}$. This corresponds to a rather small section of the stable operating window shown in Fig. 4.5, and illustrates the difficulty of achieving efficient laser energy extraction in the presence of halogen absorption. Other absorption processes are important as well, as discussed in Sect. 4.2. Their effect on efficiency is more difficult to estimate since their bleaching behavior is more complex and may affect the laser performance in other ways. For example, photodetachment and photoionization of excited species could act as a source of electrons and destabilize the discharge, although this has not been observed [4.87].

To date, the largest pulse energy which has been reported with an electron beam stabilized discharge is 50 J [4.87]. This was obtained from KrF in a 0.3 μs pulse with an enhancement ratio of the order of 2. The experiments were conducted with a broad area (0.2 × 2 m), 400 kV electron beam having a current density of a few Amperes per square centimeter. The laser volume was 30–40 l, containing two atmospheres of Ar with about 6 % Kr and 0.3 % F_2. The electric field strength was about 2–2.5 kV cm^{-1}·atm^{-1}, and the discharge current density was of the order of 15 A cm^{-1}, corresponding to an electron density of the order of 10^{14} cm^{-3}. The maximum efficiency (laser output ÷ energy deposited in the gas) observed in these experiments was 3 %. The low efficiency is attributed to poor laser extraction efficiency, η_{EX}, of the order of 10 %. Fluorescence measurements indicated a pumping efficiency ($\eta_D\eta_Q$) of the order of 20 % or more. Although the conditions are slightly different from those of Fig. 4.5, the results are in general agreement. Significantly lower intrinsic efficiency, of the order of 0.3 %, has been observed from XeF pumped by an electron-beam stabilized discharge [4.114]. In this experiment NF_3 was used, which does not absorb at the laser wavelength, but the laser was probably not saturated, which would account for a low extraction efficiency. Absorption by excited and ionized rare gas atoms and molecules are expected to be important for the saturation behavior of XeF lasers [4.56].

4.4.3 Avalanche Discharges

For many applications small pulses and somewhat lower efficiency are acceptable in the interest of simplicity. For these purposes it is possible to use unstable, fast transverse discharge, as shown schematically in Fig. 4.7. Discharges of this type have been used in the past for pulsed N_2 (337 nm) and CO_2 (10 μm) lasers. In rare gas halide lasers such discharges are fundamentally unstable, as described above. Once the gas breaks down the electrons multiply rapidly and the discharge impedance collapses. The discharge current is then inductance limited, reducing the voltage across the discharge [4.109, 116]. Moreover, spatial nonuniformities in the discharge grow with time and develop into arcs, or streamers. For these reasons it is essential to minimize the inductance in the discharge circuit and keep the current pulse short.

Many different configurations have been used for the power supply. These include Blumlein circuits with strip transmission lines [4.116–123], charged coaxial cables [4.124], lumped element transmission lines (or pulse-forming

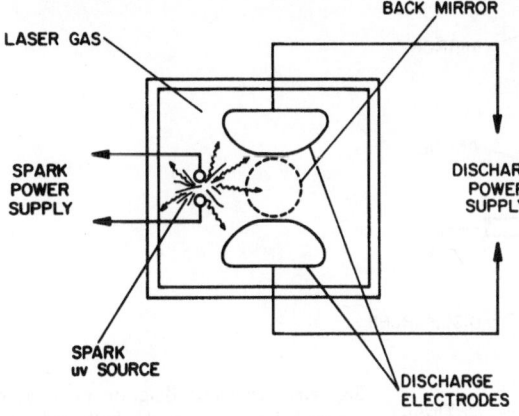

Fig. 4.7. Schematic diagram (viewed along the laser axis) of a laser pumped by a transverse discharge using uv radiation from a spark source to preionize the gas

networks) [4.125], and simple capacitors [4.46, 86, 126]. The power supply may be switched into the discharge with a low inductance spark gap [4.125], or the series inductance of the switch may be eliminated by impulse charging the power supply [4.46, 123] or allowing it to ring up to voltage (in a Blumlein or $L–C$ inversion circuit) [4.86, 115–123] and then allowing the laser gas to break down when the voltage is high enough.

Because of the development of streamers, or arcs, the useful electrical pulse length of lasers with no external stabilization is limited to about 30 ns at low pressure ($\lesssim 0.5$ atm) [4.126], and proportionately less at higher pressures [4.116–118, 120–123]. To suppress instabilities at higher pressures, the gas can be preionized. This can be done by means of a short preionizing corona discharge pulse [4.45, 117, 119], but better results have been achieved by weakly preionizing the gas with uv radiation from a series of sparks distributed along the length of the laser [4.46, 86, 115, 125], as shown schematically in Fig. 4.7. The uv radiation from the spark gaps ionizes impurities in the laser gas [4.127, 128]. Most of these electrons rapidly attach to the halogen in the gas to form negative ions. However, the unattached electrons (and possibly the negative ions [4.129, 130], start the discharge with a larger and more uniform electron density than would be the case otherwise. By reducing the number of orders of magnitude through which the electron density must avalanche, the development of arcs is forestalled, although usually not eliminated.

Somewhat longer pulses have been obtained from resistively ballasted pin electrodes, shown schematically in Fig. 4.8 [4.124, 131]. Stable glow discharge pulses as long as 150 ns have been observed [4.131], although the laser pulses were much shorter, appearing at the end of the discharge pulse. Laser pulse energies have been small, of the order of 50 μJ from KrF [4.124], but larger pulse energies should be possible.

In general, fast avalanche discharge lasers operate best with a buffer gas of He, rather than Ar, although Ne is also satisfactory [4.126]. Discharges in Ar

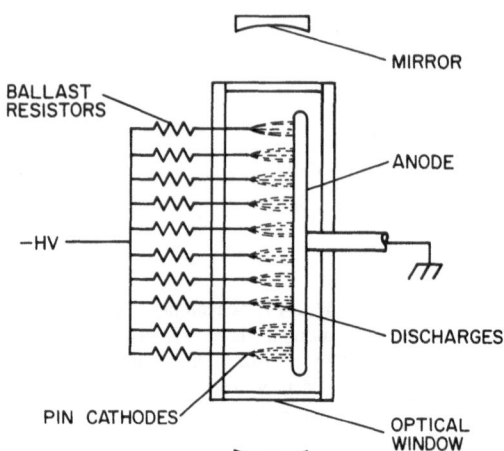

Fig. 4.8. Schematic diagram of a laser pumped by a resistively ballasted pin discharge

are observed to form arcs rather than stable glow discharges. As mentioned above, the presence of large amounts of He in the discharge leads to significant losses due to momentum transfer from the discharge electrons to the relatively light He atoms, especially at low values of the electric field. For example, at $2.5\,\mathrm{kV\,cm^{-1}\cdot atm^{-1}}$ in a mixture of He with 10 % Ar and 0.2 % F_2, the loss due to elastic collisions is about 50 % [4.109]. Typical mixtures consist of 1–2 atm of He with about 10 % Kr (or Ar) for KrF (or ArF) lasers, or about 1 % Xe for XeF lasers. The reason why smaller concentrations of Xe are best is not known, but it may be due to absorption by Xe_2^+ or Xe_2^*, as discussed in Sect. 4.2, or to the relatively large rate of quenching of XeF* by Xe (see Table 4.5). F_2 is generally used as the halogen source in ArF and KrF lasers, and NF_3 in XeF lasers, to achieve maximum performance [4.86]. However, NF_3, N_2F_4, and SF_6 have also been used in KrF lasers, F_2 and SF_6 in XeF lasers, and NF_3 and SF_6 in ArF lasers, with somewhat poorer results [4.45, 46, 123]. In general, the optimum concentration of F_2 is of the order of 0.2 %. Somewhat higher concentrations of NF_3 are optimum, probably because NF_3 has smaller cross sections than F_2 for both forming and quenching XeF in reactions analogous to reactions (4.8, 13) (see Tables 4.4 and 4.5), and because NF_3 does not absorb at the XeF laser wavelength.

The largest pulse energy which has been reported to date with an avalanche discharge is 600 mJ, obtained from KrF [4.132]. In these experiments a short, lumped-element, $0.15\,\Omega$ pulse forming network was impulse charged to more than 100 kV and switched into the discharge through an overvolted surface rail gap. This provided a very fast voltage rise-time which, together with the high voltage, made it possible to operate at as much as 6 atm total pressure. At these high pressures, a leaner mixture (He with only 1.2 % Kr and 0.07 % F_2) was found to be optimum. The laser volume was 60 cm long, 2.4 cm between electrodes, and totaled 180 cm^3, corresponding to an output energy density of

$0.6 \, \mathrm{J} \, \mathrm{l}^{-1} \cdot \mathrm{atm}^{-1}$. The mixture was preionized with uv radiation from a series of sparks and yielded a 15 ns laser pulse with no evidence of arc formation. The corresponding peak power was 40 MW. The energy stored in the impulse charging circuit was about 1 kJ, so the overall efficiency (laser energy ÷ stored energy) was only 0.06 %. The same experiment produced 250 mJ XeF laser pulses from mixtures of He with Xe and NF_3 [4.125].

Much higher efficiencies have been achieved in smaller devices with lower pulse energies. The best result reported to date is 1.4 % (laser energy ÷ stored energy) obtained in a 130 mJ pulse from KrF [4.86]. The laser used in these experiments was similar to the one described above in concept and size, but the power supply consisted of a lumped-element $L–C$ inversion doubling circuit initially charged to 25 kV. The total pressure was limited to about 2 atm, at which pressure the maximum output was obtained. The discharge was observed to be only about 0.5 cm wide, corresponding to an active volume of 60 cm^3 and an output energy density of $1 \, \mathrm{J} \, \mathrm{l}^{-1} \cdot \mathrm{atm}^{-1}$. This same apparatus produced 65 mJ XeF laser pulses and 60 mJ ArF laser pulses from He/Xe/NF$_3$ and He/Ar/F$_2$ mixtures, respectively. The largest output energy density achieved so far is about $10 \, \mathrm{J} \, \mathrm{l}^{-1} \cdot \mathrm{atm}^{-1}$, obtained in a 17 mJ pulse from XeF [4.121]. In this experiment no preionization was used and the pulse was very short (about 6 ns), although the pressure was only 0.7 atm.

The scaleability of these lasers is difficult to predict since their behavior is not well understood. The principal limitation seems to be the growth of spatial nonuniformities (arcs), which severely limits the pulse length. This problem is much more restrictive in the rare gas halide lasers than in CO_2 lasers because the attachment dominated discharges are fundamentally unstable, as discussed above. Simply extrapolating from present results, it appears that pulse energies of the order of Joules should be possible in the near future. With resonant charging of the energy storage capacitors, which can be done with more than 80 % efficiency [4.133], overall ("wall-plug") efficiencies in excess of 1 % are possible with no advance beyond present laser performance. Higher efficiencies may prove difficult to achieve since the discharge impedance is collapsing during the pulse as the electron density is exponentially growing. This makes it difficult to effectively couple the power supply to the discharge.

The biggest advances remain to be made in pulse repetition frequency. Up to the present time the highest pulse repetition rate has been 1000 Hz, with an average power of 10 W [4.134]. Ultimately, the pulse repetition frequency will be limited by: 1) the effect of the discharge on medium homogeneity and the rate at which the disturbances can be damped out and new gas flowed into the cavity, and 2) the development of high voltage, high current switching technology. Up to the present time, high pressure gas filled spark gaps have generally been used since they are simple and inexpensive, have relatively low inductance and short switching times, and can handle the large currents. However, they are generally not suitable for high repetition rates since the lifetime (typically of the order of 10^7 shots) is too short for reliable long term operation. Moreover, the recovery time of conventional gaps limits the pulse repetition frequency to hundreds of Hertz, although blowing and cooling (as in quenched gaps) can greatly extend this. Vacuum gaps recover much more quickly, but have not yet been developed for high repetition rate applications. Thyratrons have been operated at high repetition frequencies (up to 100 kHz), but they have some difficulty handling the large currents and short rise times required by fast discharge rare gas halide lasers. At the present time, several types of high power solid state switches are under active development, including thyristors, reverse blocking diode thyristors and light activated silicon switches (LASS) [4.135, 136]. Of these, the LASS appears most promising for fast rise time, high current, pulsed laser applications. Peak currents as high as 25 kA, and current rise rates as high as 760 kA μs^{-1} have been demonstrated in experimental devices. Although the voltage standoff of such devices is only of the order of kilovolts, they can be stacked in series to switch much higher voltages. Triggering (by means of a Nd:YAG laser in present devices) is a complication, but kilohertz pulse repetition frequencies do not present any fundamental problems. Thus, for the near term thyratrons offer the highest repetition frequency and reliability, but in the long term other technologies may prove superior.

4.5 Applications

The basic performance features which, if they are realized, will make rare gas halide lasers useful include large pulse energy, high average power, and high efficiency, previously not available at uv wavelengths. Already 0.6 J pulses have been achieved from tabletop fast discharge KrF lasers [4.132], and pulse energies of the order of Joules should be possible in the near future. Single pulse energies as large as 350 J have been obtained from electron-beam pumped KrF [4.87], and it appears possible to increase this by at least an order of magnitude. If problems of laser energy extraction can be overcome, electron-beam stabilized discharges should produce comparable pulse energies. The limits of direct nuclear pumping are not even known. High average power has not received much attention, yet, but it should be possible to repetitively pulse fast discharge lasers at several Joules per pulse and kilohertz frequencies, corresponding to kilowatt average powers. Higher average powers should be possible from lasers pumped by electron beams or electron-beam stabilized discharges. In such lasers the pulses will be larger and the repetition frequency lower, due to foil heating. Efficiencies as high as 1.4 % (laser energy ÷ stored energy) have been demonstrated in fast discharge lasers [4.86], and with efficient charging circuits overall efficiencies in excess of 1 % should be possible in the near future. Efficiencies as high as 0.8 % (laser energy ÷ stored energy) have been demonstrated with electron beam pumped KrF, even though the energy deposition in the laser cavity was very inefficient [4.44]. Intrinsic efficiencies in excess of 10 % (laser energy ÷ energy deposited in cavity) have been demonstrated [4.44, 105], suggesting that overall efficiencies of a few percent should be possible. Comparable or greater efficiency should be possible from electron beam stabilized discharges, depending on what laser extraction efficiency can be obtained.

For many applications the uv wavelengths at which the other rare gas halide lasers operate are also interesting. Efficient oscillation has been obtained from ArF, KrF, XeCl, and XeF in narrow bands near 193, 248, 308, and 351 nm, respectively. Weaker oscillation has been obtained from KrCl and XeBr near 222 and 282 nm, respectively. It has also been possible to tune KrF and ArF lasers over the width of the emission bands, about 2 nm [4.86, 123, 137]. The other lasers should be similarly tunable.

The wavelength capability of these lasers can be considerably extended by a number of techniques including optical pumping of other laser media, stimulated Raman scattering and parametric conversion. Optically pumped dyes are attractive because they are continuously tunable over a broad range. XeF (351 nm) laser radiation should be useful for pumping a variety of visible dyes with greater overall efficiency than N_2 (337 nm) lasers and less dye degradation than flashlamps. Shorter (uv) wavelengths have been obtained by pumping para-terphenyl in cyclohexane with KrF (248 nm) laser radiation [4.138, 139]. Broadband (untuned) output of the dye at 340 nm amounted to as much as 28 % conversion, and the emission was tunable between 323 and 364 nm. Other dyes and other pump lasers may produce tunable emission at shorter wavelengths, but absorption to higher singlet levels in the dye is expected to be a problem. Alternatively, it may be possible to achieve discretely tunable oscillation at shorter wavelengths by optically pumping simpler molecules, such as I_2, with ArF (193 nm) radiation [4.85, 140].

Recently, stimulated Raman scattering has been used to shift the wavelengths of XeF, KrF and ArF lasers. XeF (351 nm) radiation was shifted to 585 nm by near-resonant Raman scattering in Ba vapor, with a conversion efficiency of 80 % [4.141]. ArF and KrF radiation has been shifted to a variety of wavelengths in the uv and visible portions of the spectrum by nonresonant Raman scattering in high pressure H_2, D_2, and CH_4 [4.142] and liquid N_2 [4.143]. Because the incident laser radiation is not resonant with the scattering medium, the gain at the Raman shifted wavelengths is comparable to that at the initial wavelength, and successive shifts are observed. This gives rise to a number of lines from each combination of laser and scattering medium. Shifts as far as the seventh Stokes position and second anti–Stokes position have been observed from KrF scattered in H_2, although 4–wave mixing is believed to account for the anti–Stokes and higher Stokes components. Total pump conversion in excess of 50 % has been observed, with more than 25 % conversion to the first or second Stokes component.

Except for the 4–wave mixing mentioned above, parametric processes have not been used to shift the wavelengths of rare gas halide lasers, but they offer interesting possibilities, particularly for achieving shorter wavelengths, as has been done with other uv lasers. For example, Xe_2^* (171 nm) excimer laser emission has been tripled to 57 nm by 4–wave mixing in Ar [4.144], and the fifth harmonic of quadrupled Nd^{+++} (266 nm) emission has been generated in He to produce coherent radiation at 53 nm [4.145].

Both short and long pulses can be generated with rare gas halide lasers. Ultrashort pulses can be formed by mode-locking techniques. In preliminary experiments with XeF, pulses about 2 ns long were obtained using active mode-locking [4.120]. Much shorter pulses, of the order of 100 ps or less, should be possible. However, in these experiments the gain lasted only about 15 ns, enough for about four round trips of the mode-locked pulses. Shorter pulses will require longer gain times. In the absence of sufficiently long gain times for effective mode-locking, an XeF laser has been used to amplify small 200 ps pulses of tripled Nd:glass (353 nm) radiation [4.146]. Amplification factors as high as 6000 were achieved, corresponding to a small signal gain coefficient of $13\% \text{cm}^{-1}$ (no saturation effects were observed with output pulse energies of the order of 10 μJ/pulse). Because of the short lifetime of the rare gas halides, they cannot be used to store energy for times longer than about ten nanoseconds or so. Thus, they do not lend themselves to the formation of very large pulses by Q-switching or by amplification of short pulses from a master oscillator. Moreover, their high gain becomes difficult to stand off when significant amounts of energy are stored. For example, a gain of $10\% \text{cm}^{-1}$ in KrF corresponds to a stored energy of only $0.3 \text{J} \text{l}^{-1}$. However, other techniques can be used to generate giant pulses from rare gas halide laser emission, including optical pumping of suitable lasers and Raman pulse compression. Examples of optically pumped systems include the atomic I (1.3 μm) photodissociation laser and the proposed vapor phase rare earth ion lasers. In the atomic I laser, KrF radiation could be used to photolyze C_3F_7I or CF_3I to produce excited $I^*(^2P_{1/2})$ atoms [4.147]. These have a long lifetime and low gain coefficient suitable for storing large amounts of energy. In a closely related experiment laser action was observed from C_3F_7I photolyzed by XeBr (282 nm) fluorescence [4.147]. In the proposed vapor phase rare-earth lasers, vapors of rare-earth trihalogens, chelates or rare-earth transition metal trihalide complexes are used in place of a rare-earth doped glass or crystal [4.148]. Although oscillation has not been demonstrated, several transitions appear promising [4.149] and some could be pumped by rare gas halide lasers, such as $TbCl_3(AlCl_3)_n$ pumped by KrF (248 nm) [4.150]. Short, large pulses could also be obtained by stimulated Raman scattering [4.151]. To achieve pulse compression, a short pulse at the Stokes wavelength is injected into the scattering medium in the direction opposite the pump laser pulse. As it passes through the oncoming pump pulse it is amplified by stimulated backward Raman scattering.

Long pulses (perhaps even cw operation) are also possible from rare gas halide lasers. Because of the dissociative nature of the lower laser level, there is no inherent limitation on pulse length due to bottlenecking. Even in XeF, which might have proved the exception since the ground state is weakly bound, laser pulses as long as 1 μs have been observed [4.100]. Because of the short radiative lifetimes and rapid kinetics of the rare gas halides, the lasers operate in a steady-state, or quasi–cw mode on time scales longer than about 10 ns. The fundamental limitation on pulse length is common to all broadband lasers, that is, the high power per unit volume needed to achieve threshold gain. From (4.2, 51) we see that the pump power is given by the expression

$$P_{p} = \frac{hvg_0}{\eta_p \eta_Q \sigma \tau} = 4\sigma \left(\frac{\sigma}{\ln 2}\right)^{1/2} \frac{\tau_v}{\tau} \frac{hc^2 \Delta \lambda g_0}{\eta_p \eta_Q \lambda^5}. \tag{4.62}$$

For KrF with a small signal gain of $1\% \, cm^{-1}$ this corresponds to more than $10 \, kW \, cm^{-3}$. Besides necessitating a large power supply, the high threshold pumping power affects long pulse laser performance in two other ways. First of all, nonuniform heating of the gas by the pump causes density and index of refraction gradients which destroy the beam quality. It is easily shown that for a given amount of energy deposited in the gas over a time interval Δt, the density fluctuations $\Delta \varrho$ increase with the pumping time according to the approximate expression (valid for short times)

$$\Delta \varrho \approx \frac{1}{6} \left(\frac{a \Delta t}{l}\right)^2 \Delta \varrho_\infty, \tag{4.63}$$

where a is the speed of sound, l the scale size of the fluctuations, and $\Delta \varrho_\infty$ the density fluctuation which would result from adiabatic expansion of the heated gas to the ambient pressure [4.152]. Thus, we see that the fluctuations cannot grow fast enough to affect short pulse lasers, but they can be important for high power laser pulses lasting more than a few microseconds. A second restriction on pulse length is imposed by consumption of the halogen, as discussed at the end of Sect. 4.2. Thus, for a small signal gain of $1\% \, cm^{-1}$ and an initial F_2 density of $10^{17} \, cm^{-3}$, the F_2 will be consumed in a time of the order of $10 \, \mu s$.

Considering their potential advantages in terms of power and efficiency, the rare gas halide lasers appear to be suitable for a variety of applications. At a cost of $0.01/kWh for electricity and an overall efficiency of 1%, KrF laser photons can be produced for as little as $0.10/mole (plus capital costs, maintenance, laser gases, etc.!). This suggests the possibility of large scale laser chemistry, at least for expensive chemicals. Among the most expensive chemicals are rare isotopes and trace impurities (expensive to remove). Considerable attention has been given to laser isotope separation [4.153]. For example, KrF pumped p–terphenyl lasers have been proposed for separating isotopes of hydrogen by photolysis of formaldehyde [4.154, 155], and recently an ArF laser was used to separate isotopes of oxygen [4.156]. Laser purification of chemicals has received less attention, but recently an ArF laser was used to remove PH_3, AsH_3, and B_2H_6 impurities from SiH_4 by selective photolysis [4.157]. This appears to be a very promising process since high purity SiH_4 is used extensively in the electronics industry. Cheap uv photons also could be used for other types of materials processing, such as micro-machining and lithography [4.158].

The pulsed nature of the rare gas halide lasers does not make them well suited for communications on earth, where available lower power cw lasers are satisfactory and atmospheric absorption and scattering are problems for uv lasers at long ranges. However, the high brightness of the rare gas halide lasers is an advantage in space-borne applications such as long range communications and optical radars. But even on earth, short, high intensity uv laser pulses should be useful for applications such as remote sensing of the atmosphere by Raman backscatter. For the near term, of course, most of the applications of rare gas halide lasers are likely to be in research. Nevertheless, these recently discovered lasers promise to be useful in a much broader sphere than the one which gave them birth.

4.6 Recent Advances in Rare Gas Halogen Systems

Since the publication of the first edition, important refinements in the understanding and operation of rare gas halogen systems have occurred. A review of many kinetic aspects has been given by *Johnson* and *Hunter* [4.159]. In particular, the kinetics involving electron collisions with excited states and super-elastic quenching processes have received greater attention. The reaction rate system, given as an example from the work of Johnson and Hunter for KrF* in Table 4.14, provides a picture of the complexity and level of detail incorporated in current models.

Electron super-elastic collisions can cause considerable quenching of excited state species [4.187–189]. In particular, processes such as

$$e^- + KrF(B\,^2\Sigma) \rightarrow Kr + F + e^- \tag{4.64}$$

can serve to reduce the efficiency of the laser medium at sufficiently high electron densities. This general class of processes has been examined both theoretically for KrF* and XeF* [4.190], and experimentally for KrF* and ArF* [4.191]. The theoretical analyses indicated rates of $7 \times 10^{-8}\,\mathrm{cm^3/s}$ and $3 \times 10^{-8}\,\mathrm{cm^3/s}$ for XeF* and KrF*, respectively. Experimentally, however, somewhat larger values corresponding to $\sim 2 \times 10^{-7}\,\mathrm{cm^3/s}$ were found for KrF* and ArF*. Values of this latter magnitude establish an upper limit on the electron density of $\sim 10^{15}\,\mathrm{cm^{-3}}$ for efficient operation.

Preionization of rare gas halogen discharge lasers with x-rays [4.192, 193] has been successfully applied. This technique combines simplicity of design with efficient operation at large optical apertures. This development is expected to be an important feature of reliable high-power systems.

Table 4.14. KrF reaction rate system

No.	Reaction	Rate constant $[\mathrm{s}^{-1}]$	References
	Pumping, quenching, and stability reactions		
1	$Kr^* + F_2 \rightarrow KrF^* + F$	8.1×10^{-10} cm^3	[4.160]
2	$ArF^* + Kr \rightarrow KrF^* + Ar$	3.0×10^{-10} cm^3	[4.160, 161]
3	$Kr^+ + F^- \rightarrow KrF^*$	1.0×10^{-6} cm^3	[4.161, 162]
4	$Ar^* + F_2 \rightarrow ArF^* + F$	8.5×10^{-10} cm^3	[4.160]
5	$Ar^+ + F^- \rightarrow ArF^*$	1.0×10^{-6} cm^3	[4.162]
6	$Kr^{**} + F_2 \rightarrow KrF^* + F$	8.1×10^{-10} cm^3	[4.160]
7	$Ar^{**} + F_2 \rightarrow ArF^* + F$	8.5×10^{-10} cm^3	[4.160]
8	$Kr^+ + F^- \rightarrow KrF^* + Kr$	1.0×10^{-6} cm^3	[4.161–163]
9	$Ar_2^+ + F^- \rightarrow ArF^* + Ar$	1.0×10^{-6} cm^3	[4.161–163]
10	$ArKr^+ + F^- \rightarrow KrF^* + Ar$	1.0×10^{-6} cm^3	Estimated
11	$ArKr^* + F_2 \rightarrow KrF^* + Ar + F$	6.0×10^{-10} cm^3	[4.160]
12	$KrF^* + Kr \rightarrow 2Kr + F$	2.0×10^{-11} cm^3	[4.160]
13	$KrF^* + Ar \rightarrow Kr + Ar + F$	0.0 cm^3	Estimated
14	$KrF^* + 2Ar \rightarrow ArKrF^* + Ar$	8.0×10^{-32} cm^6	[4.161, 164]
15	$ArF^* + 2Ar \rightarrow Ar_2F^* + Ar$	5.0×10^{-32} cm^6	[4.160]
16	$Kr + ArKrF^* \rightarrow Kr_2F^* + Ar$	2.0×10^{-11} cm^3	[4.161]
17	$Ar + ArKrF^* \rightarrow Ar_2F^* + Kr$	2.0×10^{-11} cm^3	[4.161]
18	$KrF^* + Kr + Ar \rightarrow Kr_2F^* + Ar$	6.5×10^{-31} cm^2	[4.165]
19	$ArKrF^* + F_2 \rightarrow Ar + Kr + F + F_2$	1.0×10^{-9} cm^3	[4.160]
20	$Kr_2^* + F_2 \rightarrow Kr_2F^* + F$	3.0×10^{-10} cm^3	[4.160]
21	$Kr_2^* + F \rightarrow KrF^* + Kr$	3.0×10^{-10} cm^3	[4.160]
22	$Kr_2F^* + F_2 \rightarrow 2Kr + F + F_2$	1.0×10^{-9} cm^3	[4.160]
23	$Ar_2^* + F_2 \rightarrow Ar_2F^* + F$	2.5×10^{-10} cm^3	[4.160]
24	$Ar_2^* + F \rightarrow ArF^* + Ar$	3.0×10^{-10} cm^3	[4.160]
25	$Ar_2F^* + F_2 \rightarrow 2Ar + F + F_2$	1.0×10^{-9} cm^3	[4.160]
26	$Ar_2F^* + Kr \rightarrow ArKrF^* + Ar$	1.0×10^{-10} cm^3	[4.160]
27	$ArKr^* + Kr \rightarrow Kr_2^* + Ar$	1.0×10^{-10} cm^3	[4.160]
28	$Kr^* + 2Ar \rightarrow ArKr^* + Ar$	1.0×10^{-32} cm^6	[4.160]
29	$Ar^* + Kr + Ar \rightarrow ArKr^* + Ar$	1.0×10^{-32} cm^6	[4.160]
30	$Kr^* + Kr + Ar \rightarrow Kr_2^* + Ar$	1.0×10^{-32} cm^6	[4.160]
31	$Ar^* + 2Ar \rightarrow Ar_2^* + Ar$	1.14×10^{-32} cm^6	[4.160]
32	$Ar^* + Kr \rightarrow Kr^* + Ar$	6.2×10^{-12} cm^3	[4.160, 166]
33	$Ar_2^* + Kr \rightarrow Kr^* + 2Ar$	4.0×10^{-10} cm^3	[4.161]
34	$Ar^+ + 2Ar \rightarrow Ar_2^+ + Ar$	2.5×10^{-31} cm^6	[4.167]
35	$Ar_2^+ + Kr \rightarrow Kr^+ + 2Ar$	7.5×10^{-10} cm^3	[4.168]
36	$Ar^+ + Kr + Ar \rightarrow ArKr^+ + Ar$	1.0×10^{-31} cm^6	[4.169]
37	$Kr^+ + 2Ar \rightarrow ArKr^+ + Ar$	1.0×10^{-31} cm^6	Estimated
38	$ArKr^+ + F^- \rightarrow ArKrF^*$	1.0×10^{-6} cm^3	Estimated
39	$ArKr^+ + Kr \rightarrow Kr_2^+ + Ar$	3.2×10^{-10} cm^3	Estimated
40	$ArKr^+ + e \rightarrow Kr^{**} + Ar$	1.0×10^{-7} cm^3	Estimated
41	$Kr^+ + 2Kr \rightarrow Kr_2^+ + Kr$	2.5×10^{-31} cm^6	[4.167]
42	$Ar^+ + Kr \rightarrow Kr^+ + Ar$	3.0×10^{-11} cm^3	[4.168]
43	$Kr^+ + Kr + Ar \rightarrow Kr_2^+ + Ar$	2.5×10^{-31} cm^6	[4.161, 163]
44	$F + F + M \rightarrow F_2 + M$	1.0×10^{-33} cm^6	Estimated

Table 4.14 (continued)

	Electron reactions		
	Reaction	Typical rate constant $[s^{-1}]$	References
45	$F_2 + e \to F^- + F$	1.1×10^{-9} cm^3	[4.161, 170]
46	$Kr + e \to Kr^* + e$	4.1×10^{-11} cm^3	[4.171–173]
47	$Kr^* + e \to Kr^{**} + e$	6.4×10^{-7} cm^3	[4.174]
48	$Kr + e \to Kr^+ + 2e$	1.5×10^{-15} cm^3	[4.175, 176]
49	$Kr^* + e \to Kr^+ + 2e$	4.8×10^{-8} cm^3	[4.177]
50	$Kr^{**} + e \to Kr^* + e$	8.0×10^{-7} cm^3	[4.174]
51	$Kr^* + e \to Kr + e$	8.0×10^{-11} cm^3	[4.171–173]
52	$Kr^{**} + e + \to Kr^+ + 2e$	1.8×10^{-7} cm^3	[4.177]
53	$Ar + e \to Ar^* + e$	2.8×10^{-12} cm^3	[4.171–173]
54	$Ar^* + e \to Ar^{**} + e$	6.6×10^{-7} cm^3	[4.174]
55	$Ar + e \to Ar^+ + 2e$	1.4×10^{-19} cm^3	[4.175, 176]
56	$Ar^* + e \to Ar^+ + 2e$	2.8×10^{-8} cm^3	[4.177]
57	$Ar^{**} + e \to Ar^* + e$	9.0×10^{-7} cm^3	[4.174]
58	$Ar^* + e \to Ar + e$	5.4×10^{-12} cm^3	[4.171–173]
59	$Ar^{**} + e \to Ar^+ + 2e$	1.9×10^{-7} cm^3	[4.177]
60	$Kr_2^+ + e \to Kr^{**} + Kr$	1.1×10^{-7} cm^3	[4.178–180]
61	$Ar_2^* + e \to Ar^{**} + Ar$	7.7×10^{-8} cm^3	[4.178–180]
62	$F_2 + e \to 2F + e$	3.0×10^{-10} cm^3	Estimated
63	$F + e \to F^-$	1.0×10^{-12} cm^3	Estimated

	Radiation		
	Reaction	Rate $[s^{-1}]$	References
64	$KrF^* \to Kr + F + h\nu$	1.5×10^8	[4.160, 181]
65	$ArF^* \to Ar + F + h\nu$	2.5×10^8	[4.181]
66	$Kr_2^* \to 2Kr + h\nu$	3.3×10^6	[4.160]
67	$Ar_2^* \to 2Ar + h\nu$	3.8×10^6	[4.160]
68	$Kr_2F^* \to 2Kr + F + h\nu$	6.7×10^7	[4.160]
69	$Ar_2F^* \to 2Ar + F + h\nu$	2.0×10^8	[4.160]
70	$ArKr^* \to Ar + Kr + h\nu$	3.0×10^6	[4.160]
71	$ArKrF^* \to Ar + Kr + F + h\nu$	5.0×10^7	[4.160]

	Absorption Reaction	Cross section $[cm^2]$	References
72	$F_2 + h\nu \to F + F$	1.2×10^{-20}	[4.182]
73	$F^- + h\nu \to F + e$	5.0×10^{-18}	[4.183]
74	$Kr_2^+ + h\nu \to Kr^+ + Kr$	1.5×10^{-17}	[4.184]
75	$Ar_2^+ + h\nu \to + Ar^+ + Ar$	1.0×10^{-17}	[4.184]
76	$Ar^{**} + h\nu \to Ar^+ + e$	2.3×10^{-18}	[4.185]
77	$Kr^{**} + h\nu \to Kr^+ + e$	4.5×10^{-18}	[4.185]
78	$Ar^* + h\nu \to Ar^+ + e$	1.0×10^{-19}	[4.185]
79	$Kr^* + h\nu \to Kr^+ + e$	3.2×10^{-20}	[4.185]
80	$Kr_2F^* + h\nu \to$ products	5.0×10^{-18}	Estimated
81	$Ar_2F^* + h\nu \to$ products	1.0×10^{-18}	Estimated
82	$KrF^* + h\nu \to Kr + F + 2h\nu$	2.4×10^{-16}	[4.186]

Acknowledgements. The author would like to express his gratitude to numerous colleagues at many laboratories who graciously shared their results with him prior to publication, and gave him the benefit of numerous helpful discussions.

References

4.1   S.K.Searles, G.A.Hart: Appl. Phys. Lett. **27**, 243 (1975)
4.2   C.A.Brau, J.J.Ewing: Appl. Phys. Lett. **27**, 435 (1975)
4.3   E.R.Ault, R.S.Bradford, Jr., M.L.Bhaumik: Appl. Phys. Lett. **27**, 413 (1975)
4.4   J.J.Ewing, C.A.Brau: Appl. Phys. Lett. **27**, 350 (1975)
4.5   J.A.Mangano, J.H.Jacob: Appl. Phys. Lett. **27**, 495 (1975)
4.6   G.C.Tisone, A.K.Hays, J.M.Hoffman: Opt. Commun. **15**, 188 (1975)
4.7   L.A.Kuznetsova, Y.Y.Kuzyakov, V.A.Shpanskii, V.M.Khutoretskii: Vestn. Mosk. Univ. Ser. II Khim. **19**, (1964)
4.8   M.F.Golde, B.A.Thrush: Chem. Phys. Lett. **29**, 486 (1974)
4.9   J.E.Velazco, D.W.Setser: J. Chem. **62**, 1990 (1975)
4.10   J.J.Ewing, C.A.Brau: Phys. Rev. A **12**, 129 (1975)
4.11   M.F.Golde: J. Mol. Spectrosc. **58**, 261 (1975)
4.12   C.A.Brau, J.J.Ewing: J. Chem. Phys. **63**, 4640 (1976)
4.13   J.Tellinghuisen, J.M.Hoffman, G.C.Tisone, A.K.Hays: J. Chem. Phys. **64**, 2484 (1976)
4.14   J.Tellinghuisen, G.C.Tisone, J.M.Hoffman, A.K.Hays: J. Chem. Phys. **64**, 4796 (1976)
4.15   A.L.Smith, P.C.Kobrinsky: J. Mol. Spectrosc. **69**, 1 (1978)
4.16   P.C.Tellinghuisen, J.Tellinghuisen, J.A.Coxon, J.E.Velazco, D.W.Setser: J. Chem. Phys. **68**, 5187 (1978)
4.17   M.Krauss: J. Chem. Phys. **67**, 1712 (1977)
4.18   L.D.Landau, E.M.Lifshitz: *Quantum Mechanics* (Pergamon Press, Oxford 1958)
4.19   C.A.Brau, J.J.Ewing: Unpublished
4.20   T.H.Dunning, Jr., P.J.Hay: Appl. Phys. Lett. **28**, 649 (1976)
4.21   P.J.Hay, T.H.Dunning, Jr.: J. Chem. Phys. **66**, 1306 (1977)
4.22   J.R.Murray, H.T.Powell: Appl. Phys. Lett. **29**, 252 (1976)
4.23   J.E.Velazco, J.H.Kolts, D.W.Setser: J. Chem. Phys. **65**, 3468 (1976)
4.24   C.E.Moore: *Atomic Energy Levels*, Vol. II Natl. Bur. Stand. (U.S.) Circular 467, (1952) reissued 1971
4.25   R.Shuker: Appl. Phys. Lett. **29**, 285 (1976)
4.26   P.J.Hay, T.H.Dunning, Jr.: J. Chem. Phys. (in press)
4.27   H.H.Nakano, R.M.Hill, D.C.Lorents, D.L.Huestis, M.V.McCusker, D.J.Eckstrom: "New Electronic Transition Laser Systems", Tech. Rpt. MP 76–99, Stanford Res. Inst. (1976)
4.28   J.A.Mangano, J.H.Jacob, M.Rokni, A.Hawryluk: Appl. Phys. Lett. **31**, 26 (1977)
4.29   M.Rokni, J.H.Jacob, J.A.Mangano, R.Brochu: Appl. Phys. Lett. **31**, 79 (1977)
4.30   D.C.Lorents, D.L.Heustis, M.V.McCusker, H.H.Nakano, R.M.Hill: J. Chem. Phys. **68**, 4657 (1978)
4.31   W.R.Wadt, P.J.Hay: Appl. Phys. Lett. **30**, 573 (1977)
4.32   G.A.Hart, S.K.Searles: J. Appl. Phys. **47**, 2033 (1976)
4.33   J.G.Eden, S.K.Searles: Appl. Phys. Lett. **30**, 287 (1977)
4.34   R.Burnham, N.W.Harris: J. Chem. Phys. **66**, 2742 (1977)
4.35   J.J.Ewing, J.C.Swingle, A.Szoke: 2nd Winter Colloquium on Laser Induced Chemistry, Park City, Utah (1977)
4.36   R.Burnham, S.K.Searles: J. Chem. Phys. **67**, 5967 (1977)
4.37   J.Goodman, L.E.Bruce: J. Chem. Phys. **65**, 3808 (1976)
4.38   B.S.Ault, L.Andrews: J. Chem. Phys. **65**, 4192 (1976)
4.39   J.Tellinghuisen, A.K.Hays, J.M.Hoffman, G.C.Tisone: J. Chem. Phys. **65**, 4473 (1976)
4.40   B.A.Lengyel: *Introduction to Laser Physics* (Wiley and Sons, New York 1966)
4.41   R.O.Hunter, J.Oldenettel, C.Howton, M.V.McCusker: J. Appl. Phys. **49**, 549 (1978)

4.42   D.J.Seery, D.Britton: J. Phys. Chem. **68**, 2263 (1964)

4.43   R.K.Steunenberg, R.C.Vogel: J. Am. Chem. Soc. **78**, 901 (1956)

4.44   M.L.Bhaumik, R.S.Bradford, Jr., E.R.Ault: Appl. Phys. Lett. **28**, 23 (1976)

4.45   V.N.Ishchenko, V.N.Lisitsyn, A.M.Razhev: Appl. Phys. **12**, 55 (1977)

4.46   A.J.Andrews, A.J.Kearsley, C.E.Webb: Opt. Commun. **20**, 265 (1977)

4.47   E.J.Robinson, S.Geltman: Phys. Rev. **153**, 4 (1967)

4.48   D.E.Rothe: Phys. Rev. **177**, 93 (1969)

4.49   A.Mandl: Phys. Rev. A **3**, 251 (1971)

4.50   H.A.Hyman: Appl. Phys. Lett. **31**, 14 (1977)

4.51   J.H.Jacob, J.A.Mangano: Appl. Phys. Lett. **28**, 724 (1976)

4.52   D.C.Lorents, D.J.Eckstrom, D.Huestis: "Excimer Formation and Decay Processes in Rare Gases", Tech. Rpt. MP 72–2, Stanford Res. Inst. (1973)

4.53   D.C.Lorents: Physica **82**c, 19 (1976)

4.54   W.J.Stevens, M.Gardner, A.Karo, P.Julienne: J. Chem. Phys. **67**, 2860 (1977)

4.55   W.R.Wadt, D.C.Cartwright, J.S.Cohen: Appl. Phys. Lett. **31**, 672 (1977)

4.56   L.F.Champagne, N.W.Harris: Appl. Phys. Lett. **31**, 513 (1977)

4.57   A.M.Hawryluk, J.A.Mangano, J.H.Jacob: Appl. Phys. Lett. **31**, 164 (1977)

4.58   J.Maya, P.Davidovits: J. Chem. Phys. **59**, 3143 (1973)

4.59   L.A.Gundel, D.W.Setser, M.A.A.Clyne, J.A.Coxon, W.Nip: J. Chem. Phys. **64**, 4390 (1976)

4.60   D.L.King, L.G.Piper, D.W.Setser: J. Chem. Soc., Faraday Trans. II **73**, 177 (1977)

4.61   G.E.Caledonia: Chem. Rev. **75**, 333 (1975)

4.62   L.G.Christophorou: *Atomic and Molecular Radiation Physics* (Wiley and Sons, London 1971)

4.63   H.-L.Chen, R.E.Center, D.W.Trainor, W.I.Fyfe: Appl. Phys. Lett. **30**, 99 (1977)

4.64   K.J.Nygaard, S.R.Hunter, J.Fletcher, S.R.Foltyn: Appl. Phys. Lett. **32**, 351 (1978)

4.65   J.J.Ewing, R.Milstein, R.S.Berry: J. Chem. Phys. **54**, 1752 (1971)

4.66   J.H.Jacob, M.Rokni, J.A.Mangano, R.Brochu: Appl. Phys. Lett. **32**, 109 (1978)

4.67   J.J.Thompson: Philos. Mag. **47**, 337 (1924)

4.68   M.R.Flannery: *Atomic Processes and Applications*, ed. by P.G.Burke, B.L.Moiseiwitsch (North-Holland, Amsterdam, 1976) Chap. 12

4.69   P.Langevin: Ann. Chim. Phys. **28**, 433 (1903)

4.70   M.R.Flannery, T.P.Yang: Appl. Phys. Lett. **32**, 327, 356 (1978)

4.71   M.Rokni, J.H.Jacob, J.A.Mangano: Phys. Rev. A **16**, 2216 (1977)

4.72   C.K.Rhodes: IEEE J. QE-**10**, 153 (1974)

4.73   J.N.Bardsley, M.A.Biondi: *Advances in Atomic and Molecular Physics* (Academic Press, New York 1970) Chap. I

4.74   C.A.Brau: Appl. Phys. Lett. **29**, 7 (1976)

4.75   C.H.Chen, M.G.Payne, J.P.Judish: Private communication

4.76   A.Gedanken, J.Jortner, B.Raz, A.Szoke: J. Chem. Phys. **57**, 3456 (1972)

4.77   P.Langevin: Ann. Chim. Phys. **5**, 245 (1905)

4.78   G.Gioumousis, D.P.Stevenson: J. Chem. Phys. **29**, 294 (1958)

4.79   D.K.Bohme, N.G.Adams, M.Mosesman, D.B.Dunkin, E.E.Ferguson: J. Chem. Phys. **52**, 5094 (1970)

4.80   G.E.Veach, H.J.Oskam: Phys. Rev. A **2**, 1422 (1970)

4.81   J.H.Kolts, D.W.Setser: 7th Winter Colloquium on High Power Visible Lasers, Park City, Utah (1977)

4.82   P.K.Leichner, K.F.Palmer, J.D.Cook, M.Thieneman: Phys. Rev. A **13**, 1787 (1976)

4.83   D.W.Setser: 7th Winter Colloquium on High Power Visible Lasers, Park City, Utah (1977), and subsequent private communication. Quenching of KrF* by Xe was observed to be very fast (of the order of gas kinetic), but XeF* emission was very weak

4.84   W.S.Struve, J.R.Krenos, D.L.McFadden, D.R.Hershback: J. Chem. Phys. **62**, 404 (1975)

4.85   J.R.Murray, H.T.Powell: "Laser Experiments on KrCl and ArF $^2\Sigma - ^2\Sigma$ Transitions", in *Eelectronic Transition Lasers II*, ed. by L.E.Wilson, S.N.Suchard, J.I.Steinfeld (MIT Press, Cambridge, Mass. 1977) pp. 23–29

4.86   R.Burnham, N.Djeu: Appl. Phys. Lett. **29**, 707 (1976)

4.87   R.Hunter: 7th Winter Colloquium on High Power Visible Lasers, Park City, Utah (1977)
4.88   C.A.Brau, T.R.Loree, S.D.Rockwood, W.E.Stein: 8th Winter Colloquium on Quantum Electronics, Snowbird, Utah (1978)
4.89   R.T.Schneider, K.Thom: Nucl. Technol. **27**, 34 (1975)
4.90   D.C.Lorents, C.K.Rhodes: Opt. Commun. **18**, 14 (1976)
4.91   D.A.McArthur, P.P.Tollefsrud: Appl. Phys. Lett. **26**, 187 (1975)
4.92   H.H.Helmick, J.L.Fuller, R.T.Schneider: Appl. Phys. Lett. **26**, 327 (1975)
4.93   R.J.DeYoung, W.E.Wells, G.H.Miley, J.T.Verdeyen: Appl. Phys. Lett. **28**, 519 (1976)
4.94   N.W.Jalufka, R.J.DeYoung, F.Hohl, M.D.Williams: Appl. Phys. Lett. **29**, 188 (1976)
4.95   S.P.Bugaev, G.A.Mesyats, D.I.Proskurovskii: Sov Phys. – Dokl. **14**, 605 (1969)
4.96   C.D.Child: Phys. Rev. **32**, 492 (1911)
4.97   B.N.C.Agu, T.A.Burdett, E.Matsukawa: Proc. Phys. Soc. (London) **72**, 727 (1958)
4.98   L.Pages, E.Bertel, H.Joffre, L.Sklavenitis: Atomic Data **4**, 2 (1972)
4.99   J.G.Eden, S.K.Searles: Appl. Phys. Lett. **29**, 350 (1976)
4.100 L.F.Champagne, J.G.Eden, N.W.Harris, N.Djeu, S.K.Searles: Appl. Phys. Lett. **30**, 160 (1977)
4.101 J.M.Hoffman, A.K.Hays, G.C.Tisone: Appl. Phys. Lett. **28**, 538 (1976)
4.102 L.R.Peterson, J.E.Allen: J. Chem. Phys. **56**, 6068 (1972)
4.103 E.E.Huber, D.A.Emmons, R.M.Lerner: Opt. Commun. **11**, 155 (1974)
4.104 C.J.Elliott, A.E.Greene: J. Appl. Phys. **47**, 2946 (1976)
4.105 C.A.Brau, J.J.Ewing: "Spectroscopy, Kinetics and Performance of Rare Gas Halide Lasers", in *Electronic Transition Lasers*, ed. by J.I.Steinfeld (MIT Press, Cambridge, Mass. 1976) pp. 195–198
4.106 M.S.Livingston, J.P.Blewett: *Particle Accelerators* (McGraw-Hill, New York 1962)
4.107 C.A.Brau, H.Milde: Unpublished
4.108 O.P.Judd: J. Appl. Phys. **47**, 5297 (1976)
4.109 A.E.Greene, C.A.Brau: IEEE J. Quant. Electron. (in press)
4.110 C.H.Fisher, R.E.Center: Appl. Phys. Lett. **31**, 106 (1977)
4.111 S.C.Brown: *Basic Data of Plasma Physics* (Wiley and Sons, New York 1959)
4.112 J.D.Daugherty, J.A.Mangano, J.H.Jacob: Appl. Phys. Lett. **28**, 581 (1976)
4.113 L.F.Champagne: "Electron Beam Controlled, Neon Stabilized XeF Laser", in *High-Power Lasers and Applications*, ed. by K.-L.Kompa, H.Walter, Springer Series in Optical Sciences, Vol. 9 (Springer, Berlin, Heidelberg, New York 1978) pp. 32–36
4.114 J.A.Mangano, J.H.Jacob, J.B.Dodge: Appl. Phys. Lett. **29**, 426 (1976)
4.115 R.C.Sze, T.R.Loree: IEEE J. Quant. Electron. (in press)
4.116 V.Hasson, C.M.Lee, R.Exberger, K.W.Billman, P.D.Rowley: Appl. Phys. Lett. **31**, 167 (1977)
4.117 D.G.Sutton, S.N.Suchard, O.L.Gibb, C.P.Wang: Appl. Phys. Lett. **28**, 522 (1976)
4.118 C.P.Wang: Appl. Phys. Lett. **29**, 103 (1976)
4.119 R.Burnham, F.X.Powell, N.Djeu: Appl. Phys. Lett. **29**, 30 (1976)
4.120 C.P.Christensen, L.W.Braverman, W.H.Steier, C.Wittig: Appl. Phys. Lett. **29**, 424 (1976)
4.121 B.Godard, M.Vannier: Opt. Commun. **18**, 206 (1976), and subsequent private communication
4.122 T.J.McKee, B.P.Stoicheff, S.C.Wallace: Appl. Phys. Lett. **30**, 278 (1977)
4.123 Yu.A.Kudryavtsev, N.P.Kuzmina: Appl. Phys. **13**, 107 (1977)
4.124 R.C.Sze, P.B.Scott: J. Appl. Phys. **47**, 5492 (1976)
4.125 W.J.Sarjeant, A.J.Alcock, K.E.Leopold: Appl. Phys. Lett. **30**, 635 (1977)
4.126 R.Burnham, N.W.Harris, N.Djeu: Appl. Phys. Lett. **28**, 86 (1976)
4.127 H.J.Seguin, J.Tulip, D.C.McKen: IEEE. J. QE-**10**, 311 (1974)
4.128 R.V.Babcock, I.Liberman, W.D.Partlow: IEEE J. QE-**12**, 29 (1976)
4.129 R.C.Sze, T.R.Loree, C.A.Brau, S.D.Rockwood: 29th Gaseous Electronics Conference, Cleveland, Ohio (1976)
4.130 J.Hsia: Appl. Phys. Lett. **30**, 101 (1977)
4.131 C.A.Brau, R.V.T.Kung: Unpublished
4.132 W.J.Sarjeant, A.J.Alcock, K.E.Leopold: IEEE J. QE-**14**, 177 (1978)

4.133 R. W. McMillan: Laser Focus **13**, (2), 62 (1977)

4.134 T. S. Fahlen: J. Appl. Phys. **49**, 455 (1978)

4.135 O. S. F. Zucker, J. R. Long, V. L. Smith, D. J. Page, P. L. Hower: Appl. Phys. Lett. **29**, 261 (1976)

4.136 P. F. Pittman, D. J. Page: Int. Pulsed Power Conference, Lubbock, Texas (1976)

4.137 T. R. Loree, K. B. Butterfield, D. L. Barker: Appl. Phys. Lett. **32**, 171 (1978)

4.138 D. G. Sutton, G. A. Capelle: Appl. Phys. Lett. **29**, 563 (1976)

4.139 B. Godard, O. DeWitte: Opt. Commun. **19**, 325 (1976)

4.140 J. Tellinghuisen: Chem. Phys. Lett. **29**, 359 (1974)

4.141 N. Djeu, R. Burnham: Appl. Phys. Lett. **30**, 473 (1977)

4.142 T. R. Loree, R. C. Sze, D. L. Barker: Appl. Phys. Lett. **31**, 37 (1977)

4.143 T. R. Loree, D. L. Barker: 5th Conference on Chemical and Molecular Lasers, St. Louis, Missouri (1977)

4.144 M. H. R. Hutchinson, C. C. Ling, D. J. Bradley: Opt. Commun. **18**, 203 (1976)

4.145 J. Reintjes, R. C. Eckhardt, C. Y. She, N. E. Karangelen, R. C. Elton, R. A. Andrews: Phys. Rev. Lett. **37**, 1540 (1976)

4.146 I. V. Tomov, R. Fedosejevs, M. C. Richardson, W. J. Sarjeant, A. J. Alcock, K. E. Leopold: Appl. Phys. Lett. **30**, 146 (1977)

4.147 J. C. Swingle, C. E. Turner, Jr., J. R. Murray, E. V. George, W. F. Krupke: Appl. Phys. Lett. **28**, 387 (1976)

4.148 W. F. Krupke: "Prospects for Trivalent Rare-Earth Vapor Lasers", in *Electronic Transition Lasers*, ed. by J. I. Steinfeld (MIT Press, Cambridge, Mass. 1976) pp. 148–152

4.149 R. R. Jacobs, W. F. Krupke: Appl. Phys. Lett. **32**, 31 (1978)

4.150 J. P. Hessler, F. Wagner, Jr., C. Williams, W. T. Carnall: "Experimental Studies of Potential Rare-Earth Vapor Phase Laser System", in *Electronic Transition Lasers II*, ed. by L. E. Wilson, S. N. Suchard, J. I. Steinfeld (MIT Press, Cambridge, Mass. 1977) pp. 242–245

4.151 J. R. Murray, J. Goldhar, A. Szoke: Appl. Phys. Lett. **32**, 551 (1978)

4.152 C. A. Brau: Umpublished

4.153 R. J. Jensen, J. G. Marinuzzi, C. P. Robinson, S. D. Rockwood: Laser Focus **12**, (5), 51 (1976)

4.154 J. B. Marling: Chem. Phys. Lett. **34**, 84 (1975)

4.155 J. C. Vanderleeden: Laser Focus **13**, (6), 51 (1977)

4.156 R. K. Sander, T. R. Loree, S. D. Rockwood, S. M. Freund: Appl. Phys. Lett. **30**, 150 (1977)

4.157 J. H. Clark, R. G. Anderson: Appl. Phys. Lett. **32**, 46 (1978)

4.158 E. Spiller, R. Feder: In *X-Ray Optics*, ed. by H.-J. Queisser, Topics in Applied Physics, Vol. 22 (Springer, Berlin, Heidelberg, New York 1977)

4.159 T. H. Johnson, A. M. Hunter, II: J. Appl. Phys. **51**, 2406 (1980)

4.160 H. H. Nakano, R. M. Hill, D. C. Lorents, D. L. Huestis, M. V. McCusker: SRI Report MP-76-99 (1976)

4.161 J. A. Mangano, J. H. Jacob, M. Rokni, A. Hawryluk: Appl. Phys. Lett. **31**, 26 (1977)

4.162 J. N. Bardsley: Appl. Phys. Lett. **32**, 76 (1978)

4.163 J. H. Jacob, M. Rokni, J. A. Mangano, R. Brochu: Appl. Phys. Lett. **32**, 109 (1978)

4.164 V. H. Shui: Appl. Phys. Lett. **31**, 50 (1977)

4.165 M. Rokni, J. A. Mangano, J. H. Jacob, J. C. Hsia: IEEE J. QE-**14**, 464 (1978)

4.166 L. G. Piper, J. E. Velazco, D. W. Setser: J. Chem. Phys. **59**, 3323 (1973)

4.167 E. W. McDaniel, V. Ermak, A. Dalgarno, E. E. Ferguson, L. Friedman: *Ion Molecule Reaction* (Wiley-Interscience, New York 1970)

4.168 D. K. Bohme, N. G. Adams, M. Muselman, D. B. Donkin, E. E. Ferguson: J. Chem. Phys. **52**, 5094 (1970). A measurement of a considerably different (10^1 smaller) value of this rate constant than that estimated in this reference (as used in the code reported here (was reported by R. Johnson, J. McDonald, M. A. Bondi [J. Chem. Phys. **68**, 2991 (1978)]

4.169 W. B. Lacina, D. B. Cohn: Appl. Phys. Lett. **32**, 106 (1978)

4.170 H. Chen, R. E. Center, D. W. Trainor, W. I. Fyfe: Appl. Phys. Lett. **30**, 99 (1977)

4.171 E. Eggarter: J. Chem. Phys. **62**, 833 (1975)

4.172 J. H. Jacob, J. A. Mangano: Appl. Phys. Lett. **29**, 467 (1976)

4.173 M. Shaper, H. Scheibner: Bett. Plasma. Phys. **9**, 45 (1969)

4.174 J. H. Jacob, J. A. Mangano: Appl. Phys. Lett. **28**, 724 (1976)

4.175 L.R.Peterson, J.E.Allen, Jr.: J. Chem. Phys. **56**, 6068 (1972)
4.176 D.Rapp, P.Englander-Golden: J. Chem. Phys. **43**, 1464 (1965)
4.177 L.Vriens: Physica **31**, 395 (1965)
4.178 J.N.Bardsley, M.A.Biondi: Adv. At. Mol. Phys. **6**, 1 (1970)
4.179 C.J.Chea: Phys. Rev. **177**, 245 (1969)
4.180 M.A.Bondi: Phys. Rev. **129**, 1181 (1963)
4.181 T.H.Dunning, P.J.Hay: Appl. Phys. Lett. **28**, 649 (1976); J. Chem. Phys. **69**, 134 (1978)
4.182 J.G.Calvert, J.N.Pitts: *Photochemistry* (Wiley, New York 1966)
4.183 A.Mandl: Phys. Rev. A **3**, 251 (1971)
4.184 H.H.Michels, R.H.Hobbs, L.A.Wright: Int. J. Quantum Chem. **12**, 257 (1978)
4.185 Estimated by H.Michels (United Technology Research Center) and L.Wright (Air Force Weapons Laboratory)
4.186 A.M.Hawryluk, J.A.Mangano, J.H.Jacob: Appl. Phys. Lett. **31**, 164 (1977)
4.187 F.Roussel, P.Breger, G.Spiess, C.Manus, S.Geltman: J. Phys. B **13**, L631 (1980)
4.188 T.J.McIlrath, T.B.Lucatorto: Phys. Rev. Lett. **38**, 1390 (1977)
4.189 B.Carré, F.Roussel, P.Breger, G.Spiess: J. Phys. B **14**, 4289 (1981)
4.190 A.U.Hazi, T.N.Rescigno, A.E.Orel: Appl. Phys. Lett. **35**, 477 (1979)
4.191 D.W.Trainor, J.H.Jacob: Appl. Phys. Lett. **37**, 675 (1980)
4.192 S.-C.Lin, J.I.Levatter: Appl. Phys. Lett. **34**, 505 (1979)
4.193 B.M.Forestier, B.L.Fontaine, T.Solenne: J. Phys. Lett. **42**, L221 (1981)
4.194 W.L.Nighan, Y.Nachshon, F.K.Tittel, W.L.Wilson, Jr.: Appl. Phys. Lett. **42**, 1006 (1983)

Note Added in Proof

In a recent paper, *Nighan* et al. [4.194] reported on an e-beam pumped XeF $(C \rightarrow A)$ laser with output energy density of 0.1 J/l corresponding to an electrical-optical conversion efficiency of approximately 0.1 %.

5. Metal Vapor Excimers

A. Gallagher

With 19 Figures

In this chapter we consider diatomic excimer molecules (AB) in which a Group I, II, or III metal atom is the radiative species (A). Group II *metal atoms* can play the role of a noble gas (B), since they have $^1S_0$ ground states and have predominantly repulsive interactions with ground-state atoms. Thus Group II and VIII atoms will be included in the class of "B" atoms and we have divided the discussions of specific systems into the six cases of Group I, II, or III atoms paired with Group II or VIII atoms. In Sect. 5.2.7 we have also included a discussion of recent work on mercury halide systems, an example of a Group II/Group VII material.

These metal-based excimer molecules share several important characteristics. First, their excimer bands are on the wings of the metal-vapor lines; hence the important bands from the ground to first excited states are generally in the visible or near uv and ir wavelength regions. Next, many of the excited states AB^* associated with these bands are weakly bound (i.e., by less than $10\,kT$). With the partial exception of Hg, elevated temperatures are necessary to obtain adequate metal vapor pressures for reasonable gain coefficients and high powers. An associated engineering problem is a frequent reactivity with window and gasket materials. Finally, the lowest excited states of the metal atoms generally have excitation energies of less than half their ionization energies. This has important consequences for the electron collisional cross sections which determine the efficiencies of potential e–beam or discharge-excited high-power lasers. In particular, as noted in Sect. 5.3 below this may allow the possibility of efficient, high power, direct electric discharge lasers.

The weakly bound character of many metal-based excimers has a profound effect on their optical properties as a laser medium. It results in a low-gain coefficient per excited metal atom, homogeneous broadening across the excimer band, rapid transitions between excited atoms A^* and the associated excimer molecules AB^*, the necessity of a high "noble gas" density, and relatively stringent requirements on the fractional excitation of the metal atoms necessary for net gain. The optical properties of such weakly bound excimers can be characterized by some very simple and relatively universal relations, which are discussed in Sect. 2.12. These are also described in Sect. 5.1, which discusses the excimer vibrational-rotational population distributions, relations between stimulated and spontaneous emission and absorption, and the consequences of the classical Franck–Condon principle and homogeneous broadening. These discussions, in fact, apply with very minor modifications to many strongly bound excimers as well, and this generalization is discussed therein. The results of Sect. 5.1 are applied in Sect. 5.2, which discusses those metal-based excimers for which we have information. The last section discusses potential excitation methods and their potential efficiencies, with emphasis on a prototype example of discharge-excited Na–Xe.

The weakly bound character of many metal based excimers leads to the possibility of high power levels, resulting from the low gain coefficient and homogeneous broadening, and high pulse energies due to the high-pressure noble-gas heat sink. (Some of the more strongly bound cases such as Hg_2 also have small gain coefficients in the bands associated with the lowest excited states.) However, the low gain coefficients combined with the possibility of spatial variations in the index of refraction associated with the high pressure gas make these systems difficult to develop into successful lasers. Some promising candidates are now under study, particularly Tl–Hg, but laser action has yet to be demonstrated.

5.1 Optical Properties

5.1.1 Vibrational-Rotational Population Distribution in High-Pressure Gases

We consider one or more relatively low density molecular species in the presence of a high pressure atomic gas, with energy deposition by an electron current, a high-energy electron beam, an external light source, or some combination thereof. A major fraction of the deposited energy is presumed to go into electronic excitation, ionization, and gas heating.

Vibrational and rotational excitation also result from electron collisions and molecular fluorescence. Collisions of the excited molecules with the high-pressure gas drive excited vibrational and rotational populations toward equilibrium with the gas-kinetic temperature, so these forms of energy deposition also go ultimately into gas heating.

For light, strongly bound molecules, the vibrational separations are so large that slow atomic collisions are inefficient at relaxing bound-state vibrational populations. The present discussions will be primarily aimed at heavy, large and/or weakly bound diatomic molecules in which the vibration spacings are small and the *vibrational relaxation* probability is high. Factors which often contribute to this rapid vibrational relaxation are attractive long-range interactions with the atomic buffer gas, combined vibration-rotation transitions, and a high probability of many vibrational steps in a single heavy-particle collision once the single step probability is high [5.1]. Typical vibrational relaxation times thus correspond to only a few gas-kinetic collisions, and at the typical inert gas densities $[B]$ of 10^{20} cm^{-3} proposed for laser, this corresponds to a 10^{-9}–10^{-10} s relaxation time. By comparison radiative rates for visible atomic transitions and their associated excimer bands are generally 10^{-8} s or longer, while radiative trapping further decreases the effective radiation decay rate for resonance transitions. If excimer vibrational excitation rate coefficients by electron impact, of which we have no direct information, were similar to those of the stable molecular gases then they would typically be ~ 10 times smaller than for electronic superelastic collisions and the latter will limit the time available for gas vibrational relaxation rates for $n_e/[B] \sim 10^{-3}$. The proposed discharge excited laser systems (Sect. 5.3) typically utilize electron densities a factor of 1–100 smaller than this. Thus, in the models considered below, vibrational and rotational relaxation rates will generally be assumed to be faster than the electron collision and radiative rates with which they compete. The distribution of population within the vibrational and rotational states will be characterized by a temperature. The gas kinetic temperature will normally be used, although a somewhat higher effective temperature could be used if relaxation is not complete. This applies to the strongly bound as well as the weakly bound excimers as long as they do not involve light atoms such as H or He and the buffer gas densities are in the typical 10^{20} cm^{-3} range. On the other hand, the total bound-state population may not be in equilibrium with the associated atomic-state concentrations for strongly bound excimers, and the total populations of different electronic states will definitely not be in equilibrium at the gas temperature if net laser gain is to be achieved.

The total populations of each molecular electronic state will result from competition between gas collisional processes, radiative processes, and electron collisions in electrically excited systems. If the gas collisional dissociation rate R_d is the dominant destruction mechanism and the three-body association rate R_f (via complexes or any mechanism) the dominant formation rate for a particular state, then the populations will approximately satisfy the equilibrium relation at the gas temperature T. That is $[AB_{ij}]/[A_i][B_j] = R_f/R_d = K_{eq}(T)$, where the state AB_{ij} dissociates into $A_i + B_j$, K_{eq} is the equilibrium constant for state AB_{ij} and the brackets refer to densities. Collisional dissociation rates vary approximately exponentially with binding energy divided by kT, so this will generally be the dominant destruction mechanism only for shallow potentials, typically up to 5 or 10 kT deep depending on the size of competing rates and buffer gas density. Three-body association will often be the dominant production mechanism for such weakly bound states since most of the constituents A_i and B_j are then in the atomic form ($[AB_{ij}]/[A_i] \ll 1$ and $[AB_{ij}]/[B_i] \ll 1$). Thus, the total molecular population of such weakly bound electronic states will frequently be in equilibrium with the parent atomic components, particularly in cases where the atomic form predominates at equilibrium. In this case it is conceptually most useful to relate the absorption and stimulated emission coefficients to $[A_i][B_i]$ rather than $[AB_{ij}]$. One other advantage of the description in terms of $[A_i][B_i]$ for weakly bound states is that the radiation from free and quasi-bound collision states of AB_{ij} is naturally included, as it should be since they also contribute to the excimer spectrum. On the other hand, a formulation in terms of the molecular concentration and its equilibrium constant refers only to bound states; these adequately describe the optical properties only for strongly bound electronic states. The Hg$_2$ case described below is an example of the strong-binding limit and is discussed directly in terms of $[AB_{ij}]$.

For repulsive interatomic potentials, the free nuclear states of the collision complex replace the bound vibrational states in the above "molecular" picture. In this case the continuum distribution of vibrational states is always in equilibrium with the atomic concentrations at the gas kinetic temperature. In the discussions below we treat free and bound vibrational states in the same manner; thus the derived relations can be applied uniformly to bound–bound, bound–free and free–free transitions. In addition they are applicable from the wings of atomic lines on through the associated molecular bands.

5.1.2 Thermodynamic Relations Between Absorption and Emission

Many of the results of this section have been given previously by *Phelps* [5.2] and *York* and *Gallagher* [5.3]. We first note a generalization of the Einstein relation for atomic or molecular lines (2 is the upper and 1 the lower level)

$$N_2 B_{2\to1} = \frac{\lambda^2}{8\pi} N_2 A_{2\to1} \tag{5.1}$$

to cover continuum radiation or portions of lines. We replace $N_2 A_{2\to1}$ with a spectral distribution $S(v)$ and $N_2 B_{2\to1}$ with G_v

$$G_v = \frac{\lambda^2}{8\pi} S(v). \tag{5.2}$$

Here G_v is the stimulated emission coefficient and $S(v)dv$ the spontaneous emission rate per unit volume for $v \to v + dv$ photons. $N_2 B_{2\to1} = \int_{\text{line}} G_v dv$ and $N_2 A_{2\to1} = \int_{\text{line}} S(v) dv$. The traditional definition of the absorption coefficient k_v is the difference between pure absorption, which we label K_v, and G_v $(k_v = K_v - G_v)$. Next we recall that for a vapor contained in a vessel of temperature T and in equilibrium with the walls, the net absorption of the black-body radiation must at every v be balanced by spontaneous emission at the same v

$$(K_v - G_v)\frac{8\pi}{\lambda^2(e^{hv/kT} - 1)} = S(v). \tag{5.3}$$

Combined with (5.2) this yields

$$\frac{G_v}{K_v} = e^{-hv/kT} \tag{5.4}$$

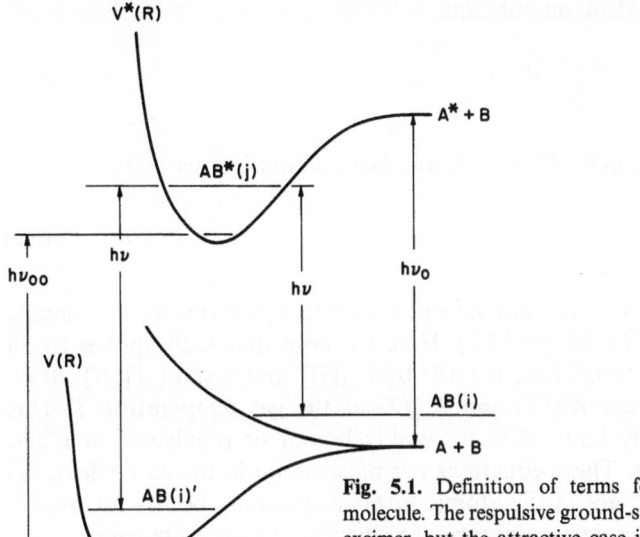

Fig. 5.1. Definition of terms for an idealized diatomic molecule. The respulsive ground-state case corresponds to an excimer, but the attractive case is also included within the framework of the theory in Sect. 5.1

for an equilibrium vapor. Also, for an equilibrium vapor, the atomic state densities are given by

$$\frac{[A^*]}{[A]} = \frac{g^*}{g} e^{-h\nu_0/kT}, \tag{5.5}$$

where we have replaced A_1 and A_2 with A and A^* for a ground and electronically excited state, and $h\nu_0$ is the excitation energy of A^* (see Fig. 5.1). Combining (5.4) and (5.5) yields

$$\frac{G_\nu}{K_\nu} = \frac{[A^*]}{[A]} \frac{g}{g^*} e^{-h(\nu - \nu_0)/kT} \quad \text{(equilibrium vapor)}. \tag{5.6}$$

We now consider a situation in which $[A^*]/[A]$ exceeds the equilibrium ratio for the gas temperature T, but the population ratios $[AB^*(j)]/[A^*][B]$ and $[AB(i)]/[A][B]$ are still in equilibrium at temperature T. Here $AB(i)$ refers to a bound or free vibrational and rotational state i in the ground electronic state, and $AB^*(j)$ to a state j in the excited electronic state (Fig. 5.1). Equation (5.6) then still applies, but with $[A^*]/[A]$ independently determined (e.g., by electron collisional or optical excitation). If one represents the ratio $[A^*]/[A]$ in terms of an effective excitation T_e

$$[A^*]/[A] = (g^*/g)\exp(-h\nu_0/kT_e). \tag{5.7}$$

Combining this with (5.6), one obtains

$$\frac{G_v}{K_v} = \exp\left[\frac{-h}{k}\left(\frac{v_0}{T_e} - \frac{v - v_0}{T}\right)\right].$$

(5.8)

The condition for net gain $(G_v/K_v > 1)$ and laser action is thus

$$\frac{v_0 - v}{v_0} > \frac{T}{T_e}.$$

(5.9)

This condition was given by *Carbone* and *Litvak* [5.4] for Hg_2 excimer lasers, and was generalized in *Phelps* [5.2]. Here we note that (5.9) applies to all molecular electronic transitions if $[AB^*]/[A^*][B]$ and $[AB]/[A][B]$ attain their equilibrium values $K_{eq}^*(T)$ and $K_{eq}(T)$ at the gas temperature T. This requirement effectively limits (5.9) to weakly bound or repulsive lower and upper state potentials. These equations are most useful in the case where the majority of A and A^* are in the atomic form, as generally occurs for weakly bound excimers. In the next paragraph we consider the opposite case.

As noted in Sect. 5.1.1, the atomic versus molecular population ratios $[AB^*(j)]/[A^*][B]$ are only expected to be in equilibrium at the gas temperature T for weakly bound or repulsive $V^*(R)$. But we expect the ratios of bound-state populations $[AB^*(j)]/[AB^*(j')]$ to be characterized by equilibrium ratios at temperature T or a somewhat elevated temperature. In the strong-binding case it is better to refer the G_v/K_v ratios to a ratio of bound-state populations, since most of the excited and ground-state species will be in the form of bound molecules. Here we will refer to the population ratio $[AB^*(0)]/[AB(0)]$, where the 0's refer to the $V=0$, $J=0$ states (Fig. 5.1). Then (5.5) is replaced by

$$\frac{[AB^*(0)]}{[AB(0)]} = \frac{g_m^*}{g_m} e^{-hv_{00}/kT},$$

(5.5a)

where hv_{00} is the energy separation of $AB^*(0)$ and $AB(0)$, and g_m^* and g_m refer to the molecular electronic states (Fig. 5.1). Then (5.6) is replaced by

$$\frac{G_v}{K_v} = \frac{[AB^*(0)]}{[AB(0)]} e^{-h(v - v_{00})/kT} \quad \text{(equilibrium vapor)}.$$

(5.6a)

Defining an effective excitation temperature similarly to (5.7)

$$\frac{[AB^*(0)]}{[AB(0)]} = \frac{g_m^*}{g_m} \exp(-hv_{00}/kT_e'),$$

(5.7a)

we obtain

$$\frac{G_v}{K_v} = \exp\left[\frac{-h}{k}\left(\frac{v_{00}}{T_e'} - \frac{v - v_{00}}{T}\right)\right].$$

(5.8a)

The net gain condition becomes

$$\frac{v_{00} - v}{v_{00}} > \frac{T}{T_e'}. \tag{5.9a}$$

In both cases, (5.6) or (5.6a), the ratio G_v/K_v increases as $\exp(-hv/kT)$ with increasing wavelength, which is a key point of this discussion. The excited state fraction necessary for net gain thus decreases exponentially as the photon energy decreases. It should be apparent that this result refers to all portions of a band, including discrete lines, portions thereof, and continua. Extending the range of states i and j to include quasi-bound resonance states and free-continuum states also causes no changes in (5.6–9) and (5.6a–9a) as long as a single temperature T is used to characterize the entire population distribution.

In some cases it is more useful to refer the $[AB^*(i)]$ to total bound-state populations, $[AB^*] = \sum_i [AB^*(i)]$, rather than $[AB^*(0)]$ as in (5.6a) and similarly for the ground-state. Using $[AB^*] = [AB^*(0)]Z^*(T)$ and $[AB] = [AB(0)]Z(T)$, where $Z(T)$ is the partition function, (5.6a) can be reduced to this form, as was done by York and Gallagher [5.3]. Equations (5.6) and (5.6a) are equivalent conditions, but (5.7) and (5.7a) are not equivalent for $T_e = T_e'$. Equations (5.6) and (5.6a) are universal equations for vapors with vibrational and rotational populations in an equilibrium distribution at temperature T, but the population ratios $[A^*]/[A]$ or $[AB^*(0)]/[AB(0)]$ in electronically excited vapors will be a consequence of the competition between various collisional and radiative processes. Estimating these population ratios required rather detailed modeling. Examples and references are given below.

5.1.3 Applications of the Classical Franck-Condon Principle

The classical Franck-Condon principle (CFCP), which is the basis of the quasistatic theory of line broadening [5.5], predicts that the internuclear separation R and nuclear kinetic energy do not change during an electronic radiative transition. Consequently, the photon energy hv equals $V_i^*(R) - V_j(R)$ for a transition that occurs at R, regardless of the initial motion. (In the interest of simplicity, we will discuss only one R coordinate, equivalent to a diatomic molecule, but generalization to polyatomic is implied.) Here $V_i^*(R)$ is the electronic energy of the ith adiabatic state which dissociates to $A^* + B$, while $V_j(R)$ dissociates to $A + B$. For excimer molecules B is in a $^1S_0$ state with statistical weight $g_B = 1$. Thus $A^* + B$ forms g_{A^*} adiabatic states and $A + B$ forms g_A states, half of which are effectively degenerate if A or A^* is not a $^1S_0$ state. The total rate $[S(v)dv]$ is, therefore, given by the total density of excited molecules with separations $R + R + dR$, where $hv = V_i^*(R) - V_j(R)$, times the transition rate $A(R)_{ij}$, i.e., if we define $[AB^*(R)_i]$ as the density of molecules in adiabatic state i with internuclear separations less than or equal to R,

$$[S(v)] = \sum_{ij} A(R)_{ij} \frac{d[AB^*(R)_i]}{dv}. \tag{5.10}$$

The stimulated emission coefficient is related to this by (5.2), and the absorption coefficient is given by g_j/g_i^* times equivalent expressions for ground-state molecules

$$G_v = \frac{\lambda^2}{8\pi} \sum_{ij} A(R)_{ij} \frac{d[AB^*(R)_i]}{dv} \tag{5.11a}$$

and

$$K_v = \sum_{ij} \frac{g_i^*}{g_j} \frac{\lambda^2}{8\pi} A(R)_{ij} \frac{d[AB(R)_j]}{dv}. \tag{5.11b}$$

Note that if the ij continuum band has a width Δv_{Band} and a mean radiative rate $\bar{A}_{ij}$ the stimulated emission coefficient can be approximated by the familiar expression

$$\frac{G_v}{[AB_i^*]} = \frac{\lambda^2}{8\pi} \frac{\bar{A}_{ij}}{\Delta v_{\text{Band}}}. \tag{5.12}$$

Molecular bands associated with dipole-allowed transitions of excimer molecules frequently have $\bar{A}_{ij} \sim A_0$, the rate for the separated atom, since the wave functions are not severely perturbed by the relatively weak interaction energies.

Since the CFCP is a classical principle, it is also appropriate to use a classical canonical phase-space distribution to describe the equilibrium distribution of states of nuclear motion [5.6]. The distribution vs R is obtained by integrating over nuclear momentum coordinates

$$d[AB(R)_j] = Z_j^{-1} h^{-3} dR^3 \int dP^3 e^{-[P^2/2 + V_j(R)]/kT}$$

$$= \frac{g_j}{g_A} [A][B] d^3R \exp[(-V_j(R) - V_j(\infty))/kT, \tag{5.12a}$$

where Z_j is the partition function. An identical expression with $[A^*]$, g_{A^*}, g_i, and V_i^* in place of $[A]$, g_A, g_j and V applies for $d[AB^*(R)_i]$. Also, for diatomic molecules $d^3R = 4\pi R^2 dR$. Combining (5.11) and (5.12a) we obtain, as in *Phelps* [5.2] and *Hedges* [5.6]

$$G_v = \frac{\lambda^2}{8\pi} \sum_{ij} A(R)_{ij} \left(\frac{d^3R}{dv}\right)_{ij} \frac{g_i^*}{g_{A^*}} [A^*][B] \exp\{-[V_i^*(R) - V_i^*(\infty)]/kT\}, \tag{5.13a}$$

and

$$K_v = \frac{\lambda^2}{8\pi} \sum_{ij} \frac{g_i^*}{g_A} A(R)_{ij} \left(\frac{d^3R}{dv}\right)_{ij} [A][B] \exp\{-[V_j(R) - V_j(\infty)]/kT\}, \quad (5.13b)$$

where d^3R/dv is often written as $4\pi R^2/|dv/dR|$. In the case $A = B$, as for Hg_2, $[A][B] \rightarrow [A]^2/2$ in (5.13a).

Thus, application of the CFCP to a classical equilibrium distribution of bound and free vibrational and rotational states leads to the absorption and stimulated emission coefficients in (5.13). The resulting ratio G_v/K_v agrees with the previous (5.6), as it must. These relations apply to both attractive and repulsive states; in cases where the equilibrium constants for $[AB]/[A][B]$ or $[AB^*]/[A^*][B]$ are less than one, most of A or A^* is in the atomic form and (5.13) are most useful. If either equilibrium constant greatly exceeds one, it may be more appropriate to refer to molecular populations, by substituting $[A][B] = [AB]/K_{eq}$ or $[A^*][B] = [AB^*]/K_{eq}^*$ in (5.13). In this form (5.13) will be correct even if $[A^*]/[AB^*]$ is not in the equilibrium ratio. These relations will be applied to the examples below.

5.1.4 Homogeneous Broadening

A characteristic of considerable importance in the generation of subnanosecond pulses is the homogeneous broadening, or proportion of the $[AB^*]$ capable of radiating at a given wavelength at one time. Three distinctly different cases occur for molecular electronic transitions, bound–bound, bound–free, and free–bound, or free.

In the case of a bound–bound molecule transition, each excited vibrational-rotational state radiates typically ~ 20 strong lines distributed across $\sim 2000 \, cm^{-1}$. If each line is collisionally broadened by the buffer gas to a few cm^{-1}, then each excited state typically radiates into a few percent of the band. The lines from different excited states are interleaved to yield a virtual continuum, particularly for heavy molecules, but at any one time only a few percent of the excited states interact with a single laser wavelength. For heavy molecules vibrational and rotational relaxation occurs with roughly gas-kinetic cross sections, or about once per 0.05 ns for $10^{20} \, cm^{-3}$ buffer gas density. But each collision has only a statistical probability of a few percent for transferring population into the states that interact with the laser wavelength, so typically ~ 30 collisions or 1.5 ns are required to stimulate most of the $[AB^*]$ to radiate. Thus, bound–bound transitions are unfavorable for efficient, subnanosecond pulse laser gain tubes.

If the ground state is repulsive, as for excimers, each bound excited state (V', J') radiates a continuum that undulates through $V' + 1$ maxima generally spread across a large portion of the band. Consequently, each excited state interacts strongly with more than half of the wavelengths in its portion of the band and a single laser wavelength in the strong portion of the band stimulates the majority of bound states to emit. Thus, only a few vibrational relaxation collisions are needed to stimulate almost the entire AB^* population to emit, a process that typically occurs in ~ 0.1 ns for 10^{20} cm^{-3} density of buffer gas.

If the excited state is repulsive, as in the Li–He case in Sect. 3.1, each free collisional vibrational state radiates an undulating continuum that fills more than half of the "band," or collisionally broadened "far-wing of the atomic line" in the language of pressure broadening. Furthermore, each excited atom in 10^{20} cm^{-3} density of buffer gas impacts a different perturber with a new collision energy about every 3 ps. Thus, it fills a new 50% portion of the continuum every 3 ps, and the resultant "homogeneous broadening time" is less than 10 ps. This result is independent of the repulsive or attractive character of the lower state.

5.2 Excimer Systems

5.2.1 Group I–VIII Excimers

The most thoroughly studied metal-noble gas excimers are the alkali-noble gases. The A–state potentials have been measured, and found to be weakly attractive; the X states are repulsive except for a weak, long range van der Waals minimum [5.7–9]. The A–X bands appear as extended continua on the red-wings of the alkali resonance lines with the strongest bands generally for Xe. These A–X bands have been suggested as potential excimer laser systems by *Phelps* [5.2] and *York* et al. [5.7] and as discharged excited excimer lasers by *Palmer* et al. [5.10], *Schlie* [5.11], and *Shuker* et al. [5.12]. In addition to these A–X bands, which are in the 600–1100 nm region for the various alkalis, excimer bands associated with the first excited ^2S and ^2D states of the alkalis occur in 400–700 nm regions. The absorption coefficients, and effectively the emission coefficients using relations in Sect. 5.1, have been measured for these shorter-wavelength excimer bands [5.13]. The bands associated with the ^2S state have been suggested as potential excimer laser systems [5.13], but have not been modeled quantitatively. Some of the potentials associated with these higher ^2S and ^2D parentage excimers have been estimated from a combination of fitting experimental band shapes and inferences drawn from approximate potential calculations [5.14]. We will discuss first the more thoroughly studied A–X bands, and then discuss the potential behavior of the shorter wavelength bands as excimer lasers.

A–X Bands

Since the A states are weakly attractive, very large noble-gas densities ($\sim 10^{21}$ cm^3) are required to shift a majority of resonance-line spontaneous emission to the A–X band [5.15]. With the exception of LiHe, and partial exception of NaHe, Xe produces the strongest, widest A–X bands and thereby optimizes the gain and inversion for a given fraction of excited alkali. For this reason modeling efforts have concentrated on alkali-Xe excimers which are described below. To obtain sufficiently large gain coefficients, noble gas densities of $\sim 10^{20}\,\mathrm{cm}^{-3}$ and alkali densities of $10^{16}-10^{17}\,\mathrm{cm}^{-3}$ are normally considered. If pure metal vapors are utilized, temperatures of 200–900 °C are required to obtain these alkali densities. Under such conditions the A states are less than $3\,kT$ deep and the molecular populations rapidly equilibrate with the atomic populations. For molecular A-state populations in equilibrium with the A^* concentrations, the collisional dissociation rate should exceed the sum of the radiative rate (typically $\sim 20\,\mathrm{ns}$) and the somewhat faster stimulated emission rate. These A-state dissociation rate constants are $\sim 3 \times 10^{-11}\,\mathrm{cm}^3\,\mathrm{s}^{-1}$ at temperatures corresponding to $[A] \cong 10^{11}\,\mathrm{cm}^{-3}$ [5.16], and can be expected to be $\sim 10^{-10}\,\mathrm{cm}^3\,\mathrm{s}^{-1}$ at the higher temperatures appropriate for excimer lasers. Thus, the equilibrium condition is well satisfied for $[B] \sim 10^{20}\,\mathrm{cm}^{-3}$, i.e., the $A^* + 2B \rightarrow AB^* + B$ process is not a "bottle neck" unless the excimer stimulated emission rate exceeds $\sim 10^{10}\,\mathrm{s}^{-1}$. Under these circumstances (5.13a) is a good approximation and (5.6–9) are directly applicable.

The feasibility of various metal-noble gas combinations for excimer laser action can be investigated in three steps. First, from knowledge of the A–X band intensity and shape we can determine the excitation fraction $[A^*]/[A]$ necessary for net gain and the alkali and noble gas densities necessary for a given gain coefficient. Next we can estimate, or calculate in some cases, the effect of other overlapping absorption bands. Finally, we consider the feasibility of obtaining the required $[A^*]/[A]$ by various methods of excitation. The latter issue is discussed in Sect. 5.3, in which the potential efficiency is also considered.

For weakly bound excimers, (5.6) or (5.13a)/(5.13b) require that $([A^*]g/[A]g^*)\exp[h(v_0 - v)/kT]$ must exceed one for net gain. This condition is most easily satisfied at the longest wavelength region of the A–X bands, recalling, of course, that the gain drops rapidly in this region. Consequently, a key feature of a good excimer laser candidate is a large red-shift. Since $hv = V^*(R) - V(R)$ this net gain criterion depends only on the difference potential, but the size of the gain coefficient depends on the size of $s(v)$. Thus, a larger stimulated emission and gain coefficient occurs if V^* is attractive at wavelengths where net gain is achievable. A strongly attractive A state and highly repulsive X state are valuable (e.g., such as found in the rare gas halogens) but by no means necessary. Conversely, lower gain coefficients allow greater energy storage before superfluorescence, and they allow higher power gain cell or laser operation if net cavity gain can be achieved. We will discuss some examples before considering other absorptions.

As an example case the potentials and excimer band of NaXe are shown in Figs. 5.2 and 5.3a [5.7]. The A state has $\sim 2\,kT$ of binding at $T = 750\,\mathrm{K}$ and optimum gain occurs at $\sim 700\,\mathrm{nm}$, compared to $\lambda_0 = 590\,\mathrm{nm}$. For $\lambda = 700\,\mathrm{nm}$ and $T = 750\,\mathrm{K}$ one obtains $(g/g^*)\exp[h(v_0 - v)/kT] = 55$ so that an excited

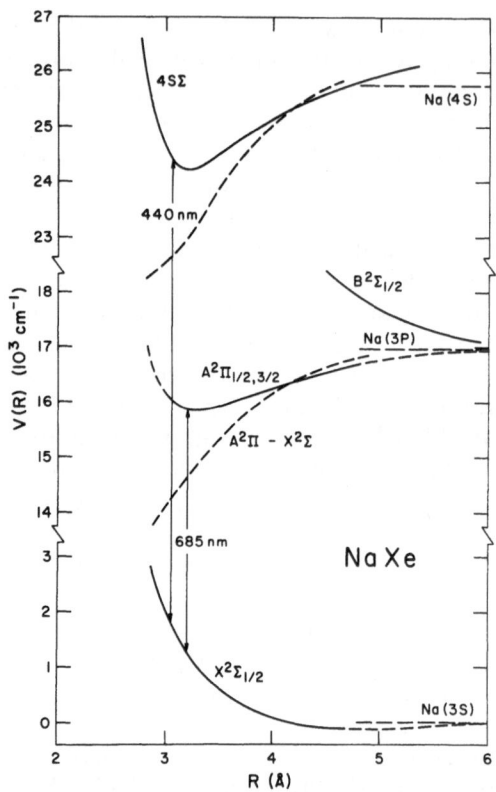

Fig. 5.2. Potential energies (——), $V^*(R) - V(R)$ differences (---), and some excimer band wavelengths for the NaXe molecule. The B and X states are theoretical [5.14], with the X state closely matched by atomic beam scattering results [5.9]. The A state is from A–X band analysis [5.8], which also agrees with the theoretical X state. The $4S$ Σ state, which is not known experimentally, is from a minor adjustment of the theoretical potential [5.9] to match some features of the spectroscopic data [5.13]

fraction $[Na^*(3P)]/[Na] > 0.018$ is required for net gain at this wavelength. Preliminary discharge modeling [5.12] yields an excited fraction 0.01–0.10, depending on various assumptions. For the purpose of showing the entire curve, the gain shown in Fig. 5.3b corresponds to the 0.10 case, corresponding to the upper limit of the estimated range. This Na excitation fraction of 0.10 corresponds to $kT_e = 0.62$ eV in (5.7). The Na_2 coefficients in Fig. 5.3b are discussed below.

The Li–He case is an example of a repulsive or very weakly attractive A state (present data cannot distinguish between these possibilities, but require that any attraction must be less than $0.2\,kT$ [5.7]). For similar densities as the Na–Xe case above, the net gain at $\lambda = 900$ nm for $[Li^*]/[Li] = 0.05$ is given in Fig. 5.4. This is almost equal to the Li–Xe gain at the wavelength (Fig. 5.4) in spite of the stronger binding ($\sim 4\,kT$) of the Li–Xe A-state. This is a result of a much more repulsive $V(R)$ for Li–He compared to Li–Xe. Since the Li–He system has very slight or no A-state attraction, the molecular state populations will be in equilibrium with the gas temperature under all feasible conditions. Furthermore, all excited atoms can deliver their energy to a very short pulse (no vibrational relaxation is required). This is a very interesting example of an "excimer" laser candidate based on free colliding atoms.

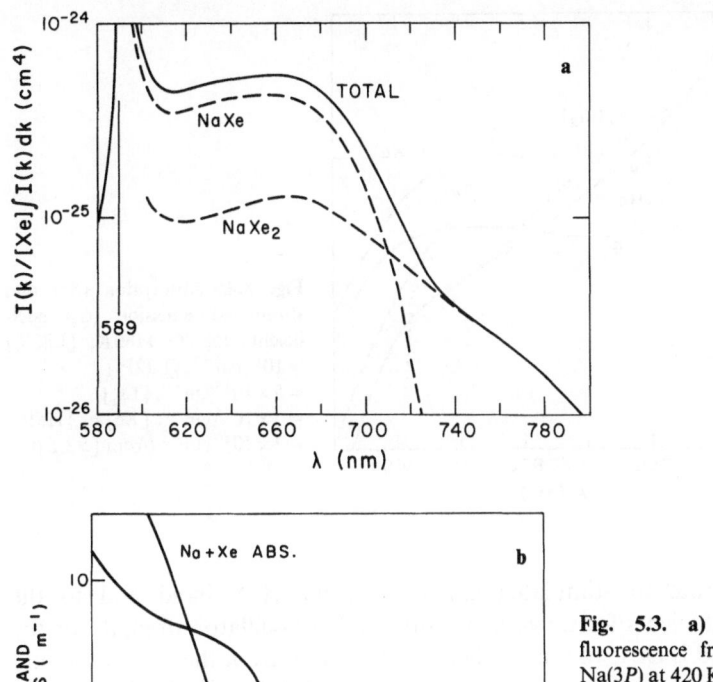

Fig. 5.3. a) The normalized fluorescence from optically thin Na(3P) at 420 K in the presence of $2 \times 10^{20}\,\mathrm{cm}^{-3}$ of Xe. The contribution from the NaXe A–X band is indicated, as is the residual intensity which is attributed primarily to the A–X band of NaXe$_2$ [5.15].
b) Absorption (K_ν) and stimulated (G_ν) coefficients for $T = 760\,\mathrm{K}$, $[\mathrm{Na}(3S)] = 3.5 \times 10^{16}\,\mathrm{cm}^{-3}$, $[\mathrm{Na}(3P)] = 3.5 \times 10^{15}\,\mathrm{cm}^{-3}$, $[\mathrm{Xe}] = 2.7 \times 10^{20}\,\mathrm{cm}^{-3}$, $[\mathrm{Na}_2(X)] = 10^{14}\,\mathrm{cm}^{-3}$, and $[\mathrm{Na}_2(A^1\Sigma)] = 2 \times 10^{13}\,\mathrm{cm}^{-3}$ (from [5.12])

The A–X bands of the other alkali–Xe cases have similar properties except that their bands are at longer wavelengths [5.6, 8, 14]. The A–X bands are weaker and less extended to the red for lighter noble gases, so that they make less viable excimer laser candidates in spite of the desirability of a smaller noble-gas index of refraction [5.6, 8]. The above Li–He case and to some extent NaHe are the only known exceptions to this rule.

We now discuss other possible absorptions. Absorption from the $[AB^*]$ to a higher electronic states $[AB^{**}]$ will generally be spread across molecular bands of width similar to the A–X band. The same initial-state populations,

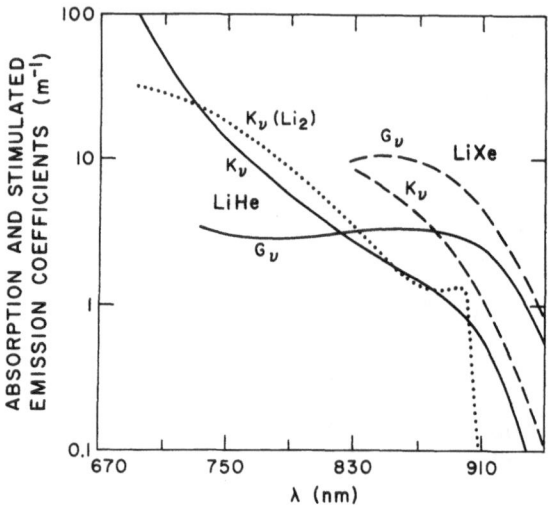

Fig. 5.4. Absorption (K_ν) and stimulated emission (G_ν) coefficients for $T = 1180$ K, $[\text{Li}(2S)] = 10^{17}$ cm^{-3}, $[\text{Li}(2P)] = 5 \times 10^{15}$ cm^{-3}, $[\text{Li}_2] = 5 \times 10^{15}$ cm^{-3}, [Xe] or [He] $= 2 \times 10^{20}$ cm^{-3} (from [5.3, 8])

[AB^*], contribute to stimulated emission in the A–X band and to the absorption coefficients for upward transitions. The oscillator strength for the alkali A–X band is close to unity, whereas absorption from the A–state should be fairly evenly distributed across many bands and the ionization continuum (based on the alkali-atom oscillator strengths and sum rules). Therefore, only the relatively rare coincidence of a particularly strong absorption band bunched into a critical wavelength range can cause difficulties.

The most serious absorption source we are aware of is the A–X band of the alkali dimers. The alkali dimers are typically bound by ~0.4 eV, their A–X bands are in the same spectral regions as the metal-noble gas A–X bands, and the dimer A–X band has a large oscillator strength. This absorption has been discussed in detail by *York* and *Gallagher* [5.3]. As an example, the Na$_2$ absorption is shown in Fig. 5.3b for [Na$_2$] calculated to occur in a high-power discharge [5.12]. Of key importance is the fact that the alkali-dimer absorption has a red edge, or satellite, beyond which absorption decreases rapidly. Thus, the net optical gain in longer wavelength portions of alkali-noble gas A–X bands is not affected. In addition to the LiHe and LiXe cases described above, with A–X bands extending past the Li satellite, the NaXe$_2$ A–X band extends far past the Na$_2$ satellite [5.15], although the gain coefficients at [Xe] $= 2 \times 10^{20}$ cm^{-3} are rather small (Fig. 5.3b). There are good reasons to expect this type of three-body excimer spectrum for many excimers. The Li$_2$ absorption spectrum in the presence of high pressure Xe is unknown, but unlikely to cause any unexpected problems as the area $\int K_\nu d\nu$ must be the same as that in Fig. 5.4.

Another important factor in the alkali-dimer absorption is that the alkali-dimers are bound by many kT (e.g., 11 kT for Na$_2$ at 750 K), so that their collisional dissociation by gas collision is relatively slow and exponentially dependent on the gas temperature. Their density and absorption thus decrease rapidly with increasing gas temperature above that corresponding to the alkali equilibrium vapor pressure. Part of this gas temperature rise could be deliberately imposed externally, but the gas heating associated with the excitation of the medium also causes it. Furthermore, if discharges or e-beam excitation is used, the electron collisions with the dimers cause dissociation and thereby depletion of the dimer concentration. As an example, the $[Na_2]$ used for Fig. 5.3a is a factor of 15 less than the equilibrium value, as was obtained from discharge modeling [5.12]. If wide-band optical pumping is used, the alkali-dimer A–X and B–X bands absorb the pumping light across many hundreds of nanometers, so that the X state will again be severely depleted. In essence, the strong dimer absorption under equilibrium conditions is due to the relatively strong binding of the A_2 molecules, but they are not formed very rapidly by $A + A + Xe \rightarrow A_2 + Xe$ collisions due to the low A density. Thus any depletion mechanism, even an excimer-laser field itself, can be effective at depleting the dimer concentration. By comparison, the AB^* excimers associated with the excimer gain are only bound by a few kT and the process $A^* + Xe + Xe \rightarrow AXe^* + Xe$ is proportional to $[Xe]^2$ rather than $[A]$ $[Xe]$ for the dimer formation process. Thus dimers are depleted long before excimers are.

Collision-Induced Excimer Bands

Collision-induced absorption and emission in the wings of forbidden atomic lines has been studied for many years. However, only in the past two years was it discovered that the strong excimer bands occur very far from the forbidden atomic lines. The Columbia group of *Tam, Moe, Happer*, and associates discovered and have now measured absorption and emission by many of the bands associated with the transition to the first excited S and D states of the alkalis perturbed by noble gases [5.13]. More recently *Sayer* et al. [5.17], *Gauthier* et al. [5.18], and *Eden* et al. [5.19] have also measured properties of these bands. The bands associated with the D state appear to the blue side of the atomic transition energy, so the exponential factor of (5.6) favors absorption over emission and they are not viable excimer-laser candidates. (This occurs because the excited state is more repulsive than the ground state.) The bands associated with the $n^2S_{1/2}$ (ground state) to $(n+1)\,^2S_{1/2}$ transitions are to the red of the atomic transition wavelength $\lambda_0 = k_0^{-1}$, so they are potential excimer-laser candidates.

For the Rb and Cs–noble gas cases, the $nS \rightleftarrows (n+1)S$ bands have a strong peak at the furthest red shift (k_s), indicating a "satellite" or that a minimum occurs in $V^*(R) - V(R)$ [5.2]. The frequency shift $hc(k_s - k_0)$ is greatest for Xe, but even in this case the $\exp[-hc(k_s - k_0)]$ factor of (5.6) that enhances G_v/K_v is only 10 at the temperatures necessary for a reasonable alkali density and gain coefficient. [The $V^*(R)$ binding is weak enough that (5.6) rather than (5.6a) is appropriate here.] Thus, net gain requires $[Cs(7^2S)]/[Cs(6^2S)] \geq 0.1$ and equivalently for Rb, and this only appears feasible with laser pumping. We conclude that the Rb and Cs–Xe bands might be used as gaseous dyes for tuneable frequency conversion in the 500–700 nm region, but they do not appear to be viable candidates for flash lamp, e-beam, or discharge pumping. The absorption coefficient at these band peaks has been measured [5.13] to be typically $K_v/[\mathrm{Xe}][A] \cong 10^{-38}\,\mathrm{cm}^5$, from which it follows that $G_v/[\mathrm{Xe}][A^*] \cong 10^{-37}\,\mathrm{cm}^5$. Thus at $[\mathrm{Xe}] = 10^{20}\,\mathrm{cm}^{-3}$, $[A] = 7 \times 10^{16}\,\mathrm{cm}^{-3}$, $[A^*] = 3 \times 10^{16}\,\mathrm{cm}^{-3}$ due to strong laser pumping, $K_v = 0.07\,\mathrm{cm}^{-1}$ and $G_v = 0.3\,\mathrm{cm}^{-1}$ might result. Some of the general properties of such "gaseous dye lasers" are discussed further in Sect. 5.3.

As the lighter alkalis are used, the $nS \rightleftarrows (n+1)S$ excimer bands broaden, indicating the absence of a satellite, and they shift further from the atomic transition energy, again with largest shifts for Xe [5.13]. Consequently, the exponential factor in (5.6) more strongly favors G_v versus K_v and these systems become viable excimer-laser candidates. In addition, the theoretical potentials of *Pascale* and *Vandeplaque* [5.14], which disagreed sharply with the Cs and Rb–noble gas data, appear to give excellent predictions for K–Ar, Kr, Xe and Na–Ar, Kr, Xe [5.13]. The Na–Xe potentials in Fig. 5.2 include the "$4S\Sigma$" state associated with $\mathrm{Na}(4\,^2S_{1/2}) + \mathrm{Xe}$. This potential is essentially that given by *Pascale* and *Vandeplaque* [5.14], except that we have lowered it 400 cm^{-1} at the minimum to conform better to the spectroscopic data of *Tam* et al. [5.13], who observed an ~ 20 nm wide emission band centered at 440 nm.

From (5.2) the spontaneous emission power reported in [5.13] can be multiplied by λ^3 to obtain the shape of $G_v/[\mathrm{Xe}][\mathrm{Na}^*]$ at the same temperature. Then from (5.6), $K_v/[\mathrm{Xe}][\mathrm{Na}]$ is obtained. Both G_v and K_v are then in the same units, but these are relative units since neither has been measured in absolute units for NaXe. At the center of the band the $\exp[-h(v-v_0)/kT]$ factor of (5.6) is ~ 200, yielding net gain for 0.5 % excitation fraction. (At this [Xe], and for the $4S\Sigma$ potential in Fig. 5.2, $K_{\mathrm{eq}}[\mathrm{Xe}] \cong 0.15$ and these relations in terms of atomic populations are accurate to ~ 10 %.) We have estimated the magnitude of G_v and K_v for Na–Xe by comparing to the Cs–Xe and Rb–Xe cases for which K_v has been measured [5.13]. For this comparison we use (5.13), assume that the $A(R)$ are proportional to the square of the alkali atomic size, and evaluate the effective dv/dR at a satellite in terms of the area under the peak [5.6]. The dominant difference is about a factor of 10 reduction of G_v from Cs–Xe to Na–Xe, since the latter band does not have a satellite and is more spread out. The result is then, at $T = 810\,\mathrm{K}$ corresponding to $[\mathrm{Na}] = 10^{17}\,\mathrm{cm}^{-3}$, $G_v(440\,\mathrm{nm})/[\mathrm{Na}(4S)][\mathrm{Xe}] \cong 10^{-38}\,\mathrm{cm}^5$. It appears that this G_v magnitude estimate should be a lower bound, since NaXe is much more strongly bound than CsXe and one might expect the induced transition moment to be also stronger than for CsXe. The resulting G_v and K_v for $[\mathrm{Na}] = 10^{17}\,\mathrm{cm}^{-3}$, $\mathrm{Na}^*(4S) = 10^{15}\,\mathrm{cm}^{-3}$, and $[\mathrm{Xe}] = 10^{20}\,\mathrm{cm}^{-3}$, are given in Fig. 5.5. The excitation fraction of $\mathrm{Na}^*(4S)/\mathrm{Na}(3S) = 0.01$ was arbitrarily chosen. The possibility of obtaining this excitation fraction in discharge, flashlamp or e–beam excitation has been neither modeled nor measured.

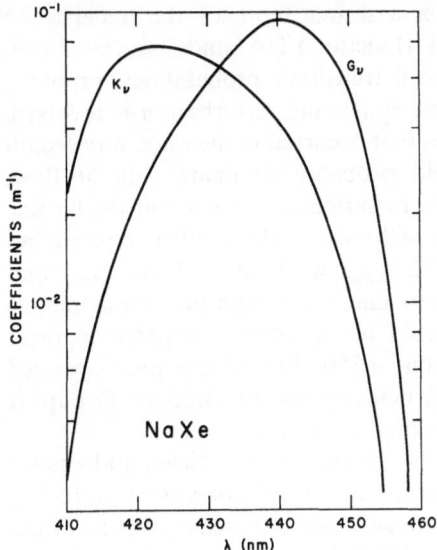

Fig. 5.5. Absorption (K_ν) and stimulated emission (G_ν) coefficients for the $4S\Sigma - X^2\Sigma$ band of NaXe, from analysis of the data in [5.13]. The absolute scale is estimated and 1% of the Na is assumed to be in the $4S$ state

It appears highly probable that the alkali dimers have absorption bands in the same spectral regions as these bands, since there will also be dimer states of Na$(4S)$ Na$(3S)$ parentage. However, the same dimer depletion arguments discussed above in connection with the A–X bands hold here; these dimer absorptions might not be a problem in a high-power system.

5.2.2 Group II–VIII Excimers

In contrast to the I–VIII excimers, for which the A–X bands and potentials of most pairs have been thoroughly characterized, the only II–VIII band that has ever been observed is that of Hg–Xe. The broadening and shift of many Group II lines, due to noble-gas collisions, have been measured [5.5], but these relate only to very long-range interactions and are of no help in predicting excimer bands. We will thus discuss only Hg–Xe below, after a few general comments applicable to all of the Group II elements.

Visible and uv excimer bands associated with the n^3P and $n^1P \to n^1S$ (ground state) transition may be viable excimer laser candidates, and even the near infrared $n^3P \to (n-1)^3D$ and $n^1P \to (n-1)^1D$ transitions might be of interest for Ca, Sr, and Ba. Generalization of the *Baylis* models [5.20] for alkali-noble gas potentials argues that Xe should produce the strongest binding to Group II excited states, but that these will still be weakly bound ($<10\,kT$). Possible rapid predissociation of n^3P or n^1P excimers to repulsive excimer states associated with $(n-1)^1D$ and $(n-1)^3D$ levels could be serious for Ca, Sr, and Ba. The rate of energy transfer via $AB^* + A \to A_2^* + B$ is expected to be 10^6–$10^7\,s^{-1}$ at typical $[A] \sim 10^{16}$–$10^{17}\,cm^{-3}$, and this could be a significant

$[AB^*(n^3P)]$ depletion mechanism. (It is a source term for the buffered A_2^* excimer bands described in the Group II–II section.) The bands associated with the $n^3P \to n^1S_0$ transition will have small transition probabilities, probably similar to the atomic values as the same spin-orbit perturbation is involved. Thus any excited-state absorption bands that occurred in the same wavelength regions as these excimer bands would probably eliminate gain at these wavelengths. The $n^1P \to n^1S_0$ atomic-line radiation is trapped, but the far line wings which include the excimer bands will escape. The result is typically an effective radiative loss rate of $0.05\text{–}0.2\,\Gamma_{Atomic} \cong 10^7\text{–}10^8\,s^{-1}$ for the total $A^*(n^1P) + AB^*(n^1P)$ population. It is clear that rapid, high-power pumping is necessary. Excited-state absorption should be negligible compared to these strongly allowed resonance-line bands, but $AB^*(n^1P)$ might be predissociated or collisionally transferred to n^3P, particularly for the heavier Group II elements for which the triplet–singlet mixing is greatest.

The Hg–Xe excimer bands as well as the fluorescence yields and excited state kinetics were studied with e–beam excitation by *Gutcheck* et al. [5.21]. Two uv bands were reported, each $\sim 30\,nm$ wide and centered at about 220 and 270 nm. The shapes of these two bands and their fluorescence yields for Hg–Ar, Kr, Xe, using e–beam excitation, have recently been reported by *Woodworth* [5.22]. In both studies the energy radiated in each band is typically $\sim 1\%$ of the e–beam energy deposited in the gas. The shorter wavelength band is on the red wing of the 185 nm $6^1P_1 - 6^1S_0$ transition and clearly is the associated $A\text{–}X$ band; the longer wavelength transition comes from the 6^3P excimer manifold. The shape of the 270 nm Hg–Xe band has also been reported by *Oldenberg* [5.23], *Freeman* et al. [5.24], *Nikoforov* et al. [5.25], and others, using 6^3P_1 optical excitation. The band has a similar but significantly different shape than that observed in the e–beam excitation experiments. The binding of the HgXe* state associated with the 270 nm band, which we will label $HgXe^*(6^3P_1)$, is estimated as $1300\,cm^{-1}$ relative to 6^3P_1 by *Gutchek* et al. [5.21], $1560\,cm^{-1}$ by *Strausz* et al. [5.26], and $600\,cm^{-1}$ by *Kielkopf* and *Miller* [5.27]. The ground-state interactions have been determined from atomic beam scattering data [5.28]. Nothing else is known of the Hg–noble gas excited-state potentials, but there should be no strongly bound nonradiative HgXe* reservoir states to compete for the excited-state population (see Hg_2^*). However, the 6^3P_0 level is $\sim 1800\,cm^{-1}$ below 6^3P_1 and could act as such a reservoir. It is discussed in the following paragraph. The data of *Gutcheck* et al. [5.21] also yield $\Gamma K_{eq} = 2 \times 10^{-14}\,cm^3\,s^{-1}$, where $K_{eq} = [HgXe^*(6^3P_1)]/[Hg(6^3P_1)]\,[Xe]$ and Γ is the effective radiative rate for $HgXe^*(6^3P_1)$ excimers. For the expected $\Gamma_1 \cong \Gamma_0 \cong 10^7\,s^{-1}$, this yields $K_{eq} = 2 \times 10^{-21}\,cm^3$, a reasonable value for an excimer bound by $\sim 1300\,cm^{-1}$ or $3\text{–}4\,kT$.

Since $kT \cong 300\,cm^{-1}$ at temperatures corresponding to [Hg] $= 10^{16} - 10^{17}\,cm^{-3}$, any of the above binding energies are less than $6\,kT$ and an equilibrated distribution of $[HgXe^*]/[Hg^*(6^3P_1)]$ is expected. Thus, the net gain criteria of Sect. 5.2 can be applied, and at 270 nm yields the requirement $[Hg(6^3P_1)]/[Hg(6^1S_0)] > 3\exp(-2400\,cm^{-1}/300\,cm^{-1}) = 10^{-3}$.

The gain coefficient at 270 nm assuming $\Gamma = 10^7 \, \text{s}^{-1}$ and 1300 cm^{-1} binding, is $G_v/[\text{Hg}(6^3P_1)][\text{Xe}] \cong 2 \times 10^{-38} \, \text{cm}^5$, yielding $K_v = 0.02 \, \text{m}^{-1}$, $G_v = 0.2 \, \text{m}^{-1}$ at $[\text{Hg}] = 10^{17} \, \text{cm}^{-3}$ $[\text{Hg}(6^3P_1)] = 10^{15} \, \text{cm}^{-3}$, $[\text{Xe}] = 10^{20}$ cm^{-3}. If $\text{Hg}(6^3P_1) + \text{Xe} \rightarrow \text{Hg}(6^3P_0) + \text{Xe}$ (excitation transfer) were a rapid process, it would be very difficult to achieve this 6^3P_1 state density as a large nonradiative 6^3P_0 population would lose excited-state energy via electron quenching or $^3P_0 - ^3P_0$ collisions. However, an upper limit to the rate coefficient of $10^{-14} \, \text{cm}^3 \, \text{s}^{-1}$ [5.29] has been measured for this excitation transfer and, therefore, in an e–beam afterglow or discharge the electrons may keep 6^3P_1 and 6^3P_0 in about a statistical ratio. Due to the shallow binding and resulting rapid Xe–induced dissociation of the $\text{HgXe}^*(6^3P_1)$, electrons will not readily change the $[\text{HgXe}^*(6^3P_1)]/[\text{Hg}(6^3P_1)]$ ratio. Consequently, we conclude that this system could be a good excimer laser candidate, probably with higher efficiency in a discharge than in the above e–beam experiment. The low gain coefficient per $\text{Hg}(6^3P_1)$ can be compensated by increasing [Hg], presumably at constant excitation fraction, until collisions between excited states represent a significant loss. Excited state absorption could, of course, destroy net medium gain at any density.

The $\text{Hg}(6^1P_1)$ excimer bands are also viable laser candidates. Assuming weakly bound $\text{Hg}(6^1P_1) - \text{Xe}$, the net gain criteria of (5.6) becomes at 215 nm $[\text{Hg}(6^1P_1)]/\text{Hg}(6^1S_0)] = 3 \exp(-7500 \, \text{cm}^{-1}/300 \, \text{cm}^{-1})$ $\cong 10^{-10}$. The 215 nm stimulated emission coefficient, from (5.13), is $\Gamma_0 = 8 \times 10^8 \, \text{s}^{-1}$ [5.30], and assuming $\sim 600 \, \text{cm}^{-1}$ binding in the 6^1P_1 state, $G_v/[\text{Hg}(6^1P_1)][\text{Xe}] = 10^{-37} \, \text{cm}^5$. For example, if $[\text{Xe}] = 10^{20} \, \text{cm}^{-3}$, $[\text{Hg}] = 10^{18} \, \text{cm}^{-3}$, and $[\text{Hg}(6^1P_1)] = 10^{15} \, \text{cm}^{-3}$, we find that $G_v \simeq 1 \, \text{m}^{-1}$ and $K_v \cong 10^{-7} \, \text{m}^{-1}$ at 215 nm. No studies of the feasibility of obtaining such 6^1P densities in e–beam, discharge, or optically pumped Hg–Xe vapor have been reported, but such a low excitation fraction is required that this appears as a fairly promising laser candiate.

5.2.3 Group III–VIII Excimers

The Group III metals attain equilibrium vapor densities appropriate for excimer lasers ($\geq 10^{16} \, \text{cm}^{-3}$) at temperatures of 800–2000 °C, but they are generally nonreactive with window and containment materials. Their resonance lines are in the 250–530 nm region so that the red-wing excimer bands occur in the 250–700 nm regions. The only III–VIII excimers that have been studied experimentally are those of Tl, the most volatile member in the group, so the remainder of this discussion will concentrate on Tl.

Thallium has a $6^2P_{1/2}$ ground state and a metastable $6^2P_{3/2}$ state separated by 1 eV fine structure. Transitions from these states to the $7^2S_{1/2}$ and 6^2D states have large oscillator strengths, so Tl has four strong resonance lines occurring at 277, 353, 378, and 535 nm. The red-wing excimer bands associated with each of these lines are viable excimer laser candidates, so they encompass much of the

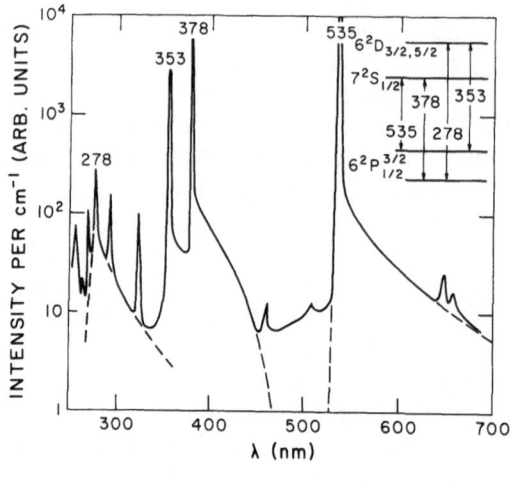

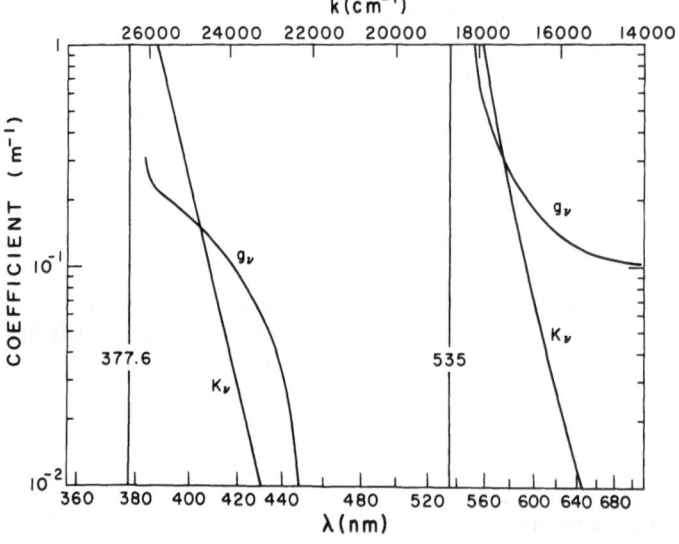

Fig. 5.6. a) Spontaneous emission from a discharge in TlXe for $T=1020\,\mathrm{K}$, $[\mathrm{Tl}]=3\times10^{15}\,\mathrm{cm}^{-3}$ and $[\mathrm{Xe}]=3\times10^{19}\,\mathrm{cm}^{-3}$ (from [5.32]). The excimer bands are indicated by dashed lines.
b) Absorption (K_ν) and stimulated emission (G_ν) coefficients for TlXe at $T=1090\,\mathrm{K}$, $[\mathrm{Tl}(6^2P_{1/2})]=[\mathrm{Tl}(6^2P_{3/2})]=5\times10^{15}\,\mathrm{cm}^{-3}$, $[\mathrm{Tl}(7^2S_{1/2})]=2\times10^{14}\,\mathrm{cm}^{-3}$, $[\mathrm{Xe}]=2.7\times10^{20}\,\mathrm{cm}^{-3}$ (from [5.31])

near uv and visible spectrum. As with the I–VIII excimers, Xe produces the strongest and widest excimer bands, and is, from the discussion above, selected as the best laser candidate [5.31]. The relevant Tl energy levels and the Tl–Xe excimer emission spectrum from a weak discharge in $[\mathrm{Tl}]\cong10^{16}\,\mathrm{cm}^{-3}$ and $[\mathrm{Xe}]\cong10^{20}\,\mathrm{cm}^{-3}$ are shown in Fig. 5.6a [5.32]. The temperature dependence of the two visible band intensities associated with the $7^2S_{1/2}-6^2P_{1/2,3/2}$ transitions has been measured, and interpreted to yield the Tl–Xe excimer

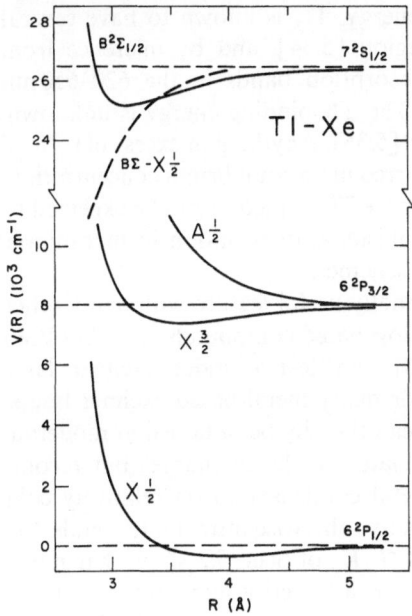

Fig. 5.7. Experimental TlXe potentials (from [5.31])

potentials [5.31] shown in Fig. 5.7. From (5.2) and (5.6), these normalized spontaneous emission intensities have been converted into the G_v and K_v in Fig. 5.6b. The conditions assumed for this figure are [Tl] and [Xe] appropriate for a high-power excimer laser, with 2% excited $Tl(7^2S)$ and the remainder divided equally between the $6^2P_{1/2} + 6^2P_{3/2}$ states. This $7^2S_{1/2}$ excitation fraction corresponds to an electron temperature $T_e \sim 1\,eV$ in (5.7). Studies of its feasibility are underway in the author's laboratories. The 278–310 and 353–378 nm bands in Fig. 5.6a are associated with the 6^2D–6^2P transitions. These have not been studied in normalized fluorescence, so the excimer potentials associated with the 6^2D states and the gain coefficient as a function of $[Tl(6^2D)]$ have not been determined. Nonetheless, based on the intensity of these bands in the discharge spectrum they are expected to be as good an excimer laser candidate as the visible bands.

Tl is essentially a one-electron atom, and the oscillator strengths of the above four resonance lines add up to 0.95 with about 1/3 of this in the $6P \rightarrow 7S$ transition [5.33]. Therefore the arguments applied to alkali-noble gas absorption from excited states apply here as well, although it is somewhat weaker for the $6P \rightarrow 7S$ bands due to their smaller oscillator strengths. It is possible for absorption in strongly allowed Tl–Xe bands out of the 7^2S or 6^2D levels to exceed the stimulated emission in the $7^2S \rightarrow 6^2P$ or $6^2D \rightarrow 6^2P$ bands, but the oscillator strengths of the latter bands are large enough that it is extremely unlikely these excited state absorptions could eliminate net gain in all regions of the bands. The other potential absorption, by Tl_2, is difficult to evaluate due to minimal knowledge of the Tl_2 binding energy. Tl_2 is known to have several absorption bands in the 400–440 nm region [5.34], and by inference from emission bands there should also be absorption bands in the 620–650 nm region from $Tl(6^2P_{3/2})$ parentage states. The Tl_2 binding energy is unknown, although it has been estimated as ~0.6 eV [5.35]. Anything in excess of 0.75 eV would cause serious absorption if Tl_2 occurred in an equilibrium concentration. However, just as in the Na_2 case above, severe Tl_2 depletion can be expected to occur due to gas heating and electron collisional dissociation in high-power discharges, e–beams, or optical pumping schemes.

The Group III elements share a potentially useful feature with many other metals; they are much more volatile as halogenated compounds. This has been used to advantage in arc lamps [5.36] and tested under excimer laser conditions [5.37]. It could be valuable for many metal-based excimer lasers. Relatively high densities of metal vapors can thereby be obtained at moderate temperatures, with the compounds dissociated in the discharges but recombining in cold outer zones to prevent metal condensation on relatively cold walls. In addition to allowing the use of many otherwise intractable metals, the exponential Boltzmann factors in G_v and G_v/K_v of Sect. 5.2 show that much larger gain coefficients and inversions can be attained if temperatures can be lowered. Other compounds besides halogens can be considered, of course, particularly as the residual halogens could have a deleterious effect on discharge stability.

5.2.4 Group I–II Excimers

The existence of low-lying ^1D and ^3D excited states of *Ca*, *Sr*, and *Ba* combined with the causticity and low vapor pressures of these elements make them poor candidates for replacing the noble gas in excimer molecules. Mg, Zn, Cd, and Hg are reasonable candidates for use with elements of equal or lesser excitation energies, but to obtain the necessary $\geq 10^{19}\,cm^{-3}$ densities requires 900–1000 °C for Mg, Zn, and Cd. Thus, due to its volatility and relatively mild chemical behavior Hg is clearly the most practical and frequently studied substitute for noble gases. In the remainder of this section and most of the following two sections we will discuss only such A–Hg type excimers.

The interaction potentials of alkali–Hg molecules have been studied by *Duren*, both experimentally through atomic beam scattering and theoretically [5.38] using the method developed by *Baylis* [5.20] for alkali-noble gas molecules. As an example these calculated Na–Hg potentials are shown in Fig. 5.8 and the G_v and K_v for the A–X band which these yield from (5.13) for typical laser-model conditions are shown in Fig. 5.9. The only spectroscopic data available for these alkali–Hg molecules are those of *Drummond* [5.39], who measured the A–X band shapes, but not in absolute units. The shape of G_v corresponding to Drummond's spontaneous-emission band shape is shown in Fig. 5.9 for comparison. The extent of the agreement between these theoretical

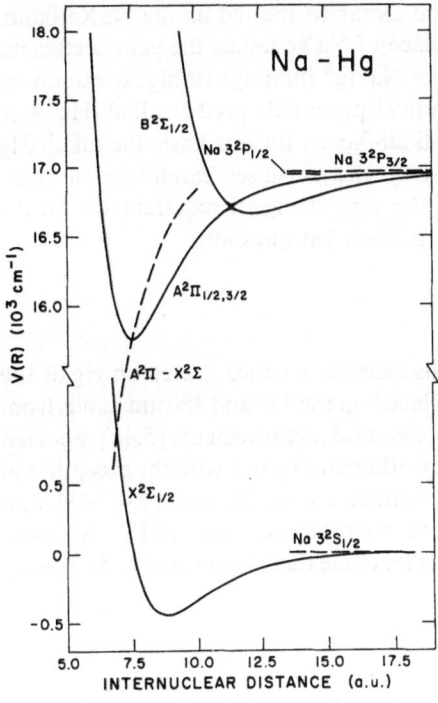

Fig. 5.8. Theoretical NaHg potentials (from [5.38])

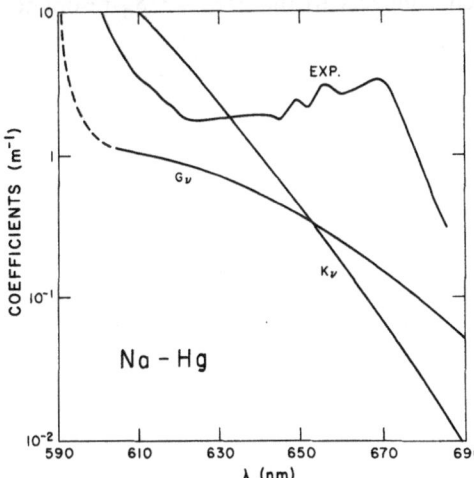

Fig. 5.9. Absorption (K_v) and stimulated emission (G_v) coefficients for NaHg, deduced from the potentials in Fig. 5.8, for conditions of $T = 690$ K, [Na(3S)] $= 10^{16}$ cm^{-3}, [Na(3P)] $= 10^{15}$ cm^{-3}, and [Hg] $= 10^{19}$ cm^{-3}. The experimental line is the measured shape of G_v obtained from the measured fluorescence spectrum of [5.39]. This experimental curve is arbitrarily normalized

and experimental band shapes is typical of that seen between measurements and Baylis-type potential calculations for the alkali-noble gas molecules [5.7, 8, 13]. The d^3R/dv term in (5.13) is very sensitive to $V^*(R) - V(R)$, so that a very accurate set of $V(R)$ is necessary to produce the A–X band intensity distribution.

As this NaHg band does not extend as far to the red as the NaXe band, there is little to recommend NaHg in place of NaXe unless the gain coefficient for NaHg is much larger (due to stronger NaHg* binding), or higher excitation efficiency were feasible. Duren's theoretical potentials predict alkali-Hg A–X bands of strength similar to those of alkali-Xe, so on this basis the alkali-Hg molecules would not be very interesting excimer-laser candidates if these potentials were correct. But in view of the very stringent requirements on the accuracy of the potentials, this is still an unsettled question.

5.2.5 Group II–II Excimers

Most excimer-laser related studies in this class have concentrated on Hg as the "noble gas", with most of this attention placed on the 330 and 480 nm bands from pure Hg vapor. The exception is some models and measurements [5.40] reported for Cd–Hg. Very little is known about any other atom pairs, with the exception of Mg_2 [5.41]. The Mg_2 A and X–state potentials are well known [5.41–43], and allow its amplification characteristics to be modeled using (5.13). As these characteristics are fairly interesting, we will include them below before discussing the Cd–Hg and pure Hg vapor work.

Mg_2

Balfour and *Douglas* [5.41] have measured the bound–bound spectrum of the $A^1\Sigma_u^+ - X^1\Sigma_g^+$ system of Mg_2, which connects to the $3^1P_1 - 3^1S_0$ resonance

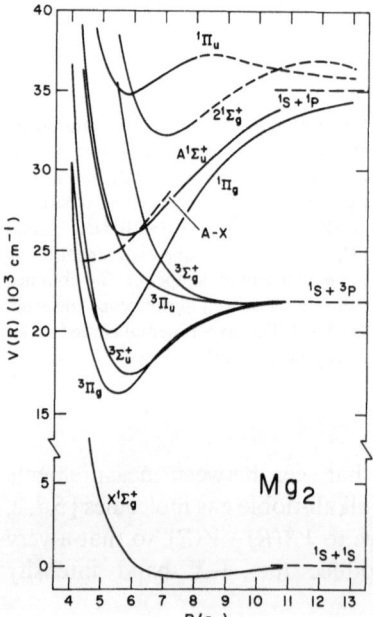

Fig. 5.10. Mg_2 potentials. The $A^1\Sigma_u^+$ and $X^1\Sigma_g^+$ potentials are from spectroscopic data [5.41] and the rest are theoretical [5.44]

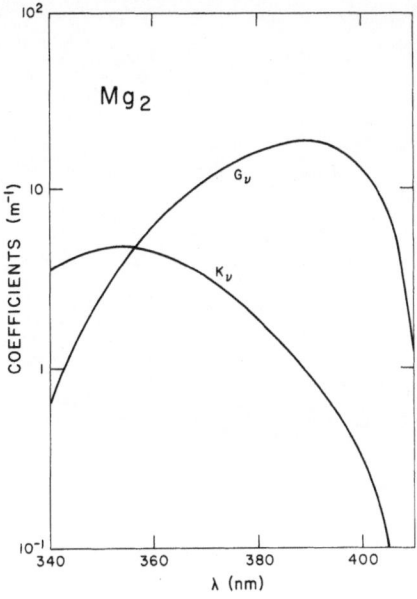

Fig. 5.11. Absorption (K_v) and stimulated emission (G_v) coefficients for Mg_2 for $T = 1150$ K, $[Mg] = 10^{18}$ cm^{-3}, $[Mg_2(A^1\Sigma_u^+)] = 10^{15}$ cm^{-3}

transition of Mg. Their initial analysis to obtain the potentials has been extended by *Li* and *Stwalley* [5.42] and by *Vidal* and *Scheingraber* [5.43] to accurately determine dissociation limits and many higher-order Dunham coefficients. These potentials [5.43] are given in Fig. 5.10, along with other potentials calculated by *Stevens* and *Krauss* [5.44]. *A*–state predissociation into the triplet manifold was also observed in [5.43], but lifetimes are not given. The difference between the *A* and *X*–state potentials is also plotted in Fig. 5.10, so that the small slope at $R \sim 5.5\, a_0$ is apparent. This helps concentrate the band and yields a large G_v at $\lambda \cong 390$ nm. The absorption and gain coefficients for $[Mg] = 10^{16}$ cm^{-3} at $T = 870$ K, $Mg_2^*(A^1\Sigma) = 10^{14}$ cm^{-3}, and $\Gamma_0 = 3.5 \times 10^8$ s^{-1} [5.45] in (5.13, 13a) are shown in Fig. 5.11. This is shown to indicate the possibilities as a high-power laser in a buffered system. However, the feasibility of obtaining a vibrationally thermalized or partially thermalized *A*–state population of this magnitude has not been investigated. In particular, predissociation to repulsive triplet states and gas collisional conversion of $A^1\Sigma$ into nonradiative $^1\Pi_g$ molecules could be very fast and eliminate the possibility of obtaining such $A^1\Sigma$ state populations. Absorptions out of large triplet and $^1\Pi_g$ populations could also be a serious problem.

Cd–Hg

The excimer bands associated with $Cd(5^3P) - Hg$, have been reported by *McGeoch* et al. [5.40] as a ~ 100 nm wide continuum centered at ~ 470 nm. *McGeoch* and *Fournier* [5.40] have also reported 0.4 % cm^{-1} gain at 488 nm in

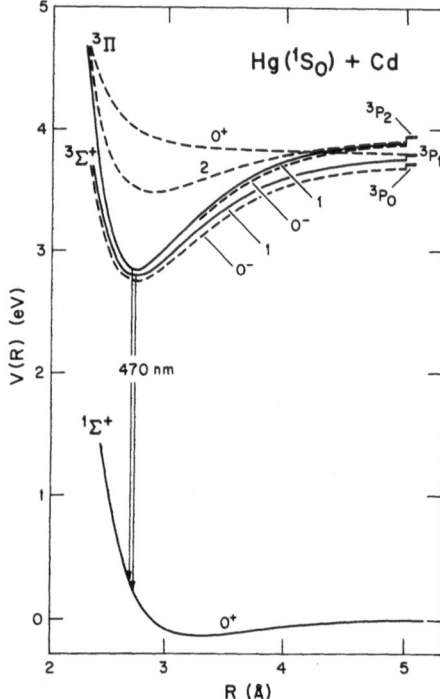

Fig. 5.12. HgCd potentials, from [5.40]. The explanation of the 470 nm band is indicated

flashlamp excited vapor of $[Cd] = 3 \times 10^{15}$ cm^{-3}, $[Hg] = 2 \times 10^{18}$ cm^{-3}. However, recent investigations at LLL have failed to show gain in a nearly equivalent experiment [5.46]. The potentials proposed by these authors are given in Fig. 5.12; they assign radiative lifetimes of $\sim 3\,\mu$s to the two levels associated with the excited ^3P states. As in the Hg$_2$ case which follows, this may be an effective radiative rate due to an excited population distributed thermally between the ^{31} and $^3O^-$ states, with the latter metastable. However, the gain coefficient per total Hg Cd* density is still given by $(\lambda^2/8\pi)\Gamma_{eff}/\Delta v$ where Γ_{eff} is this effective radiative rate and Δv is the width of the band. In view of this very small Γ_{eff} (an effective oscillator strength of $\sim 10^{-3}$), excited state absorption could be a very serious issue in this molecule.

Hg$_2$

Due to the ease of handling Hg vapor and its importance in lighting and electrical equipment, the kinetics, spectra and discharge properties of Hg vapor have been studied extensively [5.47]. In spite of this, many of the properties which determine the viability and efficiency of potential Hg$_2$ excimer lasers have only been determined very recently if at all. Studies by *Hill* et al. [5.48], and by *Schlie* et al. [5.49] indicated potential optical properties in e–beam excited Hg vapor, but only the recent very thorough experimental studies by

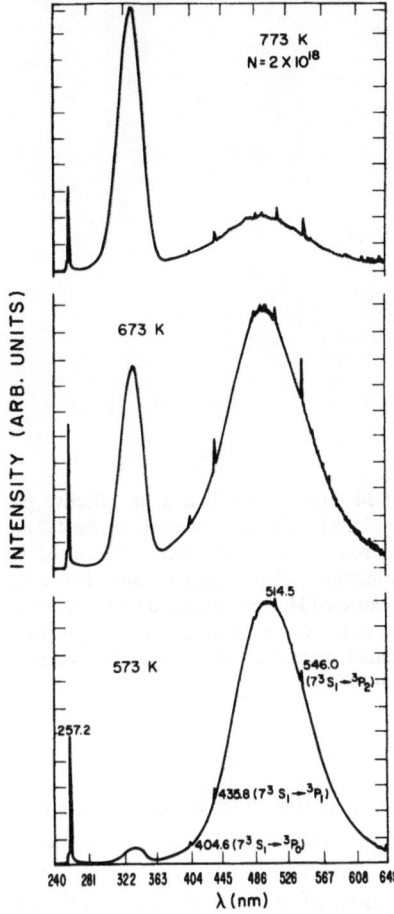

Fig. 5.13. Fluorescence spectrum from optically excited Hg vapor at $[\text{Hg}]=2\times10^{18}\,\text{cm}^{-3}$ and the indicated temperatures (from [5.50])

Drullinger et al. [5.50] combined with theoretical potentials by *Stevens* and *Krauss* [5.44] have finally yielded a fairly complete picture of the Hg_2 energy levels and radiative properties as well as the heavy particle kinetics associated with the $\text{Hg}(6^3P)$ states and connected Hg_2^* levels. Some calculations [5.51, 52] and measurements [5.4, 53] on Hg discharges have elucidated aspects of high-power discharges in high pressure Hg vapor, as appropriate for excimer lasers. These are discussed in Sect. 5.3; we summarize the *Drullinger* et al. [5.50] picture for Hg_2 potentials and Hg vapor kinetics in the following paragraphs.

Temperature dependent emission spectra of optically excited Hg vapor are shown in Fig. 5.13, and a set of Hg_2 potentials is shown in Fig. 5.14. The R scale is not well determined by the theory or data, so is left arbitrary in the figure; this uncertainty is not important for laser modeling. These potentials were derived by adjusting theoretical potentials from *Stevens* and *Krauss* to match experimental data, where available. The Hg_2 radiative transition

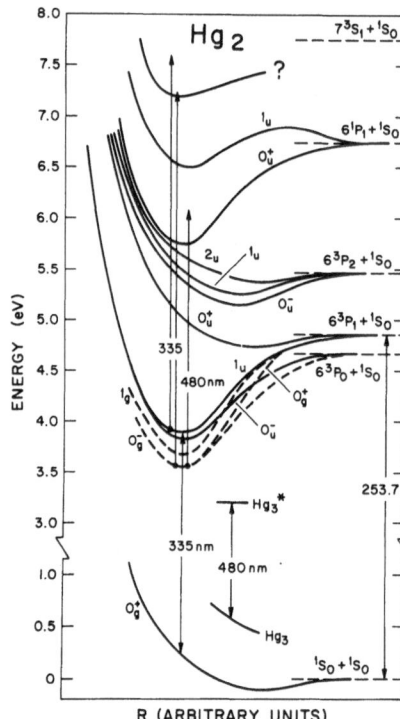

Fig. 5.14. Hg$_2$ potentials and the effective position of the Hg$_3$ levels responsible for the 480 nm band, from [5.50]. These were constructed by adjusting the results of Mg$_2$ ab initio calculations to fit expected Hg$_2$ behavior and known spectroscopic data. We have added the (?) potential associated with the 7^3S state, whose shape is unknown

associated with the 335 nm band is indicated, as is the effective energy position of the Hg$_3^*$ and Hg$_3$ levels which the authors associate with the 480 nm band. The experimental observations indicate that the nonradiative O_u^-, O_g^-, and O_g^+ states act like an effective reservoir (Res) for most of the Hg$_2^*$ population that is ~ 2700 cm^{-1} below the minimum of the radiative 1_u state, which has a band-averaged radiative rate $\Gamma_1 \cong 10^6$ s^{-1}. Another 3800 cm^{-1} below this effective Hg$_2$ reservoir is the average position of the Hg$_3^*$ population responsible for the 480 nm band. At densities [Hg] $> 10^{18}$ cm^{-3} appropriate for high-power excimer lasers, and without electron collisions, the population of this Hg$_3^*$ state is in equilibrium with the Hg$_2^*$ reservoir, so that [Hg$_3^*$]/[Hg$_2^*$(Res)] $=$[Hg]$K_3 \exp(3800$ cm$^{-1}/kT)$. Similarly, the Hg$_2^*$ populations come to an equilibrium ratio [Hg$_2^*$(1_u)]/[Hg$_2^*$(Res)] $=(Z_{1u}/Z_{Res}) \exp(-2700$ cm$^{-1}/kT)$, where the partition functions should satisfy $Z_{1u}/Z_{Res} \sim 1$. Thus if the band-average radiative rate of Hg$_3^*$ is Γ_3, the ratio of the bands in fluorescence is

$$\frac{I_{335}}{I_{485}} = \frac{\Gamma_1 \exp(-6500 \text{ cm}^{-1}/kT)}{[\text{Hg}]K_3\Gamma_3}. \tag{5.14}$$

Measurements have determined that $\Gamma_1/K_3\Gamma_3 = 4.5 \times 10^{23}$ cm^{-3}. If one estimates $K_3 \sim 10^{-22}$ cm^3, then $\Gamma_3 \sim \Gamma_1/50 \sim 2 \times 10^4$ s^{-1}, corresponding to an

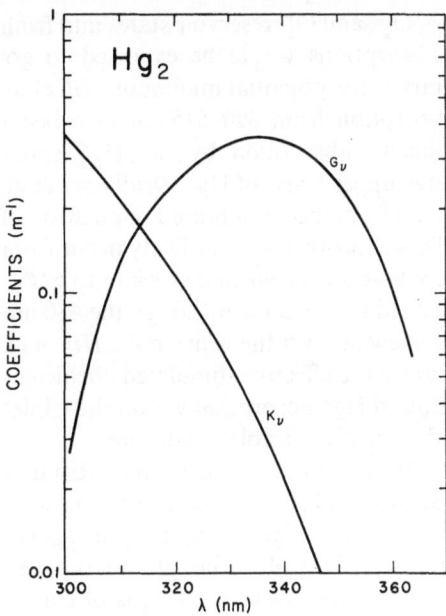

Fig. 5.15. Absorption (K_ν) and stimulated emission (G_ν) coefficients for Hg vapor at $T = 1300$ K, $[\mathrm{Hg}] = 10^{19}\,\mathrm{cm}^{-3}$, $[\mathrm{Hg}_2^*] = 1017$ cm^{-3}, where Hg_2^* refers to the 35–4.0 eV excimer states in Fig. 5.14

oscillator strength of about 10^{-4} at 480 nm. These results, the shape of the 1_u potential, and the $1_u \rightarrow O_g^+$ (ground state) transition moment variations were deduced from analysis of the temperature and time dependence of the band intensities and shapes, following optical excitation.

Due to the $\exp(-2700/kT)$ ratio of $[\mathrm{Hg}_2^*(1_u)]/[\mathrm{Hg}_2^*(\mathrm{Res})]$, the 330 nm band gain can be greatly enhanced by overheating the vapor relative to the equilibrium density. Figure 5.15 shows the 330 nm band gain and absorption for $T = 1300$ K and $[\mathrm{Hg}_2^*(\mathrm{Total})]/[\mathrm{Hg}] = 1\%$, where total refers to the 1_u and reservoir states. At this temperature about 7% of the total Hg_2^* is in the radiating state, so that raising T further could enhance $G_\nu/[\mathrm{Hg}_2^*(\mathrm{Total})]$ by as much as a factor of five. The other effects of such gas heating would be to spread the band slightly and to increase K_ν, although by a smaller factor than the G_ν increase. In a discharge or e–beam excited vapor, electron collisions will tend to raise the $[\mathrm{Hg}_2^*(1_u)]/[\mathrm{Hg}_2^*(\mathrm{Res})]$ ratio above the gas temperature, thereby increasing the gain coefficient per unit of total $[\mathrm{Hg}_2^*]$ density.

The stimulated emission coefficients for these bands are very weak when expressed as a function of the total Hg_2^* density, e.g., it is $\sim 4 \times 10^{-20}\,\mathrm{cm}^2$ at 335 nm and $T = 1300$ K (Fig. 5.15). Therefore, absorption from Hg_2^* in allowed bands, particularly from the reservoir states which contain most of the population, will be $\sim 10^3$ stronger and will destroy the medium gain if they occur in the same wavelength regions as the excimer bands. Estimates of some higher excited Hg_2^* states which could cause such absorptions, by *Stevens* and *Krauss*, are shown in Fig. 5.14. Arrows of length corresponding to 335 nm and

480 nm are drawn upwards from the O_u^-, O_g^+, and O_g^- reservoir states and from the 1_u state to indicate where these absorptions would be expected to go, although it is not necessary that they start at the potential minimum. *Hill* et al. [5.48] measured almost a constant absorption from 390–515 nm in e–beam pumped Hg vapor, and attributed this to absorption by the Hg_2^* states associated with $Hg(6^3P) - Hg$ to repulsive upper levels of Hg_2. *Drullinger* et al. [5.50] also measured absorption at 488 nm from the equilibrated population of optically excited $Hg(6^3P) - Hg$ states. These absorptions very likely occur from the reservoir states to the 1_u and O_u^- repulsive states which dissociate to 6^3P_2; these might be more repulsive than indicated in Fig. 5.14. Although the 480 nm band gain per Hg_2^* might be improved somewhat over the value in the *Hill* et al. [5.48] study (e.g., using lower temperatures), the effective stimulated coefficient per Hg_2^* is quite small and the dipole-allowed Hg_2^* absorption within the triplet manifold will probably be stronger under most realizable conditions.

Hg_2^* absorption in the region of the 335 nm band would be very strong if sufficiently attractive states associated with $Hg(7^3S) - Hg$ occur (e.g., the state with question mark in Fig. 5.14). *Stevens* [5.54] suggests that Rydberg-like states should occur about as indicated in Fig. 5.14, although no detailed calculations have been made. This absorption coefficient per Hg_2^* is of course given by (5.13) where V refer to the potentials of the Hg_2^* states and V^* to those of the $Hg(7^3S) - Hg$ state. Thus, if $V^*(R) - V(R)$ traverses the 335 nm band photon energies anywhere in the 2.7–3.5 Å region, very strong absorption bands will obscure the excimer-band gain. *York* et al. [5.53] have measured typically 10% net absorption in a discharge afterglow at a number of wavelengths in the 335 nm band, and *Byer* et al. [5.55] have measured absorption at 325 nm from strongly optically pumped Hg vapor. *Drullinger* et al. [5.50] obtained an upper limit of $\sim 3 \times 10^{-18}$ cm^2 for 325 nm absorption per total Hg_2^* in their weakly excited vapor with $[Hg^*] \cong 10^{12}$ cm^{-3}. This limit is large enough to allow the above measured absorptions to be due to $Hg_2^*(Res + 1_u)$. However, the absorptions in the strongly excited vapors could also be due to populations of Hg_2 in higher excited states with strong absorptions from these. For this absorption to show the same time dependence as fluorescence it would, of course, be necessary for these higher Hg_2^{**} state populations to follow the same decay curve as the Hg_2^* emission, but this is not unreasonable if the decay rates were predominantly due to net energy flow out of many strongly coupled levels. The coupling can occur through the $Hg_2^* - Hg_2^*$ collisions, radiative and collisional cascading following recombination, electron collisions, and other possibilities. In fact the fluorescence observed from these and e–beam excited vapors may also have major fluorescence bands due to more highly excited excimer states; the fact that an apparent single decay rate is observed is not proof that a single, closely grouped, excited-state manifold is the source of all the bands. But returning to the Hg_2 335 nm band, it requires strong excitation to obtain useful excimer gain levels, so that absorption under these conditions is the crucial test. These measurements indicate that net gain of significant size cannot be obtained on the 335 nm band with strong pumping, unless index of refraction variation or some other loss process is being misinterpreted as Hg_2^* absorption. On the other hand, if excited-state absorption could be overcome the 335 nm excimer band itself has excellent gain characteristics (Fig. 5.11). It appears reasonable to expect that a large concentration of Hg_2^* excimers might be achievable, subject to the limitations discussed in Sect. 5.3, and this is generally considered to be an attractive candidate for high-energy, subnanosecond pulses.

The excited-state absorption issue is a serious consideration for all excimer bands with low gain coefficients. Discharges and e–beams populate many high-excited levels, and there may be no net medium gain until these higher levels are largely emptied, hopefully long before the desired lowest excimer states are emptied.

5.2.6 Group III–II Excimers

In this group, Hg is the most tractable Group II replacement for the noble gas, although densities of 10^{19} cm^{-3} of Cd, Zn, or Mg are feasible at 800–1100 °C. Temperatures in this range are necessary as well to obtain appropriate Group III vapor densities. As with all the metal-based excimers, the substitution of Hg for Xe offers the possibility of stronger excited-state binding and thereby larger gain coefficients due to the exponential factor in (5.13a). But whereas this may be a vain hope for the alkali–Hg excimers, it is astonishingly successful for Tl–Hg. *Drummond* and *Schlie* [5.56] have shown that Tl–Hg has relatively narrow excimer emission bands with gain coefficients to per $[Tl^*(7^2S)]$ that are 10–100 times larger than for Tl–Xe. It is likely that similarly strong bands occur in other visible wavelength regions when other Group III elements are substituted for Tl. *Santaran* et al. [5.57] have studied the bands of *TlHg*, *TlCd*, *TlZn*, *InHg*, and *InZn*, and obtained some portions of the potentials. Since Tl–Hg has been studied as an excimer laser candidate by *Drummond* and *Schlie*, and it is a very promising candidate, we will discuss this case below.

The Tl–Hg potentials, from [5.56], are shown in Fig. 5.16. These have a similar appearance to the Tl–Xe potentials in Fig. 5.7, but there are two important and major differences. It is instructive to compare the optical consequences of these differences from the standpoint of (5.13). First the B–state of Tl–Hg is bound by 4100 cm^{-1}, compared 1300 cm^{-1} for Tl–Xe, so that the $\exp(-\Delta V^*/kT)$ population factor in (5.13a) is $\sim\exp(5.4)$ for Tl Hg and $\sim\exp(1.7)$ for Tl Xe at $T = 1100$ K corresponding to $[Tl] \sim 10^{16}$ cm^{-3}. Due to this, $G_v/[Tl^*][Hg]$ in the center of the Tl Hg emission bands is ~ 50 times greater than $G_v/[Tl^*][Xe]$ in the center of its bands. In either case, at high enough [Hg] or [Xe] most of the $[Tl^*(7^2S)]$ is in molecular form and most of the emission is in the molecular band (5.13a), but the Tl Hg band is still narrower due to the larger exponential population factor. This factor con-centrates the emission band in the wavelength region corresponding to the potential minimum. If the gain coefficient per excimer molecule is approxi-

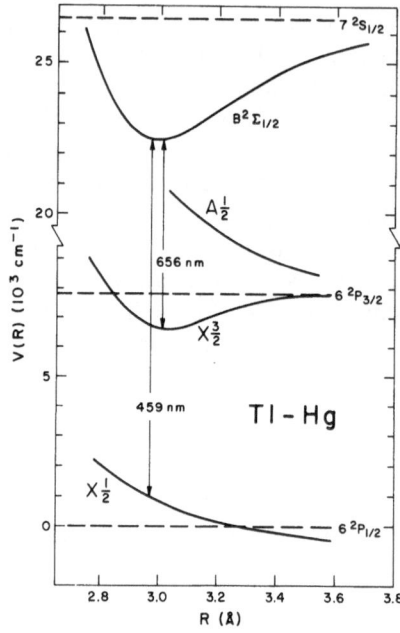

Fig. 5.16. TlHg potentials and origins of the 656 and 459 nm bands, from [5.56]. The $A_{1/2}$ state is not known, but is expected to be quite repulsive as shown

mated as in (5.12) one can see that this decreased Δv_{Band} results in a higher gain coefficient. This effect is greatly enhanced by the second important difference between the Tl Hg and Tl Xe potentials: $V^*(R) - V(R)$ goes through a minimum for Tl Hg near 3.1 Å. This results in the "head of band-heads" or "satellite" seen in the Tl–Hg bands in Fig. 5.17. (These Tl Hg potentials were in fact constructed by *Drummond* and *Schlie* [5.56] to match this measured effect.) This further narrows the effective Δv_{Band} to $\sim 500 \text{ cm}^{-1}$, compared to 5000 cm^{-1} for the Tl Xe bands. This is in fact the same effect that is responsible for the narrowness of the halogen-noble gas uv excimer bands. In the present context, it results in the large gain coefficients shown in Fig. 5.17 for the assumed Tl, Hg, and Tl* densities, which have been obtained here by applying (5.2) and (5.6) to the fluorescence data in [5.56]. Comparison to Fig. 5.6 shows that even for $[\text{Xe}]/[\text{Hg}] \cong 0$, the Tl Hg gain exceeds that of Tl Xe by a factor of ~ 5. Furthermore, as noted in [5.56] this [Hg] or lower values can be used with a He buffer so that the index of refraction effects will be less severe.

These comparisons between Tl Hg and Tl Xe have been made to demonstrate the sensitivity of the optical properties to the features of the potentials. It should be apparent that the Tl Hg potentials are characteristic of an ideal excimer-laser candidate. The excited state has enough binding ($\sim 5kT$) to concentrate the emission band at moderate "noble gas" densities, yet it is shallow enough to allow rapid relaxation from initially excited free Tl*(7^2S) into the several lowest vibrational states responsible for the band peak. (This three-body association rate constant is $3 \times 10^{-31} \text{ cm}^6 \text{ s}^{-1}$ [5.56].) A minimum occurs in $V^* - V$ at an R near the V^* potential minimum. This causes a

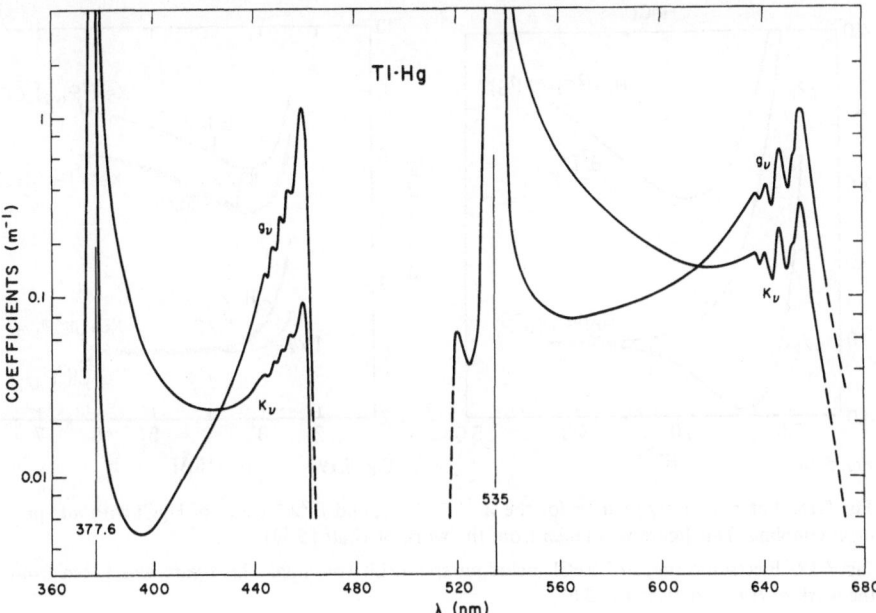

Fig. 5.17. Absorption (K_ν) and stimulated emission (G_ν) coefficients for the Tl–Hg excimer bands of Fig. 5.15, from data in [5.56]. The conditions are $[\text{Tl}(6^2P_{1/2})]=[\text{Tl}(6^2P_{3/2})]=5\times10^{15}\,\text{cm}^{-3}$, $[\text{Tl}(7^2S)]=2\times10^{14}\,\text{cm}^{-3}$, $[\text{Hg}]=3\times10^{19}\,\text{cm}^{-3}$, $T=820\,°\text{C}$

convergence of bands, so that a number of low-lying vibrational states [in $V^*(R)$] radiate predominantly at the same wavelength. The resulting high gain coefficient and homogeneous broadening make this the best known candidate for a visible metal-vapor excimer laser. The lower gain-coefficient Tl Xe system is much more difficult to achieve laser action in, although it could attain higher power and energy densities without superfluorescence losses if successful.

Hg_2 absorption is negligible in the 450 and 650 nm regions where the peak G_ν of the Tl Hg bands occur. Absorption by Tl_2 might occur in these regions, but is very unlikely to match the Tl Hg* gain at the peak G_ν wavelengths. Of course, Tl_2 would also be severely depleted in any high-power pumping scheme. Strong absorption by Tl Hg excimers to higher states is extremely unlikely to occur in the precise λ regions of 450–460 and 620–650 nm with sufficient strength to decrease these G_ν significantly.

5.2.7 Mercury Halide Excimers

Mercury monohalide molecules form a class of excimer media with properties that are similar in many respects to the corresponding rare gas halides. This is expected, since the mercury atom resembles that of a heavy rare gas atom in many of its interactions. Comparison of the Hg_2 potential curves illustrated in

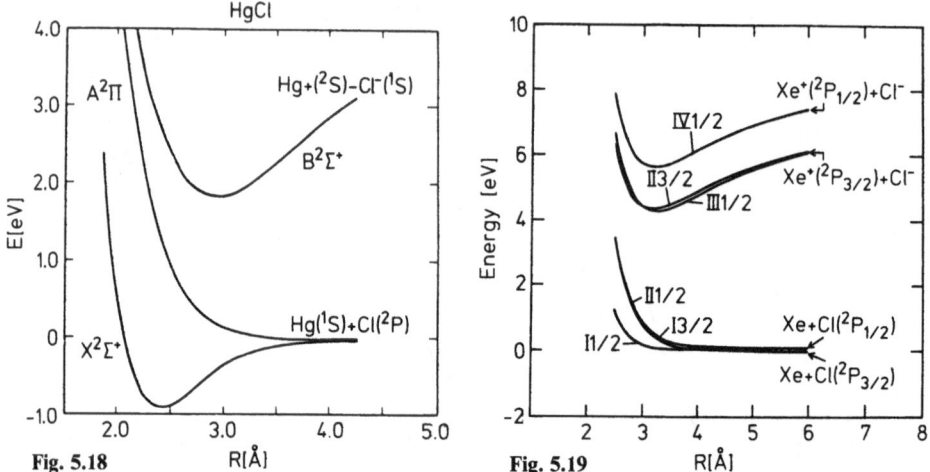

Fig. 5.18. Potential-energy curves for the $X\,^2\Sigma^+$, $A\,^2\Pi$, and $B\,^2\Sigma^+$ states of HgCl without spin-orbit coupling. This figure was taken from the work of *Wadt* [5.59]

Fig. 5.19. Electronic states of XeCl including spin-orbit coupling. This figure was taken from the work of *Hay* and *Dunning* [5.61]

Fig. 5.14 with the corresponding potential curves for Xe_2 illustrates this general point [5.58]. The ground level in both cases is weakly bound and there is a dense manifold of excited states with appreciably greater characteristic binding energies.

The strong similarity of the mercury halogen molecules with the corresponding rare gas halogen systems is immediately apparent from a comparison of the potential curves, a direct manifestation of the quasi-rare-gas behavior of the Hg atom. These similar features are clear from a comparison of the HgCl curves illustrated in Fig. 5.18 [5.59] and the corresponding curves for the xenon halides [5.60, 61]. The electronic states of XeCl are shown in Fig. 5.19. Relatively weakly bound ground states and strongly bound ionic states for excited levels are characteristic of both. Further compilations of data on the rare gas systems are compiled in [5.62].

Certain clear differences are also apparent in the comparison of rare gas halides and mercury halides. Since the lowest Hg excited state occurs at an energy considerably below that characteristic of rare gas excited levels, the mercury halides radiate at longer wavelengths, normally in the visible range of the spectrum (e.g., HgCl, 558 nm) rather than the ultraviolet. Correspondingly, the radiative lifetimes tend to be somewhat longer for the mercury halides [5.63]. It is expected that the influence of spin-orbit coupling should be appreciable in the mercury materials, particularly those involving a heavy halogen such as HgBr and HgI.

Stimulated emission has been observed [5.64] on the mercury monohalides using excitation from electron beams [5.65], discharges [5.66], and photo-

Table 5.1. Radiative properties of HgCl, HgBr, and HgI

System	Wavelength [nm]	Transition	Radiative lifetime [ns]	Stimulated emission cross section [cm²]	References
HgCl	558	$B^2\Sigma_{1/2} \rightarrow X^2\Sigma_{1/2}$	22	4.5×10^{-16}	[5.70]
HgBr	503, 498	$B^2\Sigma_{1/2} \rightarrow X^2\Sigma_{1/2}$	23	2.4×10^{-16}	[5.71, 72]
HgI	444	$B^2\Sigma_{1/2} \rightarrow X^2\Sigma_{1/2}$	27	$\sim 2 \times 10^{-16}$	[5.71]

dissociation of mercury dihalide materials [5.67]. Separate calculations of the electronic structures of $HgCl_2$, $HgBr_2$ [5.68], Hg_2F_2, and Hg_2Cl_2 [5.69] have been made.

Table 5.1 indicates some of the principal radiative characteristics of HgCl, HgBr, and HgI [5.63].

5.3 Excitation Methods and Efficiency

The potential efficiency of these molecular lasers depends on the excitation method as well as on the molecular properties discussed in Sect. 5.2. We will discuss four excitation schemes, in order of increasing potential efficiency.

These excimers can easily be used as *laser-pumped tunable lasers* if desired. Simple considerations lead to a typical single pass gain $G \sim (\lambda^2/8\pi)P/\Delta v_{\text{Band}}$ for such a laser, where P is the pump power in photons $\text{cm}^{-2}\,\text{s}^{-1}$ and Δv_{Band} is the width of the excimer emission band. For a 1000 Å wide visible excimer band and a 10 W Ar^+ pump laser with a 20 μm waist, $G \cong 0.05$. This is very similar to existing dye lasers, as indeed these considerations apply equally to them. A variety of issues such as index of refraction variations, band shapes, and flow turbulence enter into the determination of relative values for different tuneable media. These will not be discussed here, since the present work is concerned instead with efficient, high-power excimer lasers.

If flashlamps are used to excite the excimers, only radiation in the wings of the resonance lines, where the absorption coefficient is 0.1–10 cm^{-1}, is useful. These wavelength regions are usually relatively narrow, typically filling less than 1% of the visible continuum. Thus, although the laser energy might be a large fraction of the absorbed light, the latter will normally be a very small fraction of the total flashlamp radiation. The flashlamp spectral match to the absorption could be improved somewhat by filling it with the same excimer vapor species as the gain tube, but we still would not expect high efficiencies or scalable, high powers by this method. It may be easier to achieve gain in flashlamp pumped rather than electrically excited systems due to the absence of undesired electron collision processes, but their potential efficiencies appear to be much lower than if the discharge is applied directly to the laser medium.

High energy electron beam (e–beam) excitation occurs predominantly via recombination and excited state relaxation following ion-pair formation. Ion-pair energies are typically ~ 1.7 times ionization energies, and generally only one excimer is formed per ion pair. Under these circumstances the maximum efficiency is given by the excimer-band photon energies divided by the ion-pair energy. Excimer states with large ratios of excitation energy to ionization energy, as in the noble-gas dimers, are, therefore, the most efficient for e–beam pumping. The optimum e–beam excimer pumping efficiencies predicted for these cases by current models are about 40–50 % relative to the energy deposited in the gas (Chap. 3). Foil losses, beam spreading, and other problems limit the actual overall efficiencies to significantly lower values. For excimer lasers based on a small fraction of metal atoms in a high pressure Group II or Group VIII vapor, e–beam excitation would primarily ionize the major constitutent, requiring typically 10–20 times the metal-excimer photon energy. Thus maximum efficiencies would be 5–10 % relative to the energy deposited in the gas if each ionization event produced only one visible photon. No modeling or experiments have been reported and this may be an overly pessimistic assumption, but the potential efficiencies in these metal-doped noble gases do appear to be much greater for discharges. In addition to their use as the entire energy source, electron beams are of value as preionizers, sustainers, and diagnostics. Cases in which a larger metal fraction and, therefore, a larger potential efficiency is possible are Hg–noble gas and Hg_2. These have been studied with e–beam excitation; $HgXe$ by *Gutcheck* et al. [5.21] and by *Woodworth* [5.22]; Hg_2 by *Hill* et al. [5.48] and by *Schlie* et al. [5.49]. *Gutcheck* et al. [5.21] reported a maximum fluorescence energy efficiency of $\sim 0.5\%$ relative to the energy deposited in the gas for the 270 nm band associated with $Hg(6^3P) - Xe$. *Woodworth* also reported $\sim 0.5\%$ fluorescence efficiency for this band and about 3 times as much in the 200 nm band associated with $Hg(6^1P_1) - Xe$. *Hill* et al. [5.48] and *Schlie* et al. [5.49] have not reported efficiencies.

The metal-atom excitations desired for this type of excimer laser have two basic features that contribute to a potentially high laser efficiency with *discharge excitation*. First, the excited atomic state responsible for the laser transition (A^*) is the lowest excited state so that its excitation is the first inelastic process available to the accelerating electrons. [Vibrational and rotational excitation of excited molecular states occurs, but densities of these states are low and this is not a significant energy drain except at large molecular ion and electron (n_e) densities.] Second, the excitation energy is a relatively small fraction of the ionization energy. For n_e less than about 10^{14} cm^{-3} the electron energy loss in a discharge from this $A \rightarrow A^*$ excitation causes a rapid decrease in the electron energy distribution with increasing energy above the inelastic threshold. This attenuates ionization of the metal atoms and greatly reduces the excitation of the "noble gas", even though its density is typically at 10^4 greater than that of the metal vapor. Since the A^* energy is less than half of the ionization energy, the electrons (predominantly with energies below the first inelastic threshold) will generally be more efficient at exciting A to $A*$ than at further ionizing the $A*$. This situation is in contrast to the noble-gas excimers for which the excitation energy is a large fraction of the ionization energy. In this case the electrons with energies below the inelastic threshold are very effective at ionizing the noble-gas excimers and excited atoms. For $n_e > 10^{14}\,cm^{-3}$ electron-electron collisions tend to thermalize the electron energy distribution, and multiple-step ionization of $A*$ occurs rapidly. This situation is discussed below for the Na–Xe example.

Discharge modeling at power levels appropriate for excimer lasers has been carried out only for Hg vapor and alkali-noble gas mixtures. No measurements have yet been compared to these although some observations of general discharge behavior in these systems has been reported. Therefore, the discussions following will concentrate on the pure Hg and alkali-noble gas cases.

Glow and arc discharges in Hg vapor have been studied for several decades on account of their importance in lighting and electrical switching [5.47]. Neither discharge is appropriate for lasers, as they are cw discharges and the atomic level populations, gas temperature, electron density, and effective electron temperature are all approximately in equilibrium. The first attempt at a pulsed, high power discharge in Hg vapor to obtain gain was by Carbone and Litvak [5.4], who studied the Hg_3 480 nm band emission in a pulsed, longitudinal arc in $\sim 10^{18}\,cm^{-3}$ of Hg. Electrical energy deposition rates up to $\sim 10\,W\,cm^{-3}$ for $\sim 1\,\mu s$ were achieved without developing nonuniformity or instabilities, while the effective electron temperature was estimated at $\sim 0.6\,eV$. Although the 480 nm band gain calculated from the observed emission was too small to be useful, this magnitude of energy deposition is sufficient for most potential metal-vapor excimer lasers.

More recently the Boltzmann equation has been solved in Hg vapor by Rockwood [5.51], Masek and Vokaty [5.58], and by Judd [5.52]; but none of the investigations looked into the effect of excited state ionization or superelastic collisions on discharge stability. Mosburg and Witke [5.59] have measured the Hg_2 excimer vibrational temperature in Hg discharges, thereby inferring the effective vibrational excitation plus excimer quenching rates by electrons. York and Carbone [5.53] recently operated a transverse pulsed discharge in $\sim 10^{19}\,cm^{-3}$ of Hg, and obtained homogeneous discharges of $\sim 10\,Jl^{-1}$ energy deposition following uv preionization. From the excimer fluorescence intensity they inferred that $\sim 2\%$ net gain should occur on the 320 nm excimer band for their 10 cm discharge length. A measurement at 325 nm, however, indicated a net loss in transmission during the discharge pulse. Although this was reported in York et al. [5.53] as preliminary and inconclusive, later measurements at 325 nm and other, nearby wavelengths also indicated net attenuation (private communication with G. York). It has been suggested that these observations might be reconciled with the net gain at 325 nm observed with e–beam pumping [5.49] if the higher Hg_2^* states associated with $Hg(7^3S) + Hg$ had a larger population in the e–beam case. Thus, net gain might occur on these bands, associated with the $7^3S - 6^3P$ transition, for e–beam excitation while net absorption might occur in a discharge.

Discharge excitation of alkali-noble gas mixtures has also been studied for many years in relation to metal vapor lamps [5.47]. As noted above these essentially cw discharges are not very relevant to laser discharges. High power discharge excitation in cold, non-LTE gas was first modeled by *Palmer* [5.10] using a thermal electron energy distribution $f(u)$ with a limited number of metal-atom states and collision processes. His results indicated the possibility of reasonably efficiency excitation and net gain in the K–Xe excimer bands, but the effect of the expected non-Boltzmann character of $f(u)$ and of additional states and collision processes remained unclear. *Schlie* [5.60] solved the Boltzmann equation for $f(u)$ in typical laser mixtures of Na and Xe $(1:10^3 - 1:10^4)$ without excited states. His results indicated a very rapid drop in $f(u)$ above the first alkali inelastic threshold. More recently, *Shuker* et al. [5.12] investigated the effect of excited Na states on $f(u)$ in Na–Xe and self-consistently solved the Boltzmann equation for $f(u)$ and the population rate equations for the excited state and electron densities. The results of this calculation and the general characteristics of this system will now be explored.

In the absence of excited Na, about 90 % of the electrical energy goes into the desired $Na(3S) \rightarrow Na(3P)$ excitation for a wide range of E/N. The remaining energy goes primarily into Xe elastic recoil at the lower E/N values and into ionization and undesired excitation at higher E/N. The overall efficiency for discharge operation in Ne Xe mixtures is additionally limited by the typical 0.75 factor in the ratio $h\nu_{laser}/\Delta E_{exc}$, which is a fundamental consequence of the red-shift between excimer absorption and emission bands.

Quenching of $Na(3P)$ by the low-energy electrons is an important decay mechanism at the excitation fractions and electron densities appropriate for high-power discharge operation [5.2, 3, 12]. This is not a net energy loss mechanism, but it limits the [Na*]/[Na] fraction to $R_{exc}/(R_q + \Gamma_{eff})$, where R_{exc} and R_q are the electron excitation and quenching rates and Γ_{eff} is the effective Na* spontaneous radiative loss plus stimulated emission rates. The spontaneous emission loss rate is roughly equal to $\Gamma_0[Na \, Xe^*]/[Na^*]$ where Γ_0 is the natural atomic radiative rate. For a useful net gain the E/N must be high enough to keep [Na*]/[Na] above typically 0.05. This excitation fraction has not been achieved in preliminary high-n_e discharge experiments [5.32], although the models of these systems [5.12] do yield it under appropriate assumptions.

If the excited $Na(3P)$ fraction builds up to the ~ 5 % levels appropriate for appreciable optical gain, the $Na(3P)$ excitation efficiency drops due to ionization of excited states. The calculations indicate that typically 20–50 % of the electrical energy might nonetheless be extracted as laser power, with the remainder going primarily into gas heating. Most of this heating is attributed to rational and vibrational excitation of molecular ions, momentum transfer collisions with Xe, and A^* multiple-step ionization followed by recombination. These calculations also indicate that Na_2, with its associated deleterious absorption, is severely depleted by electron collisional dissociation, and would be further depleted by gas heating and laser-field bleaching. Electron densities of $10^{15}-10^{16}$ cm^{-3}, $E/[Xe] \sim 10^{-17} - 10^{-16}$ V cm^2 at typical $[Na] \cong 3 \times 10^{16}$, $[Xe] \cong 3 \times 10^{20}$ cm^{-3}, and current densities of 10–100 A cm^{-2} result from the calculations. At these power depositions of 10^2-10^3 MWl^{-1}, the gas heating rate is 10–100 °C μs^{-1}. If total gas heating of 100 °C could be tolerated during an excitation pulse, about 10^3 Jl^{-1}, would be deposited in the gas, conceivably with 20–50 % convertible into laser power. These calculations are obviously very speculative and the actual situation will be less favorable, but this at least indicates the sizes of the laser power, efficiency, and pulse energies under consideration.

One major question that is not addressed by these homogeneous discharge models is the time for growth of an arc instability in this high-pressure mixture. Various uniform preionization schemes are available, but preliminary investigations by *Drummond* [5.61] and by *Rothwell* et al. [5.32] are very encouraging, as they indicate that these metal-doped noble gas mixtures are much more stable against arcing than the pure noble gases.

Another potential major problem is cavity quality deterioration due to index of refraction variations, i.e., loss of medium homogeneity. Xe at 10^{20} cm^{-3} has an index of refraction $n-1 \cong 2.4 \times 10^{-3}$, and an expected 10 % temperature rise due to the excitation pulse will cause variations of ~ 10 % of this. If this index gradient occurs nonuniformly, it could severely distort the laser beam. As an example, if an index variation $n-1 = 2 \times 10^{-4}(y^2/y_0^2)$ occurred uniformly along a laser Z axis of length L, then in the $L < f$ limit the medium would look like a negative cylindrical lens of focal length $f = y_0^2/(4 \times 10^{-4} L)$. If, for example, $y_0 = 1$ cm and $L = 25$ cm, $f = 100$ cm results. A distribution of many small-scale density variations due to nonuniform temperatures could be even more serious. During a short excitation pulse the gas will not move very far and the index changes will be smaller. Nonetheless, it is essential to deposit the pump energy uniformly and to maintain highly isothermal initial temperatures. The temperatures gradient to room temperature at the end of heat pipes make their value for lasers highly uncertain [5.62].

We conclude this section on metal vapor systems by noting that this class of systems offers a variety of choices of excimer species which radiate in the visible and ultraviolet region.

Acknowledgements. This work was supported in part by the Air Force Weapons Laboratory under grant no. AFWL 77010, and in part by the Advanced Research Projects Agency of the Department of Defense, monitored on ONR under Contract No. N00014-76-C-0123.

References

5.1   See, for example, I.Procaccia, R.D.Levine: J. Chem. Phys. **63**, 4261 (1975) and references therein
5.2   A.V.Phelps: "Tunable Gas Lasers Utilizing Ground State Dissociation", JILA Rpt. 110, University of Colorado, Boulder, Colo. (1972)

5.3   G. York, A. Gallagher: "High Power Gas Lasers on Alkali-Dimer A–X Band Radiation", JILA Rpt. 114, University of Colorado, Boulder, Colo. (1974)

5.4   R. J. Carbone, M. M. Litvak: J. Appl. Phys. **39**, 2413 (1968)

5.5   S. Y. Chen, M. Takeo: Rev. Mod. Phys. **29**, 20 (1957)
      A. Jablonski: Phys. Rev. **68**, 78 (1945)

5.6   R. E. M. Hedges, D. L. Drummond, A. Gallagher: Phys. Rev. A**6**, 1519 (1972)

5.7   G. York, R. Scheps, A. Gallagher: J. Chem. Phys. **63**, 1052 (1975); The conditions for Fig. 11 are $[Xe] = 2 \times 10^{20}\,cm^3$, $[Na] = 10^{17}\,cm^3$, $[Na^*]/[Na] = 0.05$, $T - 810\,K$
      R. Scheps, Ch. Ottinger, G. York, A. Gallagher: J. Chem. Phys. **63**, 2581 (1975)

5.8   D. L. Drummond, A. Gallagher: J. Chem. Phys. **60**, 3426 (1974)

5.9   R. B. Bernstein, J. T. Muckerman: *Advances in Chemical Physics*, Vol. 12, ed. by J. O. Hirschfelder (Wiley and Sons, New York 1967) p. 389

5.10 A. J. Palmer: J. Appl. Phys. **47**, 3088 (1976)
      A. J. Palmer, L. D. Hess, Semiannual Rpt. #1, 2 (1975) #3 (1976) Hughes Res. Lab., Malibu, Calif.
      A. J. Palmer, L. D. Hess, Semiannual Rpt. Hughes Res. Lab., Malibu, Calif. #1, 2 (1975) #3 (1976)

5.11 L. A. Schlie: J. Appl. Phys. **47**, 1397 (1976)

5.12 R. Shuker, L. Morgan, A. Gallagher, A. V. Phelps: To be published

5.13 A. Tam, G. Moe, W. Park, W. Happer: Phys. Rev. Lett. **35**, 85 (1975)
      A. Tam, G. Moe, B. Bulos, W. Happer: Opt. Commun. **16**, 376 (1976)
      G. Moe, A. Tam, W. Happer: Phys. Rev. A**14**, 349 (1976)
      W. Happer, G. Moe, A. C. Tam: Phys. Lett. **54**A, 405 (1975)

5.14 J. Pascale, J. Vandeplanque: J. Chem. Phys. **60**, 2278 (1974)

5.15 W. P. West, P. Shuker, A. Gallagher: To be published

5.16 R. Scheps, A. Gallagher: J. Chem. Phys. **65**, 859 (1976)

5.17 B. Sayer, M. Ferray, J. Lozingot, J. Berlande: J. Phys. B**9**, L293 (1976)

5.18 J. C. Gauthier, F. Devos, J.-F. Delpech: Phys. Rev. **14**, 2346 (1976)

5.19 J. C. Eden, B. E. Cherrington, J. T. Verdeyen: IEEE J. QE-**12**, 698 (1976)

5.20 W. E. Baylis: J. Chem. Phys. **51**, 2665 (1969)

5.21 R. A. Gutcheck, R. M. Hill, D. L. Huestis, D. C. Lorents, M. V. McCusker: Stanford Res. Inst. Rpt. MP75–43, Menlo Park, Calif. (1975)

5.22 J. R. Woodworth: J. Chem. Phys. **66**, 754 (1977)

5.23 O. Oldenberg: Z. Phys. **47**, 184 (1928); **51**, 605 (1928); **55**, 1 (1929)

5.24 C. G. Freeman, M. J. McEwan, R. Claridge, L. F. Phillips: Chem. Phys. Lett. **6**, 482 (1970)

5.25 E. V. Nikoforov, L. I. Plimck, Y. B. Predtechenskii, L. D. Shcherba: Opt. Spektrosk. **41**, 339 (1976), [English transl.: Opt. Spectrosc. **41**, 195 (1976)]

5.26 O. P. Strausz, J. M. Campbell, S. De Pauli, H. S. Sandhu, H. E. Gunning: J. Am. Chem. Soc. **95**, 732 (1973)

5.27 J. T. Kielkopf, R. A. Miller: J. Chem. Phys. **61**, 3304 (1974)

5.28 J. R. Powers, R. J. Cross, Jr.: J. Chem. Phys. **56**, 3181 (1972)

5.29 J. S. Deech, J. Pitre, L. Krause: Can. J. Phys. **49**, 1976 (1971)

5.30 A. Lurio: Phys. Rev. **140**A, 1505 (1965)

5.31 B. Cheron, R. Scheps, A. Gallagher: J. Chem. Phys. **65**, 326 (1976)

5.32 H. Rothwell, D. Leep, A. Gallagher: Bull. Am. Phys. Soc. **23**, 142 (1978)

5.33 A. Gallagher, A. Lurio: Phys. Rev. **136**A, 87 (1964)

5.34 D. S. Ginter, M. L. Ginter, K. K. Innes: J. Phys. Chem. **69**, 2480 (1965)

5.35 J. Drowart, R. E. Honig: J. Phys. Chem. **61**, 980 (1957)

5.36 For example, J. F. Waymouth: Proc. IEEE **59**, 629 (1971)
      C. S. Lin, R. J. Zollweg: J. Chem. Phys. **60**, 2384 (1974)
      J. J. DeGrout, A. G. Jack: J. Phys. D**6**, 1477 (1973)

5.37 D. Drummond, L. A. Schlie: Private communication

5.38 R. Duren: Chem. Phys. Lett. **39**, 481 (1976)

5.39 D. Drummond: Private communication

5.40 M.W.McGeoch, G.R.Fournier, P.Ewart: J. Phys. B9, L121 (1976)
      G.R.Fournier, W.M.McGeoch: "Kinetics of Proposed Cd–Hg Excimer Lasers", Int. Conf. on
      Quantum Electron., Amsterdam (1976)
      M.W.McGeoch, G.R.Fournier: J. Appl. Phys. (in press)
5.41 W.Balfour, A.E.Douglas: Can. J. Phys. 48, 901 (1970)
5.42 K.C.Li, W.C.Stwalley: J. Chem. Phys. 59, 4432 (1973)
5.43 C.R.Vidal, H.Scheingraber: J. Mol. Spectrosc. 65, 46 (1977)
5.44 W.J.Stevens, M.Krauss: J. Chem. Phys. 67, 1977 (1977)
5.45 W.L.Wiese, M.W.Smith, B.M.Miles: Atomic Transition Probabilities, Vol. II Na–Ca,
      NSRDS–NBS22 (U. S. GPO, Washington, D. C. 1969)
5.46 L.Pleasance: Private communication
5.47 W.Elenbaas: The High Pressure Mercury Vapor Discharge (North-Holland, Amsterdam 1951)
      W.Elenbaas: Light Sources (McMillan, London 1972)
      J.F.Waymouth: Electric Discharge Lamps (MIT, Cambridge, Mass. 1971)
5.48 R.M.Hill, D.J.Eckstrom, D.C.Lorents, H.H.Nakano: Appl. Phys. Lett. 23, 373 (1973)
      D.Eckstrom, R.M.Hill, D.C.Lorents, H.H.Nakano: Chem. Phys. Lett. 23, 112 (1973)
5.49 L.A.Schlie, B.D.Guenther, R.D.Rathge: Appl. Phys. Lett. 28, 393 (1976)
      L.A.Schlie, B.D.Guenther, D.L.Drummond: Chem. Phys. Lett. 34, 258 (1975)
5.50 R.E.Drullinger, M.M.Hessel, E.W.Smith: NBS Monograph 143, Natl. Bur. Stand. (U.S.)
      (1974)
      M.Stock, R.E.Drullinger, M.H.Hessel: Chem. Phys. Lett. 45, 592 (1977)
      E.W.Smith, R.E.Drullinger, M.M.Hessel, J.Cooper: J. Chem. Phys. 66, 5667 (1977) (Theory)
      R.E.Drullinger, M.M.Hessel, E.W.Smith: J. Chem. Phys. 66, 5656 (1977) (Experiment)
      M.Stock, E.W.Smith, R.E.Drullinger, M.M.Hessel: J. Chem. Phys. 67, 2463 (1977)
5.51 S.D.Rockwood: Phys. Rev. A8, 2348 (1973)
5.52 O.Judd: J. Appl. Phys. 47, 5297 (1976)
5.53 G.York, J.Carbone: "Electrical and Optical Properties of a High Pressure Transverse Hg
      Discharge", Abstract in Electronic Transition Lasers II, ed. by L.Wilson, S.Suchard,
      J.Steinfeld (MIT Cambridge, Mass. 1977)
5.54 W.Stevens: Private communication
5.55 K.Komine, R.L.Byer: J. Appl. Phys. 48, 2505 (1977)
5.56 D.Drummond, L.A.Schlie: J. Chem. Phys. 65, 3454 (1976). The intensity scale of the reported
      fluorescence has been revised slightly (private communication)
5.57 C.Santaran, V.K.Vaidyan, J.G.Winans: J. Phys. B4, 133 (1971) and references therein
5.58 R.S.Mulliken: J. Chem. Phys. 52, 5170 (1970)
5.59 W.R.Wadt: Appl. Phys. Lett. 34, 658 (1979)
5.60 T.H.Dunning, Jr., P.J.Hay: J. Chem. Phys. 69, 134 (1978)
5.61 P.J.Hay, T.H.Dunning, Jr.: J. Chem. Phys. 69, 2209 (1978)
5.62 E.W.McDaniel, M.R.Flannery, H.W.Ellis, F.L.Eisele, W.Pope, T.G.Roberts: Compilation of
      Data Relevant to Rare Gas – Rare Gas and Rare Gas – Monohalide Excimer Lasers, Vols. I and
      II (U.S. Army Missile Research and Development Command, Alabama 1977)
5.63 I.S.Lakoba, S.I.Yakovlenko: Sov. J. Quantum Electron. 10, 389 (1980)
5.64 J.H.Parks: Appl. Phys. Lett. 31, 192 (1977);
      J.G.Eden: Appl. Phys. Lett. 31, 448 (1977);
      J.H.Parks: Appl. Phys. Lett. 31, 297 (1977)
5.65 W.T.Whitney: Appl. Phys. Lett. 32, 239 (1978);
      K.Y.Tang, R.O.Hunter, Jr., J.Oldenettel, C.Howton, D.Huestis, D.Eckstrom, B.Perry,
      M.McCusker: Appl. Phys. Lett. 32, 226 (1978)
5.66 E.J.Schimitschek, J.E.Celto: Opt. Lett. 2, 64 (1978)
5.67 E.J.Schimitschek, J.E.Celto, J.A.Trias: Appl. Phys. Lett. 31, 608 (1977)
5.68 W.R.Wadt: J. Chem. Phys. 72, 2469 (1980)
5.69 D.A.Kleier, W.R.Wadt: J. Am. Chem. Soc. 102, 6909 (1980)
5.70 J.G.Eden: Appl. Phys. Lett. 33, 495 (1978)
5.71 R.W.Waynant, J.G.Eden: Appl. Phys. Lett. 33, 708 (1978)
5.72 J.H.Parks: Appl. Phys. Lett. 31, 297 (1977)

6. Triatomic Rare-Gas-Halide Excimers

D. L. Huestis, G. Marowsky, and F. K. Tittel

With 25 Figures

In this chapter the relevant spectroscopic and kinetic issues including the formation, radiation, and quenching mechanisms of triatomic excimer species will be described (Sects. 6.2 and 3). Pertinent reaction rate coefficients obtained from temporal fluorescence measurements are listed. Important laser considerations, such as gain of bound-free transitions and transient absorption effects will be discussed in Sect. 6.4. This section also describes appropriate excitation techniques which include electron beam, fast discharge and optical pumping. Section 6.5 describes the laser characteristics of two trimers: Xe_2Cl and Kr_2F. Finally, the present status of triatomic rare-gas-halide excimers is summarized in Sect. 6.6.

6.1 Background

Numerous rare-gas and rare-gas-halide excimers have been reported since 1972 that are capable of generating high power laser radiation in the visible and ultraviolet spectral range (Chap. 4 and [6.1–6]). The relatively simple techniques required to pump such lasers, as well as their demonstrated high efficiency, have made them useful for many interesting applications. In recent years there has been considerable interest in exploring the broad bandwidth emission, which is observed from several diatomic and triatomic rare-gas-halide excimers, for the development of tunable laser sources [6.7], for the generation of ultrashort light pulses and for display applications [6.8]. These excimers exhibit radiative transitions in the wavelength region from 230 nm (Ar_2Cl) to 670 nm (Xe_2F).

Figure 6.1 depicts the various narrowband rare-gas excimers such as Ar_2, Kr_2, and Xe_2, the conceptual precursors of all excimer lasers, with transitions in the vuv region, and the diatomic rare-gas-halides with $B \rightarrow X$ transitions ranging from 193 nm for ArF to 351 nm for XeF*. In addition, the broadband spectral emission features from diatomic and triatomic excimers are indicated that occur on the long-wavelength side of most $B \rightarrow X$ transitions in high-pressure rare-gas-halide mixtures. For the triatomic excimers Ar_2F, Xe_2Br, and Xe_2F (shown in brackets), the wavelengths listed are at the center of the fluorescence, whereas the data given for Kr_2F and Xe_2Cl correspond to laser emission. The broadband continuum emissions are due to electronic transitions from ionically bound excited states to strongly repulsive covalent lower states, which rapidly dissociate into ground-state atoms. Hence such excimers are

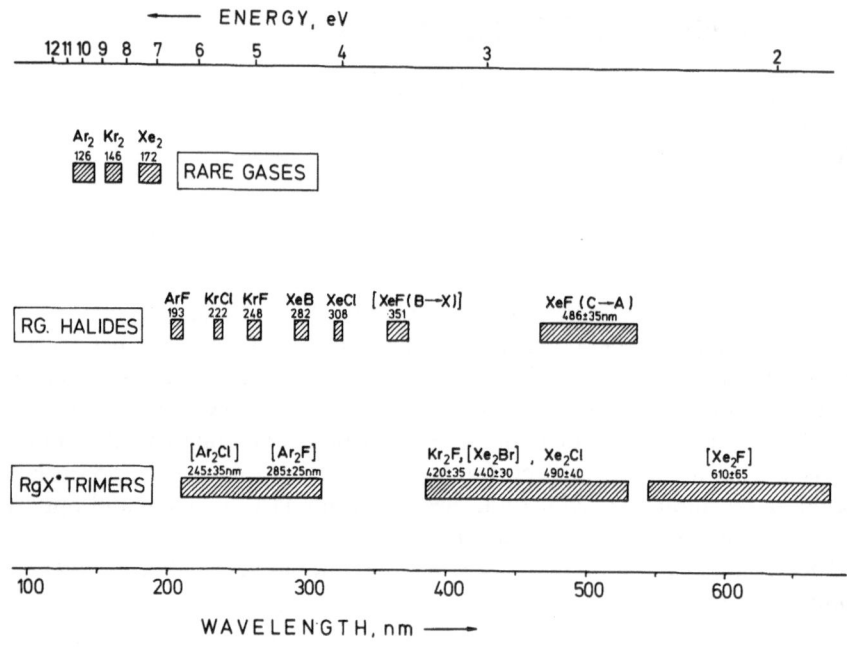

Fig. 6.1. Wavelengths and tuning ranges of the various rare-gas (Rg$_2^*$), rare-gas-halide (RgX*), broadband diatomic (RgX*) and triatomic (Rg$_2$X*) rare-gas-halide lasers

potentially capable of wide wavelength tunability (50–120 nm) as compared to the narrow spectral tuning range (~ 5 nm) of rare-gas and $B \rightarrow X$ transition-type rare-gas-halide excimers. The relatively long radiative lifetimes (~ 200 ns) of triatomic excimers imply low stimulated emission cross sections. Hence intense pumping as provided by a high-energy electron beam source is required to achieve the large excited-state population densities required for the onset of stimulated emission.

Various reaction pathways for formation of trimers have been discussed in (Chap. 4 and [6.1, 7]). The principal reaction is

$$RgX^* + 2Rg \rightarrow Rg_2X^* + Rg \tag{6.1}$$

while other pathways such as

$$Rg_2^* + X_2 \rightarrow Rg_2X^* + X \tag{6.2}$$

are possible but are apparently less important.

Electron-beam excited laser mixtures are typically composed of several atmospheres of a rare-gas buffer, several hundred Torr of another rare-gas, and a few Torr of a halogen donor. For these conditions the primary channel for the formation of Rg$_2$X* occurs via reaction (6.1). Subsequently trimer emission proceeds dissociatively:

$$Rg_2X^* \rightarrow 2\,Rg + X + h\nu. \tag{6.3}$$

This reaction was initially viewed as a loss mechanism of upper-state inversion for the diatomic rare-gas-halide lasers [6.9–12], in particular, in scaling considerations of high-power excimer lasers. However, the trimers so formed represent interesting new metastable molecules from the standpoint of their spectroscopic and kinetic properties. In 1979, it was proposed that these triatomic molecules could be employed as broadband wavelength tunable laser media [6.13]. Subsequently, electron-beam excited Xe_2Cl was shown to exhibit transient optical gain within its homogeneously broadened excimer emission band in the afterglow regime following an initial absorption [6.14]. Shortly thereafter, two triatomic lasers, Xe_2Cl^* centered at 520 nm [6.15] and Kr_2F^* centered at 435 nm [6.16, 17] were demonstrated.

6.2 Spectroscopy

The triatomic rare-gas-halides, with the general formula Rg_2X^*, were discovered in 1976 [6.9, 18–20] as broadband companion emissions from RgX^* laser media. Their identification was made through the observation that these emissions become dominant at high rare-gas densities, under which conditions clustering reactions might be expected to be important. In this section the energetics and spectroscopy of trimers will be presented.

Typical high-pressure emission spectra are shown in Fig. 6.2. These fluorescence spectra were obtained from electron-beam excited rare-gas-halide mixtures composed of several atmospheres of a buffer gas, usually argon, a few hundred Torr of the rare-gas, and a few Torr of the halogen donor. The uv RgX^* lasers are based on the intense and narrow $RgX(B1/2 \rightarrow X1/2)$ transition as shown, for example, at 308 nm for the case of XeCl (Fig. 6.2d) and at 352 nm for XeF (Fig. 6.2c). In each case there is a weaker and broader diatomic transition, $RgX(C3/2 \rightarrow A3/2)$, just to the red of the main peak (345 nm for XeCl). This weak transition is in principle a potential laser transition but has been successfully operated thus far only in XeF, near 480 nm [6.17, 21–25]. Not discernible in Fig. 6.2 are the broad $B1/2 \rightarrow A1/2$ transitions which overlap the $C \rightarrow A$ transitions for the rare-gas fluorides and chlorides but become distinct to the red side in the bromides and iodides. All of these transitions are listed in Table 6.1, and the energetics of the two most important $(B \rightarrow X)$ and $(C \rightarrow A)$ transitions for XeCl are illustrated in Fig. 6.3. Under the high-density conditions indicated in Fig. 6.2c, by far the majority of the emission is in the very broad band between 450 and 550 nm, which is assigned to Xe_2Cl. These broad trimer emissions are observed in all the rare-gas-halides at high pressure and are listed in Table 6.1. By careful choice of pressure conditions, *Brashears* et

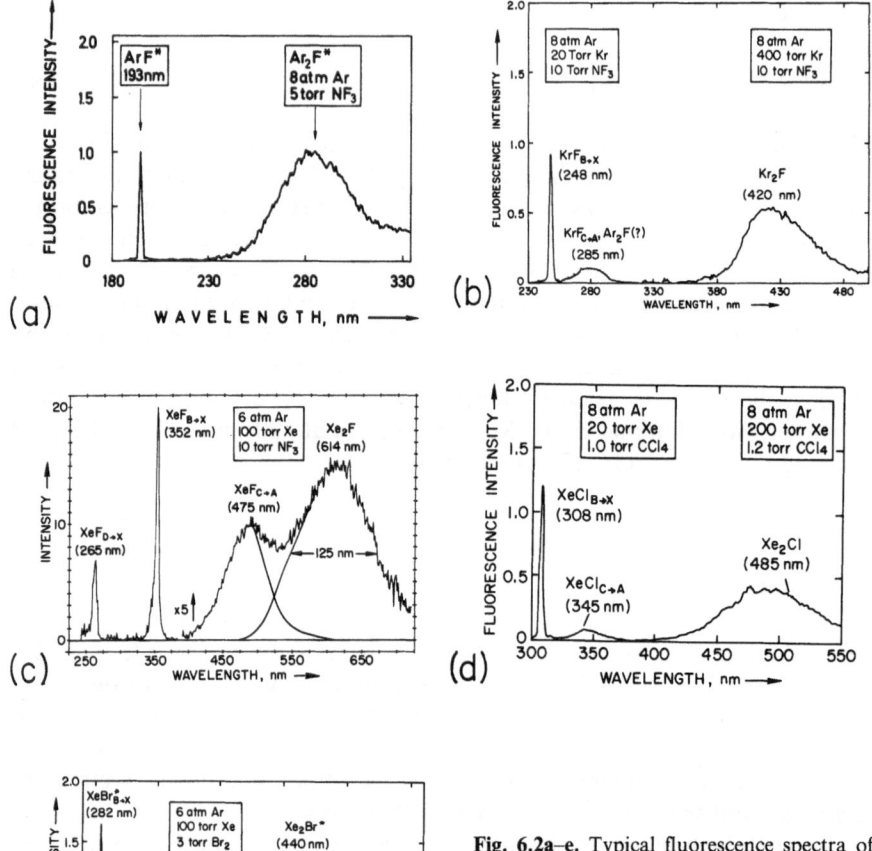

Fig. 6.2a–e. Typical fluorescence spectra of RgX* and Rg$_2$X* from e-beam-excited high-pressure mixtures of (**a**) Ar–NF$_3$, (**b**) Ar–Kr–NF$_3$, (**c**) Ar–Xe–NF$_3$, (**d**) Ar–Xe–CCl$_4$, and (**e**) Ar–Xe–Br$_2$. The appropriate rare-gas-halide-mixtures conditions are shown. For (**b**) and (**d**) low krypton and xenon pressures were used for the production of RgX*

al. [6.26] were able to identify emissions of several mixed or unsymmetrical triatomic rare-gas-halides, RgRg′X. These are listed in Table 6.2.

To explain the stability of the triatomic rare-gas-halides, *Lorents* et al. [6.19, 21] suggested an ion pair complex represented by $(Rg_2^+)(X^-)$. Such a complex is in close analogy with the ionic model for the diatomic rare-gas-halides, Rg^+X^- [6.27]. Stable molecular rare-gas ions, Rg_2^+, are known to exist with bond strengths on the order of 1 eV. Thus, Rg_2X^* should be more stable than $Rg + RgX^*$ by some fraction of the Rg_2^+ bond strength.

The observed Rg_2X emissions lie 1.5–2.2 eV to the red of the $RgX(B\rightarrow X)$ transition. The magnitude of this shift and the broadband nature of the

Table 6.1. Emission features of the rare-gas-halides

Rg	X	RgX $D1/2 \rightarrow X1/2$ [nm]	RgX $B1/2 \rightarrow X1/2$ [nm]	RgX $C3/2 \rightarrow A3/2$ [nm]	RgX $B1/2 \rightarrow A1/2$ [nm]	Rg$_2$X [nm]
Ne	F	(106)[a]	108	(110)	(111)	($\sim$145)
Ar	F	(185)	193	(203)	(204)	285 $\pm$ 25
Ar	Cl		175			245 $\pm$ 15
Kr	F	220	248	275	(272)	420 $\pm$ 35
Kr	Cl	200	222	240		325 $\pm$ 15
Kr	Br		207	222	228	$\sim$318
Xe	F	264	351	460	(410)	610 $\pm$ 65
Xe	Cl	236	308	345	(340)	490 $\pm$ 40
Xe	Br	221	282	300	320	440 $\pm$ 30
Xe	I	203	253	265	320	$\sim$375

[a] Theoretical values in parenthesis.

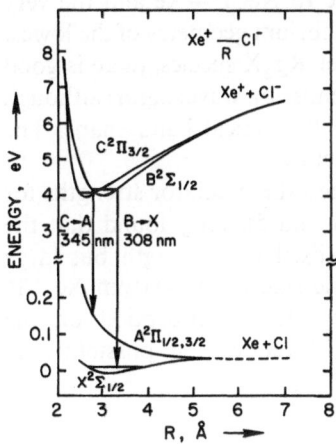

Fig. 6.3. Potential energy diagram of the XeCl* excimer. The XeCl(C) state lies below the B state 230 ± 40 cm^{-1} [6.60]

Table 6.2. Emissions of the asymmetrical triatomic rare-gas-halides [6.26]

RgRg'X	λ [nm]
ArKrF	305 $\pm$ 33
ArKrCl	270 $\pm$ 23
KrXeCl	370 $\pm$ 40
KrXeBr	330 $\pm$ 23
KrXeI	260 $\pm$ 42[a]

[a] Peak at 290 nm

Table 6.3. Calculated geometries and binding energies of the triatomic rare-gas-halides

	Ne$_2$F	Ar$_2$F	Kr$_2$F	Kr$_2$Cl	Xe$_2$F	Xe$_2$Cl
Triatomic Properties						
Rg$_2$ Distance [Å]	1.73 $\pm$ 0.02	2.47	2.78 $\pm$ 0.01	278	3.29	3.28
RgX Distance [Å]	2.18	2.54	2.67	3.14	2.78	3.32
Binding Energy [eV] relative to RgX*	0.68 $\pm$ 0.08	0.65 $\pm$ 0.08	0.58 $\pm$ 0.08	0.88	0.63	0.66
Diatomic Properties						
Rg$_2^+$ Distance [Å]	1.75	2.48	2.79	2.79	3.27	3.27
RgX(B) Distance [Å]	2.01	2.39	2.51	3.05	2.68	3.22
Rg$_2^+$ Binding Energy [eV]	1.2	1.2	1.1	1.1	1.0	1.0

emissions can be explained by the substantial repulsion expected between the ground-state atoms $(Rg + Rg + X)$ at the rather short internuclear distances in the excited state (Rg_2X^*).

Because of the unstructed continuum nature of the emissions, it is difficult to derive, from experimental data alone, more detailed information about the structure of the triatomic rare-gas-halides. Fortunately, a number of theoretical calculations have been performed [6.13, 28–31]. All of these support the simple ionic model and predict an isosceles triangular geometry with Rg – Rg distances nearly identical to those in the isolated Rg_2^+ diatomic fragments (as summarized in Table 6.3) and RgX distances 0.1–0.2 Å longer than in the corresponding RgX*. The binding energy of Rg_2X^* relative to the dissociation RgX* + Rg is calculated to be only about 65 % of the bond energy of Rg_2^+. This reduced value is due to $Rg \cdot X^-$ repulsion but is still more than an order of magnitude greater than the Rg_2 van der Waals binding.

The calculated potential energy surfaces for Xe_2Cl are shown in Figs. 6.4 and 5 [6.13]. Figure 6.4 illustrates the ionic nature of the Xe_2Cl excited states. Figure 6.5 shows the binding of Xe_2Cl^* relative to XeCl* + Xe and the very repulsive nature of the covalent states at the equilibrium geometry of the lowest ionic state $(R \sim 5.5a_0)$. As is the case for the other Rg_2X species, there is good agreement between the calculated and observed emission wavelengths although there has not yet been a detailed study of the theoretical band shape. The calculated radiative lifetimes will be discussed below.

Stevens and *Krauss* [6.31] have recently calculated the oscillator strengths for absorption by the $4^2\Gamma$ state of Xe_2Cl (Figs. 6.4 and 5). They found that the $8^2\Gamma \leftarrow 4^2\Gamma$ transition near 438 nm has a small oscillator strength but large enough to interfere with laser action on the blue side of the 490 nm Xe_2Cl^* excimer emission band. They also found very strong absorption on the $9^2\Gamma \leftarrow 4^2\Gamma$ transition near 339 nm, which could reduce the possibility of

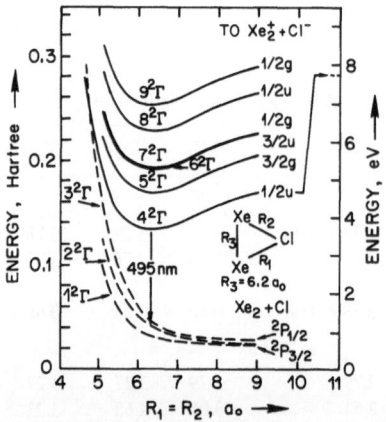
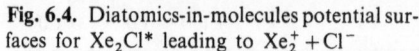

Fig. 6.4. Diatomics-in-molecules potential surfaces for Xe_2Cl^* leading to $Xe_2^+ + Cl^-$

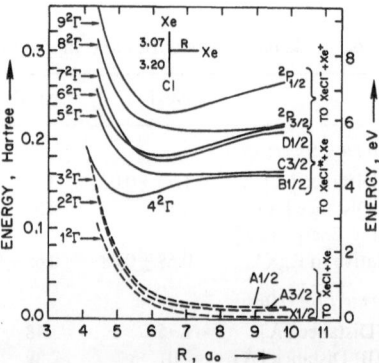

Fig. 6.5. Diatomics-in-molecules potential surfaces for Xe_2Cl^* leading to XeCl* + Xe

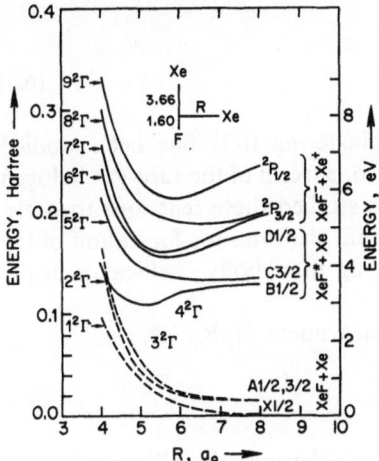

Fig. 6.6. Diatomics-in-molecules potential surfaces for Xe_2F^*

demonstrating a $XeCl(C \rightarrow A)$ laser and may interfere with the $XeCl(B \rightarrow X)$ laser at 308 nm. The corresponding absorption in Kr_2F^* has been used to explain some of the absorption observed in KrF laser media [6.32].

The calculated potential energy surfaces for Xe_2F are depicted in Fig. 6.6. The observation of fluorescence from Xe_2F^* occurred only recently [6.33]. As discussed in Sect. 6.3.3d, the quenching kinetics of XeF^* are significantly different from those of other rare-gas-halides. A major reason for these differences is shown in Fig. 6.6. The potential surface for the bound $4^2\Gamma$ state of Xe_2F is predicted to be crossed by the steeply repulsive $2^2\Gamma$ and $3^2\Gamma$ potential surfaces at an energy near the $XeF^* + Xe$ dissociation limit. This crossing explains the rapid quenching of XeF^* by Xe. Two-body rare-gas quenching of the other rare-gas-halides is quite slow.

6.3 Essential Kinetics

6.3.1 Formation Mechanism

Since their discovery, the uv diatomic rare-gas-halide excimer lasers have initiated considerable research activity devoted to characterization of the kinetic processes that control their laser performance. This literature is reviewed elsewhere [at least through 1979 in Chap. 4, and is revised for the present volume].

As mentioned previously, the Rg_2X^* emissions were identified in high-pressure RgX^* laser media and were felt to result from clustering reactions involving RgX^* or its kinetic precursors [6.9, 18–20]. The dominant reactions leading to RgX^* fall in two general classes: an ionic channel

$$Rg^+ + X^- + M \rightarrow RgX^* + M \tag{6.4}$$

and a covalent reaction channel:

$$Rg^* + X_2 \rightarrow RgX^* + X. \tag{6.5}$$

The formation of Rg_2X^* by three-body clustering (6.1) has been studied experimentally by a number of investigators for several of the rare-gas halogen combinations. *Shui* and coworkers have also studied these reactions theoretically [6.34, 35]. There is general agreement that the rates for formation of the symmetrical trimers (i.e. Rg_2X) are fast, having three-body coefficients in the range of 2×10^{-31} to $10^{-30}\,cm^6\,s^{-1}$.

Studies of formation of the unsymmetrical trimers, $RgRg'X^*$, e.g.

$$KrF^* + 2Ar \rightarrow ArKrF^* + Ar \tag{6.6}$$

have been substantially less definitive in large part because their lifetime is short as a consequence of displacement reactions of the type,

$$ArKrF^* + Kr \rightarrow Kr_2F^* + Ar. \tag{6.7}$$

Thus the formation of the mixed trimers must be inferred from deconvolution of the pressure dependence of the RgX^* decay and formation of the final Rg_2X^* product. In general, it appears that the rate coefficients for formation of the unsymmetrical triatomics in comparison with the symmetrical ones are significantly less than those of their symmetrical counterparts.

A number of studies, e.g., [6.18, 20, 36–38] have considered alternative pathways to (6.1) for Rg_2X^* formation, such as

$$Rg^+ + Rg + M \rightarrow Rg_2^+ + M, \tag{6.8}$$

$$Rg_2^+ + X^- + M \rightarrow Rg_2X^* + M, \tag{6.9}$$

$$Rg^* + Rg + M \rightarrow Rg_2^* + M, \tag{6.10}$$

and (6.2)

$$Rg_2^* + X_2 \rightarrow Rg_2X^* + X. $$

To date, the possible contributions of these reactions have not been firmly established. In some special cases, consideration of these more direct reactions is helpful in explaining the observed yields of Rg_2X^* fluorescence, but at least under the conditions typical of Rg_2X lasers, the diatomic RgX^* appears to be the major precursor.

The primary steps in a basic kinetic model for trimer formation in argon-buffered rare-gas-halide mixtures under intense e-beam excitation is shown in Fig. 6.7. Deposition of energy occurs primarily in the dense argon buffer and to

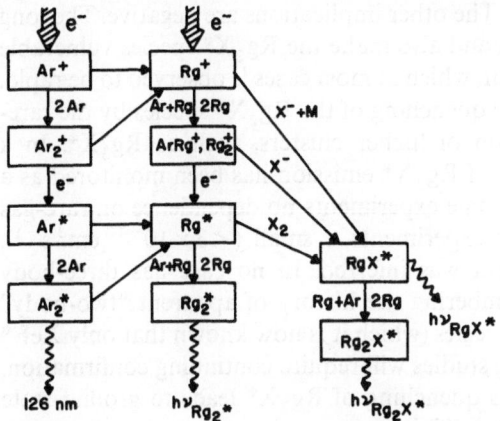

Fig. 6.7. Principal reaction pathways in high argon pressure buffered rare-gas-halide mixtures

a minor extent in the rare-gas itself. This energy is channeled into rare-gas ions and metastables which then cascade down a complex chain to eventually produce Rg_2X^*. Decomposition of the halogen donor and electron attachment leads to the required formation of halogen ions, necessary for build-up of the RgX^*-species via ionic channels, cf. (6.5). The subsequent formation of the trimer Rg_2X^* via three-body clustering, cf. (6.1), may thus be considered as an alternative decay process of the RgX^* species.

6.3.2 Radiation and Quenching

The broadband nature of the Rg_2X^* emissions is accompanied by a rather long radiative lifetime in each case that has been investigated. Lifetime values are listed in Tables 6.4–8 and range from 130 to 330 ns. This is somewhat surprising, given the short (4–16 ns) lifetimes of the $RgX^*(B \rightarrow X)$ transitions. As first pointed out by *Wadt* and *Hay* [6.30], this long lifetime can be rationalized by recognizing that the lowest excited Rg_2X^* state need not necessarily have the electronic character of the $RgX^*(B)$ state but is, in fact, much closer to the $RgX^*(C)$ state. In the triatomic complex, there are several forces at play that determine the favored electronic structure. We already know that the RgX^* B- and C-states lie close together in energy. In the other major diatomic fragment, Rg_2^+, the splittings between the electronic states are much larger. The $^2\Sigma_u^+$ state lies about 1 eV below the next state, the $^2\Pi_g$. Thus we expect the Rg_2^+ segment of Rg_2X^* will have $^2\sigma_u^+$ character, driving the RgX fragment into predominantly the $C3/2$ ($^2\Pi_{3/2}$) state because the $Rg \rightarrow X$ bonds are $\sim 60°$ relative to the $Rg–Rg$ bond. Additional factors that lengthen the radiative lifetime are the longer wavelengths (as compared to the RgX transition) and the smaller values of the transition moments that result from stretching the RgX bond.

The long radiative lifetimes of trimers have important implications for laser development. The first of these is an advantage in that a long radiative lifetime permits operation of a laser after the decay of absorption that may be dominant during the short excitation pulse. The other implications are negative. The long radiative lifetimes imply low gain and also make the Rg_2X^* species vulnerable to quenching by the halogen donor, which in most cases is observed to be rapid.

Finally, one must consider the quenching of the Rg_2X^* species by the rare-gas buffer and possible formation of higher clusters, such as Rg_3X^*. In a number of experiments the decay of Rg_2X^* emission has been monitored as a function of rare-gas pressure. In some experiments no dependence on rare-gas pressure was observed. In other experiments a small ($< 3 \times 10^{-13}\,cm^3\,s^{-1}$), two-body quenching rate constant was inferred. In no case has three-body quenching been observed. Remembering the history of apparent "two-body" rare-gas quenching of the RgX^* species (which it is now known that only XeF^* undergoes), the Rg_2X^* quenching studies will require continuing confirmation. However, it is likely that rare-gas quenching of Rg_2X^* leads to ground-state products, rather than higher clusters. This argument is supported by the much reduced binding in the Rg_3^+ species compared to Rg_2^+.

An important recent observation is that Rg_2X^* fluorescence is observed in e-beam excited liquid rare gases with small densities of fluorine donors [6.39]. This suggests that rare-gas quenching of the trimers can be very slow.

6.3.3 Specific Examples

In this subsection, the kinetic issues of the triatomic excimers Ar_2F^*, Kr_2F^*, Xe_2F^*, Xe_2Cl^*, and Xe_2Br^* are discussed as examples of well-studied trimers. Experimental and theoretical values reported in the literature for three-body reaction coefficients, spontaneous lifetime and major quenching rates are listed in Tables 6.4–9. Recently, lifetime and quenching rates of Ar_2Cl^* which radiates around 245 nm have been reported [6.118].

a) Ar_2F^*

There has been considerable interest in the triatomic exciplex, Ar_2F^*, as a potential candidate for a widely tunable uv laser [6.36–38, 40–46] due to its broad fluorescence band centered around 285 nm (Fig. 6.2a). The relevant energy levels of the $ArF–Ar_2F$ system are depicted in Fig. 6.8 showing both the bound states of ArF and Ar_2F, and the repulsive upper and lower neutral and ionic states. The addition of another argon atom into the ArF complex lowers the excitation energy from 6.5 eV (~ 193 nm) for the two-atom excimer to 4.3 eV (~ 290 nm) for the three-atom excimer.

Formation and decay processes of Ar_2F^* in excited $Ar–F_2$ and $Ar–NF_3$ mixtures were studied using short and long duration electron-beam pulses [6.42–45], short proton pulses [6.36] and fast discharge pumping [6.46]. The relevant Ar_2F rate constants reported in the literature are summarized in Table 6.4. The $ArF–Ar_2F$ kinetics essentially follow the general formation scheme shown in Fig. 6.9. However, this system is somewhat simplified as compared to most trimers due to the fact that argon acts both as the buffer gas for electron

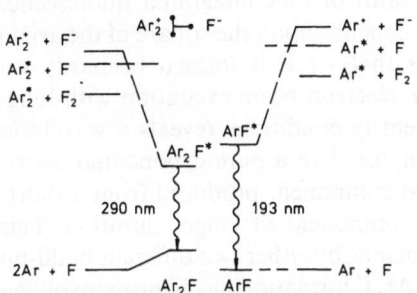

Fig. 6.8. Relevant ArF* and Ar$_2$F* energy levels showing schematically the origins of the shift of the Rg$_2$X* emission to longer wavelengths

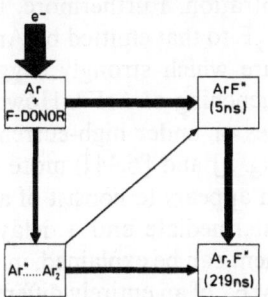

Ar$_2$F* – KINETICS

Fig. 6.9. Schematic energy-flow pathways for Ar$_2$F* formation depicting alternative formation channels

Table 6.4. Summary of Ar$_2$F* rate constants

Reaction	Rate coefficient $[s^{-1}]$	References
ArF* + Ar + Ar → Ar$_2$F* + Ar	$(1.2 \pm 0.2) \times 10^{-30}$ cm^6	[6.43]
	4×10^{-31} cm^6	[6.47]
	$(5 \pm 1) \times 10^{-31}$ cm^6	[6.36, 37, 43]
	6×10^{-31} cm^6	[6.34, 35]
Ar$_2$F* → 2Ar + F + $h\nu$(285 nm)	$(219 \pm 15)\ 10^{-9}$	[6.43]
	$(185 \pm 45)\ 10^{-9}$	[6.36, 37]
	$(230 \pm 25)\ 10^{-9}$	[6.44]
	132 (calc) 10^{-9}	[6.30]
	236×10^{-9}	[6.46]
Ar$_2$F* + F$_2$ → quenching	$(2.05 \pm 0.06) \times 10^{-10}$ cm^3	[6.43]
	2.1×10^{-10} cm^3	[6.36, 37]
	1.83×10^{-10} cm^3	[6.46]
Ar$_2$F* + NF$_3$ → quenching	$(1.23 \pm 0.05) \times 10^{-10}$ cm^3	[6.43]
	$(5.6 \pm 1.0) \times 10^{-10}$ cm^3	[6.44]
Ar$_2$F* + Ar → quenching	$< 10^{-14}$ cm^3	[6.43]
	$(2.2 \pm 0.8) \times 10^{-14}$ cm^3	[6.44]

beam energy deposition and as rare-gas for build-up of RgX* and Rg$_2$X*. The basic kinetic model for Ar$_2$F* is shown schematically in Fig. 6.9. Both experimental evidence and calculations indicate the third-order quenching reaction

$$ArF^* + 2\,Ar \rightarrow Ar_2F^* + Ar \qquad (6.11)$$

as the dominant formation process of Ar$_2$F*. The principal argument for this mechanism is based upon the observation that both the quenching behavior and fluorescence intensity of ArF* and Ar_2F show the same dependence on donor concentration. Furthermore, the ratio of time-integrated fluorescence emitted by Ar_2F to that emitted by ArF* depends upon the square of the argon partial pressure which strongly suggests that Ar_2F is formed primarily via three-body quenching of ArF*. However, electron-beam excitation with long-duration pulses or under high-current-density conditions reveals a waveform ([Ref. 6.43, Fig. 5] and [6.44]) more complex than a pure exponential decay. The waveform appears to consist of a fast component produced from a short-lived ArF* intermediate and a delayed component of longer duration. This latter component can be explained, in principle, by either two different build-up channels for ArF* of an entirely different Ar_2F formation mechanism involving long-lived excited argon neutrals such as Ar* and Ar_2^* via a reaction such as

$$Ar_2^* + F_2 \rightarrow Ar_2F^* + F. \tag{6.12}$$

Further experimental evidence of a second formation mechanism is seen in Fig. 6.10 which depicts the argon pressure dependence of both the Ar_2F^* fluorescence emitted subsequent to transverse pumping with moderate electron-beam intensities (Fig. 6.10a) and that emitted following high-current and short (< 15 ns) coaxial excitation (Fig. 6.10b). When using intense excitation densities, many secondary electrons remain unattached as the halogen donor concentration cannot be increased because of severe donor quenching. It is thought that these electrons lead to the formation of long-lived species such as Ar_2^*, which may be the precursors of the delayed Ar_2F fluorescence and the anomalous behavior of the argon buffer gas pressure dependence in Fig. 6.10b. An additional influence of the high electron density may be electron quenching of Ar_2F, which has been suggested in [6.41], Ar_2F^* experiences both severe absorption by Ar_2^+ and Ar_2^* [6.2], which spectrally overlaps the trimer fluorescence and strong self-absorption. This latter absorption can lead to

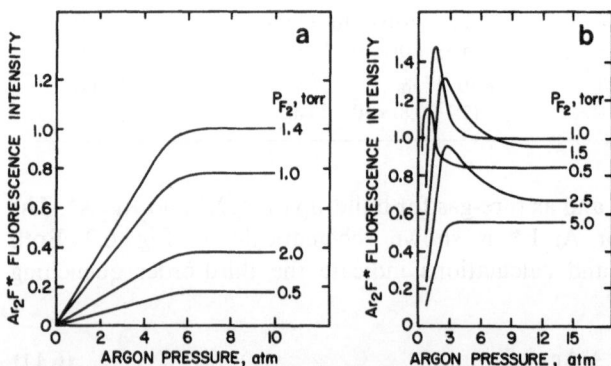

Fig. 6.10a, b. Dependence of Ar_2F^* fluorescence upon argon pressure of Ar–F_2 mixtures at various F_2 pressures for (a) transverse e-beam excitation and (b) coaxial e-beam excitations (for details see [6.43])

Ar$_2$F dissociation into high lying states of argon and fluorine, as shown in Fig. 6.8.

b) Kr$_2$F

Broad continuum emission of Kr$_2$F centered at 420 nm (Fig. 6.2b) is observed in electron-beam-pumped high-pressure mixtures containing Kr and a fluorine donor such as F$_2$ or NF$_3$ (6.9, 38, 48, 49]. The dominant formation kinetics of the triatomic excimer Kr$_2$F were discussed in [6.11]. Three-body quenching of KrF(B) by Kr leads to trimer formation

$$KrF^* + Kr + M \rightarrow Kr_2F^* + M; \quad M = Ne, \text{ Ar or Kr} \tag{6.13}$$

with subsequent blue Kr$_2$F* emission. Good agreement between theory [6.34, 35] and experiment [6.47, 48] seems to support this reaction. The pertinent reactions and rate constants for Kr$_2$F are summarized in Table 6.5. The possibility of direct production of Kr$_2$F* from Kr$_2^*$ instead of KrF* has also been reported [6.50]. It has been experimentally observed that the halogen donor NF$_3$ is a less severe quencher of the trimer fluorescence than F$_2$ by nearly one order of magnitude [6.7]. Due to this low quenching rate, NF$_3$ donor pressures of up to 10 Torr may be used effectively in the formation of Kr$_2$F. Furthermore, the transfer of excitation from KrF* to Kr$_2$F* via an

Table 6.5. Formation and decay processes of Kr$_2$F*

Reaction	Rate constant [s^{-1}]	References
Kr* + F$_2$ → Kr$_2$F* + F	7.2×10^{-10} cm^3	[6.65]
KrF* + 2Kr → Kr$_2$F* + Kr	2.9×10^{-31} cm^3	[6.38]
	9.7×10^{-31} cm^6	[6.48]
	5×10^{-31} cm^6	[6.34]
	6.4×10^{-31} cm^6	[6.47]
Kr$_2$F* + F$_2$ → products	$4.3 \pm 0.4 \times 10^{-10}$ cm^3	[6.38]
	$4.1 \pm 0.5 \times 10^{-10}$ cm^3	[6.48]
Kr$_2$F* + NF$_3$ → products	4×10^{-11} cm^3	[6.7]
	6×10^{-10} cm^3	[6.50]
Kr$_2$F* + Kr → products	$< 2.0 \times 10^{-14}$ cm^3	[6.38]
Kr$_2$F* + Ar → products	4×10^{-14} cm^3	[6.66]
	2×10^{-14} cm^3	[6.48]
	7×10^{-14} cm^3	[6.50]
Kr$_2$F* → 2Kr + F + $h\nu$(420 mm)	$(181 \pm 12) 10^{-9}$	[6.38]
	$(145 \pm 25) 10^{-9}$	[6.66]
	150×10^{-9}	[6.48]
	132 (calc) 10^{-9}	[6.30]
	$(185 \pm 20) 10^{-9}$	[6.50]

XENON   HALOGEN   EXCIPLEXES

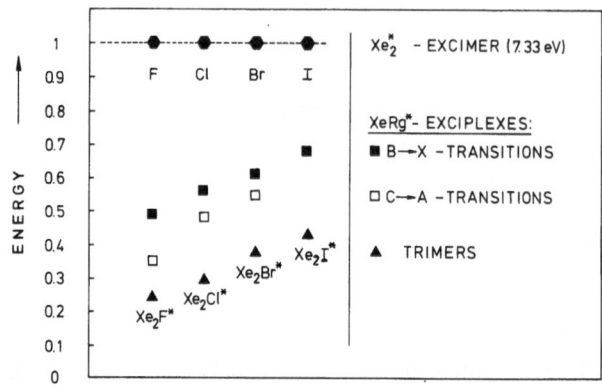

Fig. 6.11. Compilation of the transition energies of the xenon halogen exciplexes normalized to dissociation energy of Xe_2^*

intermediate excimer KrN_2F^* has been established in [6. 51] as a result of photolysis of KrF_2 in mixtures of He, CF_4, and N_2. The fluorescent intensity was approximately 50 times higher than that for a $KrF_2/CF_4/Ar$ mixture at the same pressure (2 atm) in which case Kr_2F^* is formed by reaction (6.13).

c) Xenon Halogen Trimers

The formation and quenching kinetics of the xenon-based triatomic excimers, in particular Xe_2F, Xe_2Cl, and Xe_2Br has been studied extensively under electron-beam excitation. The trimer Xe_2I has not been definitely identified, although it may have appeared near 375 nm in an early spectrum by *Brau* and *Ewing* [6.52]. Figure 6.11 depicts a compilation of the transition energies of the xenon halogen complexes normalized against the energy of the Xe_2^* excimer (7.33 eV). From this figure it is apparent that the transition energy increases monotonically with F, Cl, Br, and I. A similar relation exists for the other rare-gas halogen complexes of Ne, Kr, and Ar [6.7], and results from the increasing size of the halide ion and larger RgX internuclear separation.

d) Xe_2F

The XeF^*–Xe_2F^* excimer system was found to possess unique spectroscopic features, as compared to other rare-gas-halide molecules [6.53–61]. Besides the high efficiency XeF laser operating on the $B \to X$ transition at 351 and 353 nm, laser action is also achieved on the $C \to A$ transition of XeF around 480 nm under electron beam, photolytic, and discharge excitation [6.21–24]. Broadband emission from Xe_2F around 614 nm, the product of a three-body quenching process of $XeF(B, C)$ by xenon as well as xenon and argon, has been reported recently [6.33, 62]. The observation of Xe_2F^* emission raises questions about the previously assumed total predissociation of Xe_2F^* by a curve crossing between the lowest ionic state and one of the repulsive covalent states [6.27]. Formation of Xe_2F^* requires conditions which minimize the two-body quenching of XeF by Xe, give Boltzmann vibrational distribution of XeF and minimize quenching of Xe_2F^* [6.63]. This suggests high pressure of Ar to enhance Xe_2F production. Hence the absence of Xe_2F emission to date in experiments utilizing photolytic excitation [6.59, 60] can be explained by low argon-buffer and xenon-pressure conditions. The reaction kinetics of XeF^*–Xe_2F systems are complicated by the fact that the excited XeF(C) state lies below the XeF(B) state by $775\ cm^{-1}$ [6.62–64]. Such a large energy gap between these two states means that they have separately observable kinetic properties and are subject to collisional coupling effects that may depend on such parameters as the buffer gas, buffer gas pressure, electron density or the amount of vibrational relaxation in each electronic state [6.59]. Figure 6.2c shows the fluorescence spectra of an electron-beam excited Ar–Xe–NF_3 mixture. The relative intensities have been corrected for the OMA 1 spectral sensitivity. The XeF $(D-X)$ and $B-X$ transitions occur at 265 and 352 nm, respectively. A broad continuum is observed in the spectral region from 400 to beyond 720 nm which consists of two components. The short-wavelength peak can be assigned to the $XeF(C \rightarrow A)$ emission whereas the long-wavelength peak is due to Xe_2F. The maximum of the Xe_2F emission was determined to lie at (614 ± 5) nm with a spectral bandwidth (FWHM) of (125 ± 5) nm [6.62]. This is in good agreement with values determined from Xe–NF_3 mixtures at high xenon pressures and the potential surface calculations for Xe_2F [6.27]. Spectra similar to those obtained using Ar–Xe–NF_3 mixtures were also observed when using Ne–Xe–NF_3 mixtures. Mixed tri-atomic excimers such as ArXeF* can be excluded because the red continuum also appeared in a Xe–NF_3 mixture.

Observations of the temporal behavior of the XeF(B, C) and Xe_2F intensities as well as measurements of the fluorescence yield ratios for the different species allow the evaluation of the various rate constants. The temporal fluorescence behavior of $XeF(B \rightarrow X)$, $XeF(C \rightarrow A)$ and the Xe_2F fluorescence are shown in Fig. 6.12. The $B \rightarrow X$ transition occurs first with a

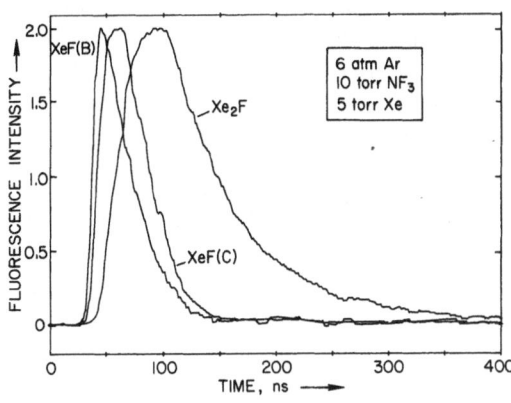

Fig. 6.12. Time-resolved fluorescence from $XeF(B \rightarrow X)$, $XeF(C \rightarrow A)$ and Xe_2F^*. The fluorescence pulses are normalized to the same relative intensity. The red argon fluorescence was subtracted from the Xe_2F^* spontaneous emission signal

steep risetime, whereas the $C \rightarrow A$ fluorescence appears somewhat delayed. This behavior seems to indicate that for certain e-beam excitation conditions, XeF(B) is formed predominantly, followed by an increase in the C-level population due to collisional mixing of the B and C states, which is similar to observations reported for photolytic excitation [6.61]. To ensure good vibrational relaxation in the B and the C states, the experiments were performed at high argon pressures (>6 atm). Below 1 atm argon pressure, decoupling of the two states can be observed. In Xe–NF_3 mixtures the $B \rightarrow X$ and the $C \rightarrow A$ transitions showed exactly the same time dependence if the xenon pressure was

Table 6.6. Summary of XeF* and Xe_2F* rate constants

Reaction	Rate coefficient $[s^{-1}]$	References
XeF(C) + NF_3 → products	$(1.5 \pm 0.3) \times 10^{-11}$ cm^3	[6.62]
	$(1.4 \pm 0.6) \times 10^{-11}$ cm^3	[6.60]
XeF(C) + Ar → products	$< 1 \times 10^{-14}$ cm^3	[6.62]
	$(7 \pm 7) \times 10^{-15}$ cm^3	[6.59]
	$(5 \pm 2) \times 10^{-14}$ cm^3	[6.60]
	$(9 \pm 0.1) \times 10^{-14}$ cm^3	[6.116]
XeF(C) + Xe → products	$(1.2 \pm 0.3) \times 10^{-10}$ cm^3	[6.33, 62]
	1.9×10^{-10} cm^3	[6.59, 83]
	$(1.8 \pm 0.5) \times 10^{-11}$ cm^3	[6.60]
	$(1 \pm 0.5) \times 10^{-11}$ cm^3	[6.116]
XeF(B) + Ar → XeF(C) + Ar	$(1.5 \pm 0.3) \times 10^{-11}$ cm^3	[6.62]
	$(1.4 \pm 0.2) \times 10^{-11}$ cm^3	[6.58, 59]
	$(0.86 \pm 0.11) \times 10^{-11}$ cm^3	[6.60]
XeF(C) → Xe + F + $h\nu$(350 nm)	$(97 \pm 5) \, 10^{-9}$	[6.33, 62]
	$(101 \pm 2) \, 10^{-9}$	[6.58, 59]
	$(95 \pm 7) \, 10^{-9}$	[6.116]
XeF(C) + Ar + Xe → products	$(3.9 \pm 0.5) \times 10^{-31}$ cm^6	[6.62]
XeF(B) + 2Ar → products	$< 3 \times 10^{-33}$ cm^6	[6.62]
XeF(C) + 2Ar → products	$< 3 \times 10^{-33}$ cm^6	[6.62]
XeF(C) + 2Xe → products	$(4 \pm 1) \times 10^{-32}$ cm^6	[6.62]
	$< 3 \times 10^{-31}$ cm^6	[6.60]
XeF(C) + 2Xe → $Xe_2F* + $ Xe	$(2.7 \pm 0.6) \times 10^{-32}$ cm^6	[6.33, 62]
	$< 3 \times 10^{-31}$ cm^6	[6.60]
XeF(C) + Xe + Ar → $Xe_2F* + $ Ar	$(1.1 \pm 0.3) \times 10^{-31}$ cm^6	[6.33, 62]
$Xe_2F* + NF_3$ → products	$(7.8 \pm 1.2) \times 10^{-13}$ cm^3	[6.33, 62]
$Xe_2F* + $ Xe → products	$(1.0 \pm 0.2) \times 10^{-13}$ cm^3	[6.33, 62]
$Xe_2F* + $ Ar → products	$(2.8 \pm 0.9) \times 10^{-14}$ cm^3	[6.33, 62]
Xe_2F* → 2Xe + F + $h\nu$(614 nm)	$(152^{+19}_{-10}) \, 10^{-9}$	[6.33, 62]

greater than 100 Torr. This is due to rapid quenching and collisional mixing between the two states by xenon. The onset of the Xe_2F pulse is on the same time scale as the width of the XeF pulses, which is due to be expected if Xe_2F is formed from XeF by fast three-body quenching processes. Xe_2F^* fluorescence typically decays considerably slower than the fluorescence of the two XeF* transitions, indicating a comparatively long radiative lifetime. Because of the rapid relaxation of $XeF(B) \rightarrow XeF(C)$, Xe_2F is produced predominantly via three-body collisions from the $XeF(C)$ state. The formation and quenching rate constants are summarized in Table 6.6.

Like the triatomic rare-gas-halides, Xe_2F is a possible candidate for a tunable excimer laser. Its cross-section for stimulated emission is high and transient absorption effects are considerably less at $\sim 600\,nm$ than at shorter wavelengths. However, the production efficiency of Xe_2F is considerably lower than for other trimers because of the strong xenon quenching of the precursor $XeF(C)$.

e) Xe_2Cl^*

Intense broadband fluorescence from electron-beam excited mixtures of argon under high pressure (typically 6–10 atm), xenon and various chlorine donors have been observed at 345 nm and in the blue-green spectral range around 490 nm (Fig. 6.2c). These broadband emissions have been assigned to the diatomic $XeCl(C \rightarrow A)$ and triatomic Xe_2Cl excimer transitions, respectively. Of all the trimer species investigated so far, Xe_2Cl emits the most intense laser radiation [6.7]. Both the formation and quenching kinetics of the $XeCl$–Xe_2Cl systems have been discussed in [6.67–69, 80]. Table 6.7 summarizes the relevant formation and quenching data for this system. Xe_2Cl is also formed primarily in a three-body clustering reaction involving XeCl*, as in the case of other Rg_2X^* species. The essential kinetic pathways involved in the decay of the electron-beam produced precursor XeCl*, and formation and decay of the trimer Xe_2Cl^* are depicted in Fig. 6.13. To facilitate the comparison of the various processes, all rates have been converted into decay times, using the set of rate coefficients of Table 6.7 and a typical rare-gas-halide mixture (6 atm Ar, 100 Torr Xe, and 1.5 Torr CCl_4). From the work reported in [6.67, 68] it appears that only about 30% of the XeCl* precursors lead to production of Xe_2Cl^* (at least in $Ar/Xe/CCl_4$ mixtures), and that the rapid quenching of XeCl* by Xe leads in part to ground state atoms. In addition to this "branching-ratio" problem, Xe_2Cl^* is rapidly quenched by all chlorine donors. For efficient laser performance, it is desirable to devise excitation schemes that overcome such losses in the kinetic pathways.

To date CCl_4 has been found to be the optimum halogen donor for Xe_2Cl^* formation in terms of fluorescence yield and quenching behavior. This is clearly evident from Fig. 6.14. Fig. 6.14a depicts electron-beam excited Xe_2Cl fluorescence spectra for optimum concentrations of three chlorine donors CCl_4, Cl_2, and HCl. The trend exhibited by these halogen donors correlates well with their

Table 6.7. Summary of rate constants for reactions quenching XeCl* and Xe_2Cl*

$XeCl* + Ar + Xe \rightarrow products$	$(3.8 \pm 0.2) \times 10^{-30}$ cm^6	[6.68]
$XeCl* + Ar + Xe \rightarrow Xe_2Cl* + Ar$	$(1.5 \pm 0.5) \times 10^{-31}$ cm^6	[6.67]
$XeCl* + Ar + Ar \rightarrow products$	$<3 \times 10^{-33}$ cm^6	[6.68]
$XeCl* + Xe + Xe \rightarrow products$	7.3×10^{-31} cm^6	[6.69]
$XeCl* + CCl_4 \rightarrow products$	$(4.6 \pm 0.2) \times 10^{-10}$ cm^3	[6.68]
	$(4.5 \pm 0.4) \times 10^{-10}$ cm^3	[6.69]
$XeCl* + Cl_2 \rightarrow products$	8.8×10^{-11} cm^3	[6.70]
$XeCl* + Ar \rightarrow products$	$<2 \times 10^{-13}$ cm^3	[6.68]
$XeCl* + Xe \rightarrow products$	1×10^{-11} cm^3	[6.70]
$XeCl* \rightarrow Xe + Cl + h\nu(308$ nm$)$	$(41 \pm 3) \, 10^{-9}$	[6.68]
	11×10^{-9}	[6.71]
	$(20 \pm 16) \, 10^{-9}$	[6.69]
	27×10^{-9}	[6.70]
	10×10^{-9}	[6.117]
$XeCl* \rightarrow Xe + Cl + h\nu(345$ nm$)$	53×10^{-9} for C state	[6.70]
	$(133 \pm 10) \, 10^{-9}$	[6.72]
	130×10^{-9}	[6.117]
$Xe_2Cl* \rightarrow 2Xe + Cl + h\nu(500$ nm$)$	$(135^{+70}_{-60}) \, 10^{-9}$	[6.67]
	$(210 \pm 25) \, 10^{-9}$	[6.73]
	210×10^{-9}	[6.69]
	185×10^{-9}	[6.70]
	$(120 \pm 30) \, 10^{-9}$	[6.14]
	330×10^{-9}	[6.31]
$Xe_2Cl* + CCl_4 \rightarrow quenching$	$(6 \pm 1) \times 10^{-10}$ cm^3	[6.67]
	$5.4 \pm 0.5 \times 10^{-10}$ cm^3	[6.73]
$Xe_2Cl* + Cl_2 \rightarrow quenching$	4.5×10^{-10} cm^3	[6.69]
	2.6×10^{-10} cm^3	[6.70]
	$(2.2 \pm 0.2) \times 10^{-10}$ cm^3	[6.73]
$Xe_2Cl* + Ar \rightarrow quenching$	$(3 \pm 1) \times 10^{-14}$ cm^3	[6.67]
$Xe_2Cl* + Xe \rightarrow quenching$	$<5 \times 10^{-13}$ cm^3	[6.67]
	$<4 \times 10^{-14}$ cm^3	[6.69]
	$<6 \times 10^{-15}$ cm^3	[6.70]

rate coefficients for dissociative attachment (i.e., Cl$^-$ production), as shown in Fig. 6.14b [6.74, 75].

The termolecular formation mechanism may be tested in several ways, as discussed in [6.67]. For example, Fig. 6.15 shows the good agreement between the calculated and experimentally measured temporal Xe_2Cl fluorescence profiles for a mixture of 2 atm Ar, 400 Torr Xe, and 2 Torr CCl$_4$ assuming that XeCl* is the main precursor for Xe_2Cl* formation. The spontaneous decay time of Xe_2Cl* together with the quenching rates of CCl$_4$, xenon and argon (Table 6.7), and theoretical fits to waveforms obtained for different rare-gas

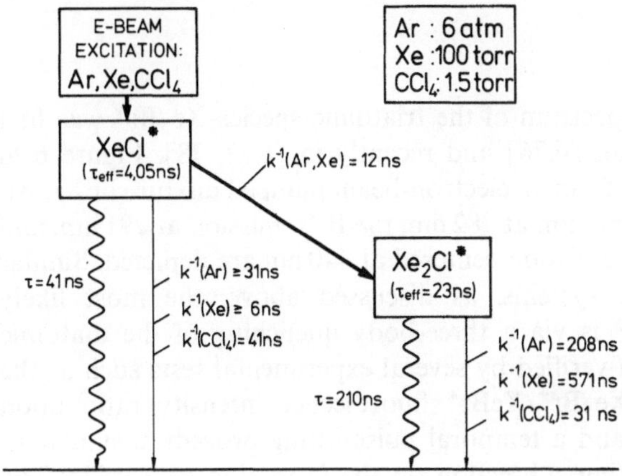

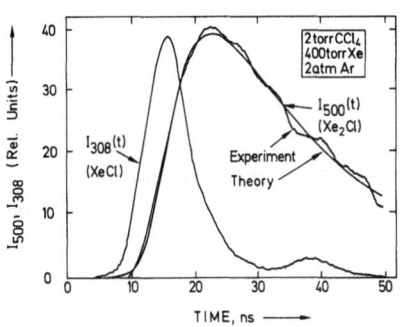

Fig. 6.13. XeCl*–Xe$_2$Cl* kinetics scheme depicting decay times for an optimized electron beam excited Ar–Xe–CCl$_4$ mixture [6.67, 68]

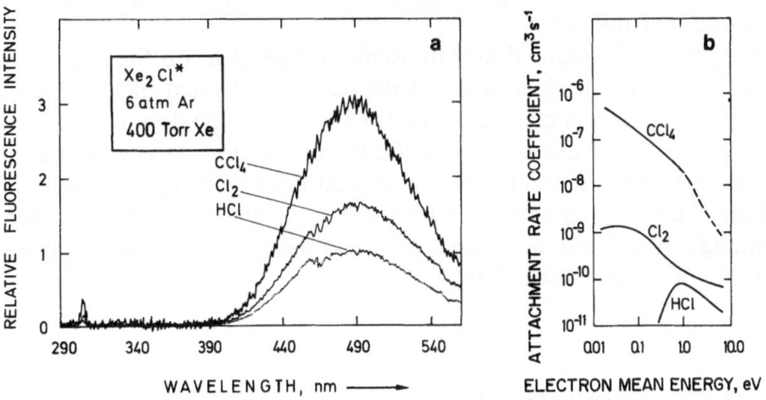

Fig. 6.14a, b. Comparison of Xe$_2$Cl* fluorescence for the chlorine donors CCl$_4$ (1.4 Torr), Cl$_2$ (1 Torr), and HCl (3 Torr) (**a**) and their respective electron attachment rate coefficients as a function of the mean energy (**b**)

Fig. 6.15. Comparison of experimentally observed Xe$_2$Cl* fluorescence pulses and calculated pulses based upon the kinetics model of [6.67], assuming that XeCl* is only the precursor of Xe$_2$Cl*

pressures yielded a determination of the three-body formation rate of $1.5 \times 10^{-31} \, \text{cm}^6 \, \text{s}^{-1}$ [6.67].

f) Xe_2Br^*

The fluorescence emission spectrum of the triatomic species Xe_2Br^* was first reported by *Konovalow* et al. [6.76] and recently in [6.77, 78]. Figure 6.2d illustrates a typical spectrum from an electron-beam pumped mixture of Xe, Ar, and Br_2. The $XeBr(B \rightarrow X)$ emission at 282 nm, the Br_2^* emission at 291 nm, and the very broadband Xe_2Br^* emission centered at 440 nm are depicted. Similar to other triatomic excimer systems, as discussed above, the most likely formation route for Xe_2Br^* is via a three-body quenching of the diatomic $XeBr^*$ [6.78]. This has been verified by several experimental tests such as the linear dependence of the $Xe_2Br^*/XeBr^*$ fluorescence intensity ratio upon xenon and argon pressure, and a temporal pulse-fitting procedure similar to that discussed in the preceding subsection on the formation of Xe_2Cl^*. The most important reactions leading to the formation and quenching of Xe_2Br^* are summarized in Table 6.8.

Quenching by the bromine donor molecule is dependent on the particular donor species used. Those donors that have been investigated include CBr_4, $CHBr_3$, HBr, and Br_2. Quenching constants for several of the donors are given in Table 6.9. Carbon tetrabromide has the highest fluorescence yield but is difficult to handle since it is solid at room temperature with a vapor pressure of only 0.5 Torr. Hence, for practical reasons $CHBr_3$ is chosen as the optimum donor, although it exhibits a slightly lower fluorescent yield and a higher quenching rate of Xe_2Br^* than CBr_4.

6.4 Basic Laser Considerations

In this section typical characteristics of gain and absorption of broadband excimers will be described together with a discussion of the various effects and appropriate excitation techniques.

6.4.1 Optical Gain

Gain is an essential parameter for the characterization of any laser system. Laser threshold, output coupling of the resonator, peak power output, efficiency, saturation behavior, and spectral narrowing are all functions of the gain properties of the active medium. In the presence of e-beam induced atomic and molecular absorbers, the small-signal gain coefficient of an excimer system is defined by

$$g_{\text{eff}}(t) = \sigma_e N^*(t) - \sigma_a N_a(t) \tag{6.14}$$

Table 6.8. Summary of rate constants for reactions quenching Xe_2Br^* [6.78]

Reaction	Rate coefficient $[s^{-1}]$
$XeBr^* \rightarrow Xe + Br + h\nu(282\ nm)$	17.5×10^{-9} [6.79]
$XeBr^* + RBr \rightarrow products$	$8 \times 10^{-10}\ cm^3$ for donor Br_2 [6.79]
$XeBr^* + 2Xe \rightarrow products$	$(2.7 \pm 0.9) \times 10^{-31}\ cm^6$
$XeBr^* + 2Xe \rightarrow Xe_2Br^* + Xe$	$2.7 \times 10^{-31}\ cm^6$
$XeBr^* + Ar + Xe \rightarrow products$	$(3.0 \pm 0.8) \times 10^{-31}\ cm^6$
$XeBr^* + Ar + Xe \rightarrow Xe_2Br^* + Ar$	$(3.2 \pm 0.4) \times 10^{-31}\ cm^6$
$XeBr^* + Ar + Xe \rightarrow other\ products$	$\leq 1 \times 10^{-31}\ cm^6$
$Xe_2Br^* \rightarrow 2Xe + Br + h\nu(440\ nm)$	$(245 \pm 30)\ 10^{-9}$
$Xe_2Br^* + RBr \rightarrow products$	See Table 6.9
$Xe_2Br^* + Ar \rightarrow products$	$\leq 2 \times 10^{-14}\ cm^3$
$Xe_2Br^* + Xe \rightarrow products$	$(2.8 \pm 0.9) \times 10^{-13}\ cm^3$

Table 6.9. Xe_2Br^* quenching rates of bromine donors [6.78]

Donor	Quenching rate, k_q $[cm^3\ s^{-1}]$
CBr_4	3.2×10^{-10}
$CHBr_3$	5.6×10^{-10}
HBr	6.5×10^{-11}
Br_2	3.3×10^{-10}
CH_3Br	8.0×10^{-10}
C_2H_5Br	7.4×10^{-10}

with σ_e and σ_a the respective cross section for stimulated emission and absorption, $N^*(t)$ the time-varying excited state population density in the upper laser level, assuming a four-level system, and $N_a(t)$ the density of absorbing species. Hence the absorption effects must be minimized in order to achieve an effective optical gain. Precise predictions of the quantities N^* and σ_e require a detailed knowledge of the spectroscopy of the laser molecule and the kinetic processes leading to the population and decay of the particular upper laser level.

For a homogeneously broadened line, the stimulated-emission cross section σ_e at the maximum of the emission of a laser transition of central fluorescence wavelength λ, bandwidth $\Delta\lambda$, spontaneous decay time τ and speed of light c is approximately given by (Chap. 4 and [6.2])

$$\sigma_e \approx \frac{1}{8\pi\tau c} \cdot \frac{\lambda^4}{\Delta\lambda}. \qquad (6.15)$$

For a trimer laser such as Xe_2Cl [6.7] with $\tau = 200\ ns$, $\lambda = 490\ nm$, and $\Delta\lambda = 60\ nm$, one obtains $\sigma_e \approx 5 \times 10^{-18}\ cm^2$, a cross section approximately one

Table 6.10. Stimulated emission cross sections for broadband transitions

Rg$_2$X*	$\sigma_e (10^{-18}$ cm$^2)$
Ar$_2$F*	0.95
KrF*	3.28
Xe$_2$Br*	2.65
Xe$_2$Cl*	4.56
Xe$_2$F*	14.05

to two orders of magnitude smaller than the comparable values of diatomic exciplexes. The long spontaneous lifetime and the broad bandwidth of the trimer transitions are responsible for a relatively low effective gain. Due to inherent kinetic limitations, it is difficult to increase the excited-state density sufficiently to counterbalance these spectral properties of trimers.

A compilation of the cross sections for stimulated emission σ_e is given in Table 6.10, calculated by means of (6.15) using the spectroscopic data of Table 6.1. The range of cross sections σ_e correspond to the λ^4 law of (6.15) because the other parameters contained in this equation are approximately equal for all trimers. For an experimentally determined Xe$_2$Cl* gain coefficient of $\sim g = 0.02$ cm^{-1} [6.7, 14], an upper state density $N^* = 5.5 \times 10^{15}$ cm^{-3} can be derived for transverse excitation of a high-pressure Ar-buffered Xe-halogen mixture with a Pulserad 110 e-beam accelerator whose beam characteristics are described in Sect. 6.4.3a. In the case of Ar/NF$_3$ mixtures the quantity $N^*(\text{Ar}_2\text{F}^*)$ will not exceed the 10^{16} cm^{-3} range and thus, this will be insufficient to observe Ar$_2$F* laser emission. In addition, there are several absorptions in this spectral range. Xe$_2$F because of its large cross section for stimulated emission may be a candidate for a broadband tunable excimer laser, as mentioned above. Its potential wavelength range of operation is spectrally separated from molecular absorptions of high pressure buffer gases in the near uv. The diatomic precursor XeF(C) is, however, strongly quenched by xenon which will limit the maximum excited state density which can be achieved.

For pulsed systems with a gain duration $\Delta t < \tau_{\text{eff}}$ (which may be required for the decay of transient absorptions) the following relation holds for the lower gain limit g_{min}, in order to clearly differentiate between emission due to stimulated emission and spontaneous fluorescence noise [6.23]:

$$\Delta t \cdot g_{\text{min}} \cdot c_0 \approx 15. \tag{6.16}$$

If c_0 is the velocity of light and with $\Delta t = 50$ ns, this yields $g_{\text{min}} = 0.01$ cm^{-1} for a typical trimer system. So far, the calculated and measured gain data of Xe$_2$Cl* [6.7, 15] and Kr$_2$F* [6.15] exceed this lower limit, and Xe$_2$Br* [6.78] appears to be close to g_{min}.

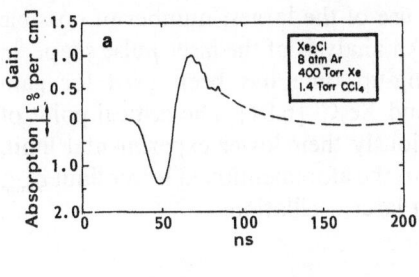

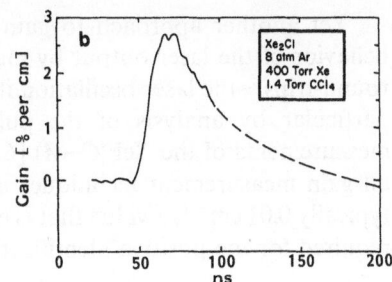

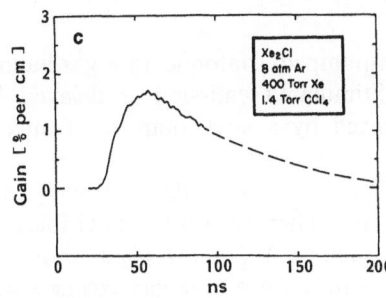

Fig. 6.16a–c. Temporal behavior of Xe_2Cl^* small signal gain and absorption versus the optical axis position (**a**) 1.5 cm, (**b**) 2.5 cm, and (**c**) 3.5 cm from the e-beam diode anode foil

Gain measurements using various experimental techniques have been performed on trimers for the prediction of laser performance and the selection of appropriate optical-cavity components. Gain studies are relatively straightforward for high-gain systems by application of amplified spontaneous emission (ASE) techniques [6.81]. Medium- and low-gain systems require more refined procedures such as multiple-pass probing of gain or loss by means of a laser beam. Paper [6.82] describes a comparison of both approaches for the broadband $C \rightarrow A$ transition of XeF. Because of the availability of several Ar^+ laser lines in the Xe_2Cl excimer emission band, and the possibility of increasing the sensitivity of the gain measurement by several horizontal or vertical multiple transits (e.g., 3–5 passes) through the excimer medium, the direct probe method is particularly useful in gain studies of Xe_2Cl [6.7]. This method makes it possible to measure conveniently both the magnitude and the temporal behavior of the effective optical gain. Figure 6.16 shows that the net gain varies strongly as a function of the excitation density. The density was varied by measuring the gain at different distances from the anode foil. The current density decreases with increasing distance. Near the anode foil, a strong initial absorption in the immediate afterglow of the electron beam pump pulse is evident, followed by a weak gain of $1\% \, cm^{-1}$ (Fig. 6.16a). The intermediate case, at the optic axis, shows gain and absorption canceling initially, followed by a gain of $2.7\% \, cm^{-1}$ (Fig. 6.16b). The lack of initial absorption was unexpected, although the delay in the gain peak corresponds well to the delay in the laser output. At the lowest excitation density both the delay and amplitude of the gain were reduced (Fig. 6.16c).

Yet another approach to gain measurements is to analyze the temporal behavior of the laser output by making use of the largest number of possible round-trips – the laser oscillation itself. An analysis of the laser pulse shape, in particular by analysis of the pulse ring-up time has been used for gain measurements of the XeF($C{\rightarrow}A$) [6.83] and Xe_2Cl [6.84]. The critical point of all gain measurement techniques is obviously their lower experimental limit, typically 0.01 cm^{-1}, a value that is equal to the aforementioned lower limit g_{min}, required for the positive identification of laser oscillation.

6.4.2 Excimer Absorption Effects

Various absorption processes in e-beam-pumped triatomic rare-gas-halides limit the performance of such lasers. Stimulated emission is delayed by broadband absorbing species and influenced by a large number of intense absorption lines.

The broadband absorbers have been identified as molecular in origin [6.85] with typical cross-sections, $\sigma_a \lesssim 5 \times 10^{-17} \text{ cm}^2$. They absorb in the visible and uv principally due to positive ionic states such as Ar_2^+ and Kr_2^+, as shown in Fig. 6.17 [6.86]. Absorption due to photo-ionization of rare-gas excimers was also recognized early, and absorptions by Ar_2^*, Kr_2^*, and Xe_2^* were reported [6.2, 87]. The absorbing excited molecules are created at the onset of e-beam excitation and have both a short (10–20 ns) and long lifetime (> 100 ns) components. This absorption seriously limits the laser efficiency even though the excited-state production efficiency is very high. Thus, laser action must occur in the afterglow regime of the e-beam pulse. The transient nature of the molecular absorption is clearly shown in Fig. 6.16. The absorption at 515 nm is of the order of 3 cm^{-1} at 8 atm of Ar.

The other absorption that has been identified involves long-lived excited atomic species, notably the rare-gas metastables, e.g. Xe* or Kr* which survive for hundreds of nanoseconds. These atomic metastables can give rise to discrete absorption lines involving transitions to excited Rydberg states [6.7, 88, 89].

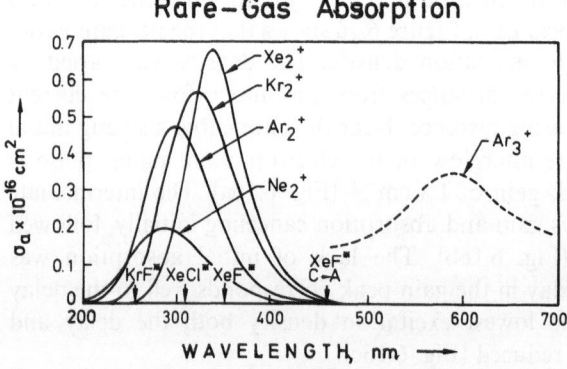

Fig. 6.17. Calculated dimer ion absorption coefficients versus wavelength [6.86]

Such absorption lines reduce the total laser output power and limit its continuous wavelength tunability. In principle, it should be possible to minimize these absorptions by introducing a quencher such as nitrogen for the xenon metastables in the rare-gas-halide mixtures [6.74, 90].

6.4.3 Excitation Techniques

Electron-beam and optical pumping have so far been the two most effective techniques in both fluorescence and laser studies of triatomic excimers [6.7]. Electron-beam excitation relies primarily on the ionization of the buffer gas and subsequent release of stored energy into the rare-gas-halide system to produce an effective excited state population. In addition, it should be possible to utilize high pressure, fast electrical discharge [6.91], microwave interaction [9.92, 93], photolytic excitation [6.25], and linear-accelerator injection [6.94]. Spectroscopic and kinetic studies can be conducted conveniently with a low-current electron-beam [6.95]synchrotron radiation [6.58] or short-pulse uv excimer laser excitation [6.70].

a) Electron-Beam Pumping

Direct e-beam excitation is one of the most widely used techniques for pumping excimer lasers on account of the high peak powers available, broad spectral applicability, scalability, and proven technology [6.1–3]. Efficiencies for conversion of electrical energy into excited states of the rare-gas buffers can be close to 50%. Pulsed e-beam generators are available in the 0.2–2 MeV range with current densities from 1 to 1000 A/cm^2. A generally useful e-beam generator employed in a major portion of this work has been a Physics International Pulserad 110 accelerator. Such a generator can deliver pulses of 1 MeV electrons at a current of 15 kA and a pulse duration of 8 ns. At the optical axis of a transversely pumped intracell optical resonator, 2.5 cm from the anode foil, the current density is ≥ 100 A/cm^2 at a typical buffer-gas pressure of 6 atm of argon. Despite the very efficient deposition of electron-beam energy into the dense rare-gas buffer, the excited-state population density is limited to 10^{16} cm^{-3} as established from gain data. Figure 6.18 shows three different geometries for coupling the electron-beam into the active medium. The amount of beam energy deposited within the laser volume depends on the excitation geometry. Because of straightforward design considerations, the one- or two-sided transverse pump geometry is most frequently employed. The electron-beam produced by a field emission diode enters the laser medium after passing through a metal foil (e.g., 25 µm thick titanium) supported by a stainless steel structure, typically an array of slots with $\geq 80\%$ transmission. Efficient beam energy utilization and penetration require an appropriate rare-gas host buffer-gas pressure, optimized anode-cathode spacing and an impedance matched pulse forming network. The coaxial or radial pumping scheme

TRANSVERSE EXCITATION

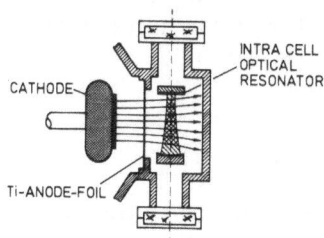

LONGITUDINAL EXCITATION

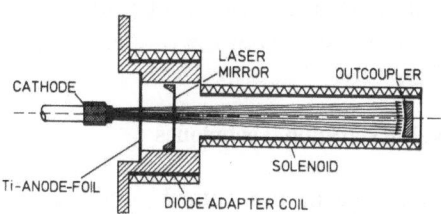

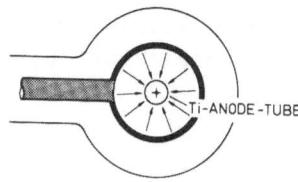

Fig. 6.18. Transverse, longitudinal and coaxial excitation geometries used in e-beam-pumped excimer lasers

described in [6.96] lends itself to the production of very high and uniform pump power densities ($> 10^2 \, Mw/cm^3$). Improved coupling of e-beam energy can be obtained by longitudinal excitation, but this requires the use of intense, pulsed magnetic guide fields [6.97].

The most direct consequence of increasing e-beam current is an increase in the electron density. The e-beam pumped rare-gas-halide mixture results in an attachment dominated plasma. The more intense the electron-beam the higher the ionization rate and the larger the electron density. This can result in a significant increase in the deactivation rate of both the excimer and trimer due to electron quenching. Furthermore, the higher ionization rates will also be accompanied by a higher production rate of rare-gas-excited states. Depending on the magnitude of their photoionization cross section, these species contribute to the total cavity absorption thereby lowering the effective optical gain.

b) Electric Discharge Pumping

Unlike the bulky apparatus associated with e-beam pumping, discharge excitation is relatively compact and easy to operate at high repetition rates. The primary requirement for efficient discharge pumping is the initiation and maintenance of a spatially uniform, stable, high-voltage discharge in rare-gas-halide mixtures with pressures of several atmospheres. So far only the broadband $C \rightarrow A$ laser transition of XeF has been successfully operated with a self-sustained discharge in mixtures of He, Xe, and NF_3 [6.23, 24]. However, it should also be possible to pump broadband trimers with discharge devices, as described in [6.46, 91, 98], provided the fast uv-preionized discharge can be maintained long enough to enable the relatively slow build-up of stable cavity oscillations of these low-gain systems.

uv-laser radiation [6.99, 100] and soft x-rays [6.100–102] can also provide uniform preionization at high buffer-gas pressures. Output energies up to 100 mJ have been obtained with a modified commercial XeCl-excimer laser upon preionization with only 5 mJ from another commercial KrF-excimer laser [6.103]. Both lasers have been appropriately synchronized to achieve XeCl laser output power optimization. Typically, the KrF-preionization laser had to be fired 80 ns in advance. The XeCl laser plasma showed a very homogeneous discharge behavior, even for partial illumination of the discharge volume. This technique of laser-induced uv preionization is scalable to large cavity volumes and provides an effective means of controlling the homogeneity of the preionization at high buffer-gas pressures.

c) Optical Pumping

Optical pumping involves the efficient conversion of e-beam energy to meta-stables of a high-pressure rare-gas (e.g., Ar, Kr or Xe) to produce intense narrowband uv fluorescence. This fluorescence, in turn, may be used to pump the active medium optically. This method of pumping minimizes optical absorption effects and quenching by electrons which occur as a result of direct e-beam excitation. *Eckstrom* et al. have demonstrated in [6.25] the potential of a photolytic pumping scheme to produce a multijoule output from the $XeF(C \rightarrow A)$ laser. Excitation is achieved in this technique by e-beam pumping of xenon to produce Xe_2^* fluorescence at 172 nm which can subsequently be transferred with high efficiency into the surrounding laser gas mixture. An unsuccessful attempt was made to achieve Xe_2Cl^* oscillations based on optical pumping by Kr_2^* radiation [6.104].

Direct optical excitation of broadband emitting complexes is also possible by using an open high-current discharge technique. In contrast to e-beam excitation, this technique allows homogeneous excitation of large mode volumes. A number of trimer systems have been studied using this type of pumping [6.17, 51, 105–107].

6.5 Laser Experiments

Of the 10 trimers listed in Table 6.1, laser operation has so far been demonstrated for Xe_2Cl^* [6.15] and Kr_2F^* [6.16, 17]. The basic experimental arrangement used in these experiments is shown in Fig. 6.19. It consists of the e-beam excitation source, a high-pressure cell with an externally adjustable optical resonator and various electrical and optical diagnostic instrumentation. Details of such a setup and the techniques employed are given in [6.7].

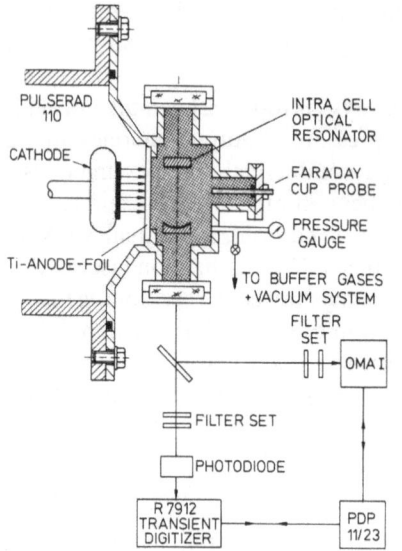

Fig. 6.19. Schematic diagram of transverse e-beam-pumped laser cell and associated instrumentation

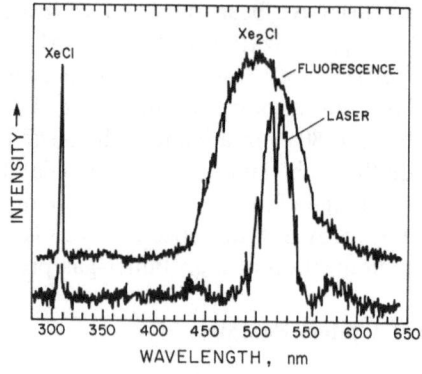

Fig. 6.20. Fluorescence and laser spectra of Xe_2Cl^* and $XeCl^*$ $(B \rightarrow X)$ of an $Ar/Xe/CCl_4$ mixture composed of 6.5 atm, 200 Torr, and 1 Torr, respectively

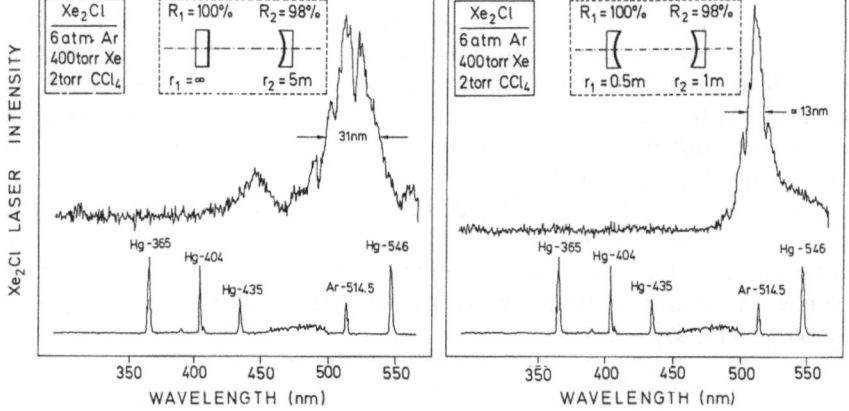

Fig. 6.21. Influence of the cavity configuration on the bandwidth of Xe_2Cl^* laser output spectrum

A typical Xe_2Cl laser spectrum is shown in Fig. 6.20. The diatomic $XeCl(B \rightarrow X)$ transition at 308 nm, the Xe_2Cl^* laser output spectrum, and, for comparison, Xe_2Cl^* spontaneous fluorescence are depicted. The laser spectrum which has a center wavelength of 520 nm shows significant spectral narrowing from 70 nm (spontaneous emission) to 32 nm. In addition, greatly enhanced intracavity absorption features can be observed. The laser output spectrum is

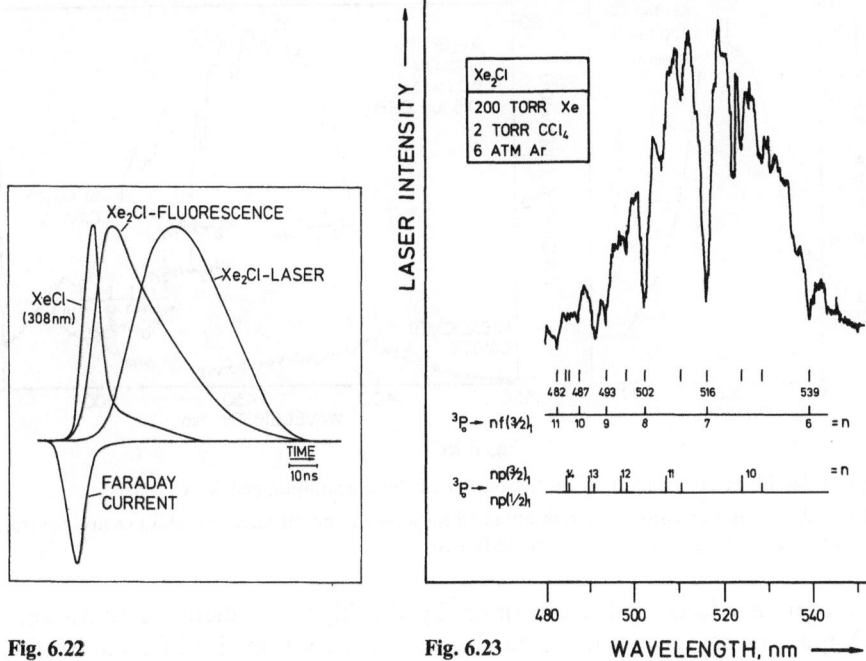

Fig. 6.22   **Fig. 6.23**   WAVELENGTH, nm ⟶

Fig. 6.22. Normalized pulses showing temporal characteristics of Xe_2Cl fluorescence and laser emission, $XeCl(B \rightarrow X)$ fluorescence and the e-beam pump pulse

Fig. 6.23. High-resolution Xe_2Cl^* laser spectrum showing intracavity absorption features due to transitions between $^3P_0$ xenon metastables and high-lying Rydberg states [6.7]

red-shifted relative to the fluorescence spectrum by approximately 20 nm. This shift is due to both the λ^4 dependence of the stimulated emission cross section and to the long-wavelength tail of the various rare-gas absorbers, such as Xe_2^+ or Ar_2^+. Figure 6.21 shows the influence of the cavity configuration on the laser bandwidth. A reduction in bandwidth from 31 nm (FWHM) to only 13 nm occurs as a consequence of tighter focusing cavity reflectors. The reduction of background noise for a high Q-cavity configuration and the resulting increase in output power by a factor of 10 should be noted. Figure 6.22 shows a typical normalized temporal composite of uv and visible fluorescence and Xe_2Cl laser output from a high-pressure $Ar–Xe–CCl_4$ mixture. The 40 ns long laser pulse is considerably delayed with respect to the e-beam pump pulse due to induced buffer-gas absorption discussed below. Furthermore, the Xe_2Cl^* emission occurs delayed with respect to the uv fluorescence pulse, and the laser output reaches its maximum peak intensity later than the Xe_2Cl fluorescence peak as a consequence of the long ring-up time of cavity oscillations.

So far, the output from the broadband excimer lasers is low due to the small active mode volume available from a stable resonator, the limited time that is available for build up of the optical pulse, their low inherent gain, transient

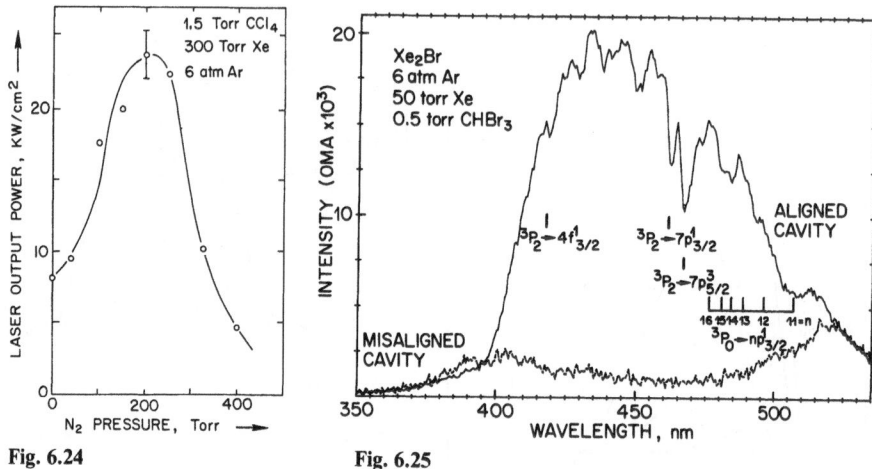

Fig. 6.24   **Fig. 6.25**

Fig. 6.24. Effect of N_2 addition on the output of the e-beam-pumped Xe_2Cl^* laser

Fig. 6.25. Xe_2Br^* emission spectrum obtained for aligned and misaligned high-Q cavity. Several absorptive Rydberg transitions are also indicated

absorption effects, and competition by the high-gain diatomic transitions. Output power densities up to $50\,kW/cm^2$ have been obtained for the Xe_2Cl^* laser. This requires optimization of both the gas mixture in order to achieve highest possible gains and the extraction efficiency [6.108, 109] by adjusting the output coupler transmission. Tuning experiments with intracavity dispersive tuning elements (i.e., a prism or grating), as demonstrated for the broadband diatomic $XeF(C \rightarrow A)$ excimer [6.97], were unsuccessful. However, broadband tunability (over a 40 nm spectral range) could be demonstrated for Xe_2Cl by using two reflectors with different center wavelengths [6.110, 111].

Figure 6.23 depicts a high-resolution emission spectrum of Xe_2Cl^* with numerous identified Rydberg-state absorption lines originating from the $6s\,^3P_0$ Xe* metastable state located at 9.5 eV. Recent work reported in [6.90] suggests that the atomic metastable absorptions may be quenched by rapid transfer to N_2 when using photolytic excitation at 172 nm with an Xe_2^* radiation source. Apparently the absorption features vanish with N_2 as buffer. Various efforts have been made to study the role of Xe* metastables on laser output characteristics for e-beam excited gas mixtures. In a recent study [6.74] the influence of N_2 addition on Xe_2Cl^* fluorescence and laser behavior has been investigated. Considerable improvement of the Xe_2Cl laser output was observed when nitrogen was added to the gas mixture, as shown in Fig. 6.24. A significant increase of the laser output power by a factor up to three was achieved by the addition of 200 Torr of nitrogen.

The role of N_2 as an additive to an $Ar/Xe/CCl_4$ mixture to enhance the Xe_2Cl laser performance can be best understood by consideration of the influence of N_2 on the production and loss by the primary absorbing species Xe*, Xe_2^*, and Xe_2^+. Assuming that the measured quenching coefficient of

1.9×10^{-11} cm^3 s^{-1} for Xe($^3P_2$) by N$_2$ is also representative of the other Xe(6s) levels [6.65], a nitrogen pressure of 100 Torr results in Xe* quenching time constant of about 20 ns, which is close to the observed delay for the onset of lasing. Since the dimer Xe$_2^*$ is produced from Xe*, it follows that the presence of N$_2$ will reduce both discrete and broadband absorption resulting from Xe* and Xe$_2^*$.

Kr$_2$F showed all the characteristics expected of stimulated emission behavior [6.7, 16, 17]. The peak laser output was considerably weaker than for the Xe$_2$Cl laser, due to severe background absorptions [6.48, 112]. The Kr$_2$F laser output occurred with 10 Torr NF$_3$ as the halogen donor, and showed no maximum with increasing Kr and Ar pressures. Atomic absorptions by krypton metastables were clearly evident in the laser spectrum.

For Xe$_2$Br, which has an estimated gain of 1 % cm^{-1} [6.78], no stable cavity oscillations with well-defined spatial emission characteristics have been observed. However, absorption controlled and cavity-alignment sensitive emission spectra have been obtained, as shown in Fig. 6.25 [6.77]. A few absorption lines, due to Rydberg transitions, have been identified as in the case of the Xe$_2$Cl laser.

6.6 Summary

Electron-beam-pumped broadband triatomic excimers, Rg$_2$X*, have been discussed in terms of their underlying spectroscopic and kinetic characteristics, emphasizing those that are relevant to laser performance. Because the triatomic rare-gas halogens have a steeply repulsive potential energy curve in the ground state, the fluorescence is inherently broadband as compared to the narrow spectral linewidth of the diatomic excimer.

The energy transfer and excited-state kinetics for trimers are complex. The detailed kinetic route by which the excimer is formed depends upon the particular trimer, the halogen donor, the excitation conditions and the partial pressures of the constituent gases. Trimers are produced principally by three-body clustering collisions involving the diatomic RgX* as precursors. Most of the three-body reaction rates are of the order of 10^{-31} to 10^{-30} cm^6 s^{-1}.

Unfortunately from the point of view of laser development, the broadband trimer emission is accompanied by long radiative lifetimes which implies low optical gain for these excimers. To date, two triatomic lasers, Xe$_2$Cl centered at 520 nm and Kr$_2$F centered at 435 nm, have been demonstrated. Electron-beam pump-induced transient atomic and molecular absorption effects were found to severely limit the potential laser efficiency and usefulness of triatomic excimers. Efficiencies are also small because of the small active volume defined by a stable high-Q optical resonator geometry. Experiments using a longitudinally pumped configuration with a long cavity have proved unsuccessful because of severe competition by the companion diatomic $B \rightarrow X$ and $C \rightarrow A$ transitions [6.111]. Other trimers with wide fluorescence bandwidths – Ar$_2$F at 290 nm,

Xe_2Br at 420 nm, and Xe_2F at 614 nm – exhibited gains which were too small to achieve laser action in the 10 cm transversely pumped laser cavity used so far in these experiments.

Optimization of gas mixtures [6.74, 119] and novel pumping schemes such as radiation transfer pumping [6.25, 90] have so far been partially successful in minimizing quenching collisions of excited species with rare-gases and competing $B \to X$ transitions. Supersonic-flow e-beam stabilized discharge excitation has been proposed to lower excimer quenching and absorption losses [6.113]. An important remaining technical challenge is the realization of a fast-discharge excited triatomic excimer laser. Such a development will require stable discharge conditions in a high pressure gas mixture which can be obtained with x-ray [6.100–102, 114], uv-laser [6.115] or e-beam preionization [6.1].

References

6.1   J.J.Ewing: In *Lasers Handbook*, Vol. 3 (North-Holland, Amsterdam 1979)
6.2   M.J.Shaw: Progr. Quantum Electron. **6**, 3–54 (Pergamon, London 1979)
6.3   M.H.R.Hutchinson: Appl. Phys. **21**, 95 (1980)
6.4   I.S.Lakoba, S.I.Yakovlenko: Sov. J. Quantum Electron. **10**, 389 (1980)
6.5   A.V.Eletskii: Sov. Phys. Usp. **21**, 502 (1978)
6.6   C.A.Brau: In *Excimer Lasers – 1983*, ed. by C.K.Rhodes, H.Egger, H.Pummer (Am. Ins. Phys., New York 1983)
6.7   F.K.Tittel, G.Marowsky, W.L.Wilson, M.C.Smayling: IEEE J. QE-**17**, 1488 (1981)
6.8   W.L.Nighan, C.M.Ferrar: Appl. Phys. Lett. **40**, 223 (1982)
6.9   J.A.Mangano, J.H.Jacob, M.Rokni, A.Hawryluk: Appl. Phys. Lett. **31**, 26 (1977)
6.10   M.Rokni, J.H.Jacob, J.A.Mangano, R.Brodus: Appl. Phys. Lett. **30**, 458 (1977)
6.11   T.H.Johnson, A.M.Hunter: J. Appl. Phys. **51**, 2406 (1980)
6.12   F.Kannari, M.Obara, T.Fujioka: J. Appl. Phys. **53**, 135 (1982)
6.13   D.L.Huestis: Presented at the 1979 Topical Meeting on Excimer Lasers, Charleston, SC (1979)
6.14   K.Y.Tang, D.C.Lorents, D.L.Huestis: Appl. Phys. Lett. **36**, 347 (1980)
6.15   F.K.Tittel, W.L.Wilson, R.E.Stickel, G.Marowsky, W.E.Ernst: Appl. Phys. Lett. **36**, 405 (1980)
6.16   F.K.Tittel, G.Marowsky, M.C.Smayling, W.L.Wilson: Appl. Phys. Lett. **37**, 862 (1980)
6.17   N.G.Basov, V.S.Zuev, A.V.Kanaev, L.D.Mikheev, D.B.Shavrovskii: Sov. J. Quantum Electron. **10**, 1561 (1980)
6.18   D.C.Lorents, R.M.Hill, D.L.Huestis, M.V.McCusker, H.H.Nakano: In *Electronic Transition Lasers II*, ed. by L.E.Wilson, S.N.Suchard, J.I.Steinfeld (MIT, Cambridge, MA 1977)
6.19   N.G.Basov, V.A.Danilychev, O.M.Kerimov, V.S.Lebedev, A.G.Molchanov: JETP Lett. **26**, 16 (1977)
6.20   D.C.Lorents, D.L.Huestis, M.V.McCusker, H.H.Nakano, R.M.Hill: J. Chem. Phys. **68**, 4657 (1978)
6.21   W.K.Bischel, H.H.Nakano, D.J.Eckstrom, R.M.Hill, D.L.Huestis, D.C.Lorents: Appl. Phys. Lett. **34**, 565 (1979)
6.22   W.E.Ernst, F.K.Tittel: Appl. Phys. Lett. **35**, 36 (1979)
6.23   R.Burnham: Appl. Phys. Lett. **35**, 48 (1979)
6.24   C.H.Fisher, R.E.Center, G.J.Mullaney, J.P.McDaniel: Appl. Phys. Lett. **35**, 26 (1979)

6.25   D.C.Eckstrom, H.C.Walker: J. IEEE J. QE-**18**, 176 (1982)
6.26   H.C.Brashears, D.W.Setser, Y.C.Yu: J. Chem. Phys. **74**, 10 (1981)
6.27   D.L.Huestis, N.E.Schlotter: J. Chem. Phys. **69**, 3100 (1978)
6.28   W.R.Wadt, P.J.Hay: Appl. Phys. Lett. **30**, 573 (1977)
6.29   C.F.Bender, H.F.Schaefer, III: Chem. Phys. Lett. **53**, 27 (1978)
6.30   W.R.Wadt, P.J.Hay: J. Chem. Phys. **68**, 3850 (1978)
6.31   W.J.Stevens, M.Krauss: Appl. Phys. Lett. **41**, 301 (1982)
6.32   A.M.Hawryluk, J.A.Mangano: Appl. Phys. Lett. **31**, 169 (1977)
6.33   W.Walter, R.Sauerbrey, F.K.Tittel, W.L.Wilson, Jr.: Appl. Phys. Lett. **41**, 387 (1982)
6.34   V.H.Shui: Appl. Phys. Lett **31**, 50 (1977); **34**, 203 (1979)
6.35   V.H.Shui, C.Duzy: Appl. Phys. Lett. **36**, 135 (1980)
6.36   C.H.Chen, M.G.Payne: Appl. Phys. Lett. **32**, 358 (1978); IEEE J. QE-**15**, 149 (1979)
6.37   C.H.Chen, M.G.Payne, J.P.Judish: J. Chem. Phys. **69**, 1626 (1978)
6.38   G.P.Quigley, W.M.Hughes: Appl. Phys. Lett. **32**, 649 (1978)
6.39   P.Fisher, W.M.Hughes: Private Communication
       H.Jara, H.Pummer, H.Egger, C.K.Rhodes: Submitted to Phys. Rev. A
6.40   H.H.Michels, R.H.Hobbs, L.A.Wright: Chem. Phys. Lett. **48**, 158 (1977)
6.41   H.T.Powell, K.S.Jancaitis: Paper presented at 34th Annual Gas. Electron. Conf., Boston, MA (1981)
6.42   H.H.Nakano, R.M.Hill, D.C.Lorents, D.L.Huestis, M.V.McCusker: Stanford Research Institute Technical Report No. MP76-99 (1976)
6.43   G.Marowsky, G.P.Glass, F.K.Tittel, K.Hohla, W.L.Wilson, Jr., H.Weber: IEEE J. QE-**18**, 898 (1982)
6.44   N.Boewering, R.Sauerbrey, H.Langhoff: J. Chem. Phys. **76**, 3524 (1982)
6.45   G.Marowsky, K.Hohla, F.K.Tittel, H.Weber: Proc. Intern. Conf. Lasers '81, New Orleans, LA (1981) pp. 712–714
6.46   G.Zhi-yu, W.Shao-yin, D.Proch, F.Rebentrost, H.Weber, K.L.Kompa: Presented at Intern. Conf. Lasers, Guangzhou, China (1983)
6.47   M.Rokni, J.H.Jacob, J.A.Mangano: Phys. Rev. **16**A, 2216 (1977)
6.48   J.F.Eden, R.S.F.Chang, L.J.Palumbo: IEEEJ. QE-**15**, 1146 (1979)
6.49   N.G.Basov, V.A.Danilychev, V.A.Dolgikh, O.M.Kerimov, V.S.Kekedev, A.G.Molchanov: Sov. J. Quantum Electron. **9**, 593 (1979)
6.50   A.Luches, V.Nassisi, A.Perrone, M.R.Perrone: Opt. Commun. **44**, 109 (1982)
6.51   V.S.Zuev, A.V.Kanaev, L.D.Mikheev, D.B.Stavrovskii: Sov. J. Quant. Electron. **11**, 1330 (1981); also Bull. Academy of Sciences USSR (Phys. Series) **46**, 1510 (Russian); **8**, 62 (English) (1982)
6.52   C.A.Brau, J.J.Ewing: J. Chem. Phys. **63**, 4640 (1975)
6.53   J.G.Eden, R.W.Waynant: J. Chem. Phys. **68**, 2850 (1978)
6.54   J.G.Eden, R.W.Waynant: Opt. Lett. **2**, 13 (1978)
6.55   M.Rokni, J.H.Jacob, J.C.Hsia, D.W.Trainor: Appl. Phys. Lett. **35**, 729 (1979)
6.56   D.W.Trainor, J.H.Jacob, M.Rokni: J. Chem. Phys. **72**, 3646 (1980)
6.57   M.Rokni, J.H.Jacob, J.A.Mangano, R.Brochu: Appl. Phys. Lett. **32**, 223 (1978)
6.58   G.Black, R.L.Sharpless, D.C.Lorents, D.L.Huestis, R.A.Gutcheck, T.D.Bonifield, D.A.Helms, G.K.Walters: J. Chem. Phys. **75**, 4840 (1981)
6.59   D.C.Lorents: Intern. Conf. Lasers '79, Orlando, FL (1979); Proc. Intern. Conf. Lasers '81, New Orleans, LA (1981)
6.60   H.C.Brashears, Jr., D.W.Setser: J. Chem. Phys. **76**, 4932 (1982)
6.61   H.C.Brashears, Jr., D.W.Setser: Appl. Phys. Lett. **33**, 821 (1978)
6.62   R.Sauerbrey, W.Walter, F.K.Tittel, W.L.Wilson, Jr.: J. Chem. Phys. **78**, 735 (1983)
6.63   Y.C.Yu, D.W.Setser, H.Horiguchi: J. Chem. Phys. **87**, 2199 (1983)
6.64   D.Kliger, H.H.Nakano, D.L.Huestis, W.K.Bischel, R.M.Hill, C.K.Rhodes: Appl. Phys. Lett. **33**, 39 (1978)
       H.Helm, D.L.Huestis, M.J.Dyer, D.C.Lorents: J. Chem. Phys. **79**, 3220 (1983)
6.65   J.E.Velasco, J.H.Kolts, D.W.Setser: J. Chem. Phys. **69**, 4357 (1978)
6.66   M.C.Smayling: Ph.D. Thesis, Rice University (1981)

6.67   G.Marowsky, G.P.Glass, M.Smayling, F.K.Tittel, W.L.Wilson: J. Chem. Phys. **75**, 1153 (1981)

6.68   G.P.Glass, F.K.Tittel, W.L.Wilson, M.C.Smayling, G.Marowsky: Chem. Phys. Lett. **83**, 585 (1981)

6.69   K.Y.Tang, D.C.Lorents, R.L.Sharpless, D.L.Huestis, D.Helms, M.Durrett, G.K.Walters: 33rd Gas Electronics Conference, Norman, OK (1980); also Lasers '80, New Orleans, LA (1980)

6.70   H.P.Grieneisen, Hu Xue Jing, K.L.Kompa: Chem. Phys. Lett. **82**, 421 (1981)

6.71   P.J.Hay, T.H.Dunning: J. Chem. Phys. **69**, 2218 (1978)

6.72   J.Thone, J.K.Ku, D.W.Setser: To be published in J. Chem. Phys.

6.73   G.Marowsky, R.Sauerbrey, F.K.Tittel, W.L.Wilson, Jr.: Chem. Phys. Lett. **98**, 167 (1983)

6.74   R.Sauerbrey, F.K.Tittel, W.L.Wilson, W.L.Nighan: IEEE J. QE-**18**, 1336 (1982)

6.75   P.J.Chantry: In *Applied Atomic Collision Physics*, Vol. 3: *Gas Lasers*, ed. by H.S.W.Massey, E.W.McDaniel, B.Bederson (Academic, New York 1982) pp. 35–71

6.76   I.N.Konovalov, V.F.Losev, V.V.Ryzhov, V.F.Tarasenko, A.G.Tostremskii: Opt. Spectrosc. (USSR) **47**, 137 (1979)

6.77   F.K.Tittel, R.W.Williams, W.L.Wilson, Jr., G.Marowsky: Appl. Phys. B **28**, 126 (1982)

6.78   W.L.Wilson, Jr., R.A.Williams, R.Sauerbrey, F.K.Tittel, G.Marowsky: J. Chem. Phys. **77**, 1830 (1982)

6.79   C.A.Hart, S.K.Searles: J. Appl. Phys. **47**, 2033 (1976)

6.80   G.Marowsky, G.P.Glass, F.K.Tittel, W.L.Wilson, Jr.: Proc. Intern. Conf. Lasers '81 (1981) p. 15

6.81   G.Marowsky, F.K.Tittel, W.L.Wilson, E.Frenkel: Appl. Opt. **19**, 138 (1980)

6.82   W.E.Ernst, F.K.Tittel: IEEE J. QE-**16**, 945 (1980)

6.83   W.K.Bischel, D.J.Eckstrom, D.L.Huestis, D.C.Lorents: Paper presented at Lasers '79, Orlando, Fl (1979)

6.84   G.Marowsky, M.Munz, F.K.Tittel: IEEE J. QE-**17**, 1281 (1981)

6.85   E.Zamir, D.L.Huestis, H.H.Nakano, R.M.Hill, D.C.Lorents: IEEE J. QE-**15**, 281 (1979); C.F.Champagne: In *Applied Atomic Collision Physics*, Vol. 3: *Gas Lasers*, ed. by H.S.W.Massey, E.W.McDaniel, B.Robinson (Academic, New York 1982) pp. 346–386; A.J.Kearsley, M.C.Gower, C.E.Webb: In *Laser Advances and Applications*, ed. by B.S.Wherrett (Wiley, New York 1980) pp. 61–64

6.86   H.H.Michels, R.H.Hobbs, L.A.Wright: J. Chem. Phys. **71**, 5053 (1979)

6.87   J.A.Vanderhof: J. Chem. Phys. **68**, 3311 (1978)

6.88   R.F.Stebbings, C.J.Latimer, W.P.West, F.B.Dunning, T.B.Cook: Phys. Rev. A **12**, 1453–1458 (1975)

6.89   R.D.Rundel, F.B.Dunning, H.C.Goldwire, Jr., R.F.Stebbings: J. Opt. Soc. Am. **65**, 628–633 (1975)

6.90   W.K.Bischel, D.J.Eckstrom, H.C.Walker, R.A.Tilton: J. Appl. Phys. **52**, 4429 (1981)

6.91   M.Diegelmann, K.Hohla, K.L.Kompa: Opt. Commun. **29**, 334 (1979); also **28**, 104 (1979)

6.92   P.J.K.Wisoff, A.J.Mendelsohn, S.E.Harris, J.F.Young: IEEE J. QE-**18**, 1839 (1982)

6.93   R.W.Waynant, C.P.Christensen, W.M.Bollen, Jr.: In *Excimer Lasers – 1983*, ed. by C.K.Rhodes, H.Egger, H.Pummer (Am. Ins. Phys., New York 1983) pp. 6–11

6.94   K.L.Hohla, T.R.Loree, C.A.Brau, W.E.Stein: Appl. Phys. **25**, 329 (1981)

6.95   F.Collier, G.Thiell, P.Cottin: Appl. Phys. Lett. **32**, 739 (1978)

6.96   D.J.Bradley, D.R.Hull, M.H.R.Hutchinson, M.W.McGeoch: Opt. Commun. **11**, 335 (1974)

6.97   J.Liegel, F.K.Tittel, W.L.Wilson, G.Marowsky: Appl. Phys. Lett. **39**, 369 (1981)

6.98   M.Diegelmann, H.P.Grieneisen, K.Hohla, Xue-Jing Hu, J.Krasinski, K.L.Kompa: Appl. Phys. **23**, 283 (1980)

6.99   R.S.Taylor, A.J.Alcock, K.E.Leopold: Opt. Lett. **5**, 216 (1980)

6.100   S.Sumida, M.Obara, T.Fujioka: Appl. Phys. Lett. **33**, 913 (1978)

6.101   J.I.Levatter, Zaizguang Li: Rev. Sci. Instrum. **52**, 2651 (1981)

6.102   Lin Shao-Chi, Bao Zhi-xiang, Gong Guang-yuan, Huo Yun-Sheng, Shu Ju-ping, Tang Shi-qing, Wei Yun-rong, Zheng Cheng-en: Appl. Phys. Lett. **38**, 328 (1981); also Lasers '82, New Orleans, LA (1982)

6.103 M.Mittelstein: Diploma Thesis, University of Hamburg (1982)
6.104 K.Y.Tang, D.L.Huestis, H.T.Powell: Unpublished
6.105 V.S.Zuev, A.V.Kanaev, L.D.Mikheev, D.B.Stavrovskii: Sov. J. Quant. Electron. **10**, 898 (1980)
6.106 N.G.Basov, V.S.Zuev, A.V.Kanaev, L.D.Mikheev, D.B.Stavrovskii: Sov. J. Quant. Electron. **9**, 629 (1979)
6.107 A.Luches, A.Perrone, A.Giannattasio: Submitted to Opt. Commun.
6.108 J.K.Rice, G.C.Tisone, E.L.Paterson: IEEE J. QE-**16**, 347 (1980)
6.109 G.M.Schindler: IEEE J. QE-**16**, 546 (1980)
6.110 F.K.Tittel, J.Liegel, W.L.Wilson, Jr., G.Zhenhua, G.Marowsky: Proc. Intern. Conf. Lasers '81, New Orleans, LA (1981) p. 8
6.111 W.L.Wilson, F.K.Tittel, W.Walter, R.Sauerbrey, G.Marowsky: Proc. Intern. Conf. Lasers '82, New Orleans (1982)
6.112 R.O.Hunter, J.Oldenettel, C.Howton, M.V.McCusker: J. Appl. Phys. **49**, 549 (1978)
6.113 B.Forestier, B.Fontaine, P.Gross: J. Physique (Paris) **41C9**, 455 (1980)
6.114 H.Shields, A.J.Alcock: Opt. Commun. **42**, 128 (1982)
6.115 R.S.Taylor, A.J.Alcock, K.E.Leopold: Opt. Lett. **5**, 216 (1980)
6.116 R.W.Waynant: Appl. Phys. Lett. **36**, 493 (1980)
6.117 G.Inone, J.K.Ku, D.W.Setser: J. Chem. Phys. **76**, 733 (1982)
6.118 J.Liegel, H.Spiegel, R.Sauerbrey, H.Langhoff: J. Chem. Phys. **79**, 247 (1983)
6.119 W.L.Nighan, Y.Nachshon, F.K.Tittel, W.L.Wilson, Jr.: Appl. Phys. Lett. **42**, 1006 (1983)

7. High-Spectral-Brightness Excimer Systems

H. Pummer, H. Egger, and Ch. K. Rhodes

With 6 Figures

Among the most significant advances made over the past few years are the refinements of the rare-gas-halogen-laser sources in the ultraviolet. This class of ultraviolet lasers exhibits an output in the ultraviolet range, high power, and energy-efficient operation, a combination of three very desirable properties which was not available in any previously known laser system. In addition, it has been experimentally shown that for this class of laser media, the fundamental parameter describing the quality of radiation, known as spectral brightness, can be increased essentially to the maximum theoretically possible. The experimentally demonstrated enhancement in spectral brightness, which is a quantitative measure of the power per unit frequency interval and per unit solid angle, was a factor of approximately 10^{10}. Overall, the discovery of this class of excimer laser media, together with the subsequent technical improvements made in performance, represents a dramatic furtherance of the technology of sources of coherent ultraviolet radiation. Those aspects are reviewed in this chapter. We shall see in Chap. 8 that these high brightness ultraviolet systems have a range of very important applications to physical studies.

7.1 Spectral, Spatial, and Temporal Control of Excimer Laser Radiation

The spectral and spatial qualities of a light source can be expressed by its spectral brightness, a quantity which represents the amount of radiated energy per unit time, surface, solid angle and frequency interval at a given wavelength. For a fixed energy, this quantity has an upper limit because the temporal duration of a pulse and its bandwidth as well as the radiating aperture and the solid angle are connected by well known uncertainty relationships. To give an example, we assume that a relative spectral brightness of 1 is obtained if a laser discharge of $600\,\mathrm{cm^2}$ surface area and 20 ns duration emits 1 J of excimer radiation in the absence of a laser resonator. In this case, the bandwidth of the radiation will be 200–400 cm^{-1} and the radiation will be emitted into $4\pi\,\mathrm{rad^2}$. If the same discharge is brought into a stable resonator, about 0.5 J of radiation can be extracted through a $3\,\mathrm{cm^2}$ front surface of the discharge volume. The pulse duration will be reduced to 10 ns (mainly due to the build up time of laser oscillation), the bandwidth narrows to $100 \, \text{cm}^{-1}$, and the divergence is $\sim(5 \times 10^{-3})^2 \, \text{rad}^2$. The relative spectral brightness is now $\sim 3 \times 10^8$. If, however, the bandwidth and divergence of this laser radiation can be reduced to its fundamental limits, namely $\sim 10^{-2} \, \text{cm}^{-1}$ and $\sim(5 \times 10^{-6})^2 \, \text{rad}^2$, the relative spectral brightness increases to $\sim 3 \times 10^{18}$. This illustrates the tremendous increase in performance associated with appropriate control of the laser radiation.

Approaches to control excimer radiation have been made using two techniques. In the first technique, the laser resonator is modified. The divergence of the radiation can be reduced using unstable resonators or spatial filtering. Narrowing of the laser bandwidth and tunability can be achieved by combinations of intracavity gratings, prisms, etalons, and apertures. The first excimer laser to be tuned in this way was an *e*-beam-pumped prism tuned Xe_2^* laser. *Bradley* et al. [7.1] reported a linewidth of 2.5 Å and a tuning range of ~ 20 Å. In a similar way, *Wrobel* et al. [7.2] have tuned an Ar_2^* laser over a range of 4 nm to generate radiation at L_α using either a grating or a prism. They stated a bandwidth of 1.16 nm. *Loree* et al. [7.3] and *Bokor* et al. [7.4] reported tuning of an ArF* laser with intracavity prisms which yields a bandwidth of ~ 1 Å and a tuning range of 1.3 nm. *Liegel* et al. [7.5] have tuned a $XeF^*(C \rightarrow A)$ laser from 455–529 nm and report a bandwidth of ~ 1 nm using a grating. Narrower bandwidths have been reported by *Andrews* et al. [7.6] for KrF* and XeCl*. Using a grating at grazing incidence, they achieved a bandwidth < 0.1 Å. *Hargrove* et al. [7.7] reported a bandwidth of $0.25 \, \text{cm}^{-1}$ in a nearly diffraction limited beam for ArF*. Their resonator contained two etalons and two prisms.

Control of the laser pulse duration using intracavity devices has been attempted by several groups. *Efthimiopoulos* et al. [7.8] reported passive mode locking of KrF* resulting in pulses of ~ 2 ns duration. *Christensen* et al. [7.9] mentioned active mode locking of XeF* with similar pulse duration. In both cases, the observed modulation was not complete. Using a new method of switching an intracavity Pockel's cell, *Reksten* et al. [7.10] have generated XeCl* laser pulses of ~ 300 ps duration. Such fast Pockel's cells could find use in extracavity pulse slicing, as stated by *Pacala* et al. [7.11].

These examples show that substantial tunability, spectral narrowing, and short pulses can be achieved by using intracavity spectral, spatial, and temporal filters. This general technique, however, has two major disadvantages. Due to the very limited number of cavity round-trips which are possible while the gain in an excimer laser is present, the desired radiation generally has insufficient time to build up and dominate the stimulated emission in the cavity. A narrow linewidth can, therefore, only be achieved with the use of several narrow bandwidth elements and the penalty for this is a severe loss of energy. Furthermore, the alignment and tuning of a multi-element cavity can become very involved, especially when an unstable resonator for divergence reduction is used. In cases, however, for which substantial energy extraction in a low divergence beam without special requirements for laser wavelength and bandwidth is needed, an unstable resonator may be a good choice. Similarly, an intracavity prism or grating may be the best choice whenever tunability is needed and moderate bandwidths (several 10^{-1} Å) and moderate energy extraction can be tolerated. It should be noted that experiments which are aimed at lengthening excimer-laser pulses have resulted in a 100 ns microwave discharge pumped XeCl* laser [7.12]. If further progress is made in this direction, intracavity techniques will certainly become of more importance.

The second technique tends to be more complex and requires considerably more apparatus. However, it provides the possibility to extract the full energy stored in an excimer laser in a laser pulse which is tunable, diffraction and bandwidth limited, and whose absolute wavelength can be measured to better than 1 part in 10^7. In addition, laser pulses of several ps duration are currently being produced and the production of sub-ps pulses can be expected soon. The different approaches all have in common the feature that radiation of the desired spectral and spatial properties is first produced by some other means and then used to control the stimulated processes in the excimer medium. This can be done by either injecting it into the resonator of the excimer laser or by amplifying it in one or more passes through the excimer medium. The power requirement for the primary radiation is simply that it has to exceed the excimer noise (spontaneous emission) into the resonator modes (injection locking) or its own spatial modes (amplification). On general grounds [7.13], this minimum power P_0 can be shown to be quite low with $P_0 \sim 8\hbar\omega\,\Delta\omega$, in which $\hbar\omega$ is the quantum energy and $\Delta\omega$ is the bandwidth available for amplification. For cases of interest, $P_0 < 10^{-3}$ W.

There are several ways to prepare the primary radiation. We briefly summarize the various approaches and then describe certain systems in more detail. Radiation used to lock excimer amplifiers or oscillators can be generated by controlling the emission of a small excimer oscillator with the techniques described above. As the radiation is not used directly, this can be done at the cost of energy losses and even extracavity filtering with etalons may to a certain degree be acceptable. *Goldhar* et al. [7.14] and *Murray* et al. [7.15] used an intracavity line-narrowed oscillator to injection lock a high energy unstable resonator oscillator. These researchers reported linewidths of $\sim 10^{-1}$ cm^{-1} in a beam with a divergence of approximately twice the diffraction limit. *Bigio* et al. [7.16] have injection locked an unstable resonator KrF* laser. They reported tunable pulses of <0.1 Å bandwidth with divergence close to the diffraction limit and an energy of 150 mJ. The same authors mentioned locking of an unstable resonator XeF* laser with the 3511 Å line of an Ar$^+$ laser [7.17]. They obtained diffraction-limited pulses of >1 MW power and ~ 50 MHz bandwidth. *Tomov* et al. [7.18] have tripled Nd:glass laser pulses of 200 ps duration and amplified them in XeF*.

Clearly the most complete and most versatile control of excimer radiation is obtained if dye lasers are used to generate the primary radiation. Dye lasers allow the highest degree of spectral, spatial, and temporal control of all existing lasers. If these properties can be transferred to the high power excimer systems, a unique light source in the vuv is made possible. The first successful application of this method was reported by *Hawkins* et al. [7.19].

7.2 KrF* System (248 nm)

As noted above, rare-gas-halogen sources have generally been limited by (i) a broad ($\sim 100\,\mathrm{cm}^{-1}$) emission profile, (ii) the absence of a convenient, accurate, and reliable tuning system for control of the output wavelength, and (iii) an output beam divergence on the order of one hundred times the diffraction limit. Overall, an enhancement of many orders of magnitude in spectral brightness is achievable if the output parameters of these sources could be made to conform to the most stringent limits fundamentally possible.

Early studies [7.14, 15] showed that substantial improvement in source properties can be obtained by using the output of a tunable, intracavity etalon line-narrowed, discharge-excited oscillator to injection lock a high-energy, unstable resonator oscillator. The best linewidth reported was $\sim 10^{-1}\,\mathrm{cm}^{-1}$ in a beam with a divergence approximately twice the diffraction limit [7.15].

More recently, the properties of a KrF* excimer source with performance parameters which closely approach the fundamental limits governing spectral width, beam divergence, and absolute wavelength control were achieved. The basic laser system is illustrated schematically in Fig. 7.1. The output of a frequency-stabilized, cw dye laser (Coherent 599–21, $\Delta v < 5\,\mathrm{MHz}$, $\sim 30\,\mathrm{mW}$ at

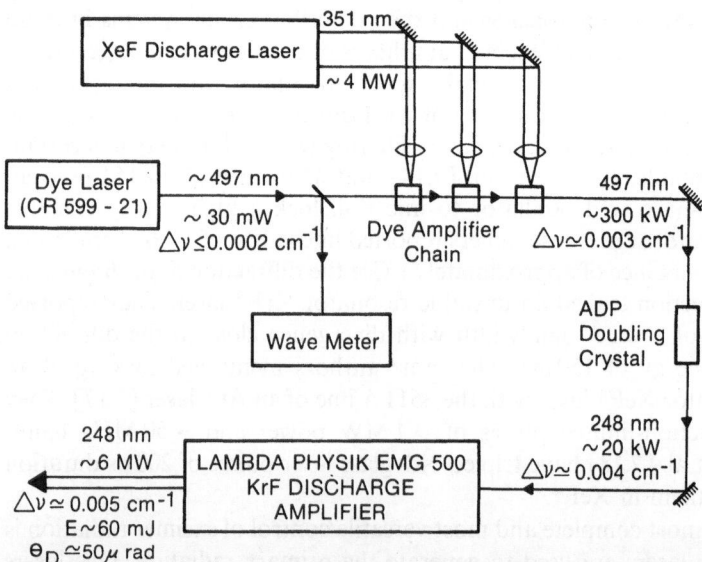

Fig. 7.1. Ultraviolet laser system illustrating the cw dye laser, the dye amplifier chain, the ADP doubling crystal, and the final KrF* amplifier stage

500 nm) is pulse amplified in a 3 stage XeF* pumped (30–50 mJ at 351 nm) dye amplifier, producing a 7-ns visible pulse with an energy of ~ 3 mJ at a repetition rate up to $1 \, \text{s}^{-1}$ (repetition rates up to ~ 1 kHz should be feasible). The linewidth of this visible radiation was examined with a scanning interferometer (Tropel 240) and was found to be 85 ± 10 MHz. Frequency-doubled radiation corresponding to any wavelength within the KrF* gain profile may readily be generated in a temperature-tuned, 90° phase-matched ADP crystal, producing ~ 5 ns second-harmonic pulses with energy $> 100 \, \mu$J. This spectrally narrow (120 ± 20 MHz FWHM, Fourier transform limited) second-harmonic radiation is subsequently amplified in a single pass through a discharge-pumped KrF* amplifier (Lambda Physik EMG 500) to produce output pulses of ~ 10-ns duration, and energies up to 60 mJ.

A study of the spatial properties of the 2×0.5-cm-output beam reveals it to have a diffraction limited divergence of $\sim 50 \, \mu$rad. It is noted that at the peak measured power, ~ 6 MW, focal intensities at 248 nm in excess of $10^{15} \, \text{W/cm}^2$ can be generated with f1 optics.

Furthermore, since the frequency of the cw dye laser can be electronically scanned over $1 \, \text{cm}^{-1}$, continuously tunable coverage of the ultraviolet radiation over a 2-cm^{-1} interval is readily accomplished. In addition, the dye laser frequency is known to within 300 MHz by interferometric comparison with a stable HeNe laser source [7.20]. In these experiments, amplification was observed at wavelengths λ_{KrF} in the range 248.2 nm $< \lambda_{\text{KrF}} <$ 250.3 nm. In this range, the output pulse energy was observed to monotonically decrease with increasing wavelength. In contrast to previous reports of a strong absorption centered at ~ 248.8 nm in both e-beam pumped KrF* amplifiers [7.21] and discharge-pumped tunable KrF* lasers [7.3, 14], no reduction in output energy was observed at that wavelength in our apparatus.

In order to establish the linewidth of the source, direct interferometry was used. A 1 GHz FSR Fabry-Perot interferometer was constructed with two dielectric coated mirrors of 98 % reflectivity, resulting in a finesse of ~ 20 for our typical 10-ns pulse duration. With this interferometer, the amplified ultraviolet linewidth was measured to be $\Delta \nu_{\text{KrF}} = 150 \pm 30$ MHz. This linewidth is equivalent, within the experimental uncertainty, to the Fourier-transform-limited linewidth $\Delta \nu_{\text{SHG}} = 120 \pm 20$ MHz of the second-harmonic output of the ADP crystal.

In order to further verify the narrow linewidth of the laser output, a preliminary study of the xenon $(6p[1/2]_0 \leftarrow {}^1 S)$ two-photon absorption at 249.63 nm was also performed under Doppler-free conditions through detection of the resulting $(6p[1/2]_0 \rightarrow 6s[3/2]_1)$ fluorescence at 828 nm. Upon scanning the KrF* source frequency across the resonance, a single feature was observed, centered at $80{,}118.73 \pm 0.10 \, \text{cm}^{-1}$, with a width of ~ 450 MHz (FWHM, at the KrF* frequeney), as shown in Fig. 7.2. These measurements were made at a xenon pressure of 0.25 Torr and an ultraviolet intensity of $\sim 5 \times 10^5 \, \text{W/cm}^2$. Although this width is a factor of 3 less than the Doppler width, the resonance is still a factor of ~ 3 broader than would be expected with

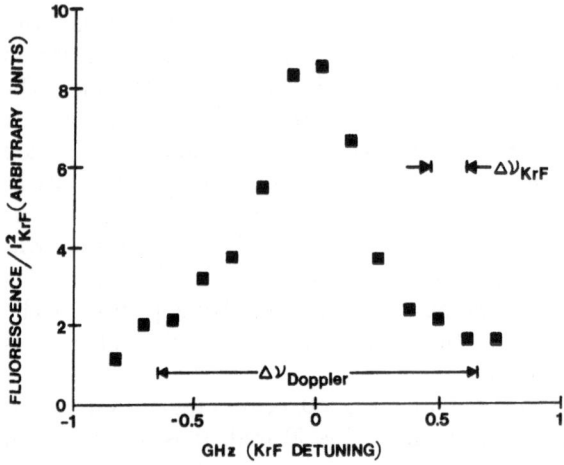

Fig. 7.2. Observed fluorescence at 828 nm following two-photon absorption in Xe($6p[1/2]_0 \rightarrow {}^1 S$) at 249.63 nm, 0.25 Torr Xe pressure, and $\sim 5 \times 10^5$ W/cm² KrF* intensity (unfocussed)

a source linewidth of ~ 150 MHz. Since the experimentally determined pressure broadening coefficient of this xenon two-quantum transition was determined to be 25 ± 15 MHz/Torr, which gives a minor contribution to the linewidth at 0.25 Torr, it was concluded that unresolved isotope splittings (five xenon isotopes with natural abundances $> 8 \%$) lead to the observed width. An estimate [7.22] of the isotope shift for such a ($5p^5 6p \leftarrow 5p^6$) transition yields a relative shift of ~ 50–100 MHz for two Xe isotopes whose masses differ by 1 amu. This estimate agrees well with both the linewidth of the observed absorption, and the absence of additional resonances within ± 6 GHz of the one observed.

These results demonstrated that a KrF* excimer source, continuously tunable over the KrF* band, and operating at the fundamental limits of spectral brightness, could be constructed [7.19]. Although this demonstration was made at a power of ~ 6 MW, scaling to far higher power is clearly implied. This aspect will be raised below in Sect. 7.4 concerning picosecond systems.

7.3 ArF* System (193 nm)

Ultrahigh spectral brightness operation can also be achieved [7.23] with the ArF* system at 193 nm. The laser system developed to produce a spectral brightness close to the fundamental limits is illustrated schematically in Fig. 7.3. The output of a frequency-stabilized, cw dye laser (Coherent 599–21, $\Delta \nu < 5$ MHz, ~ 50 mW at 580 nm) was pulse amplified in a three-stage XeF* pumped dye amplifier, producing a ~ 10-ns visible pulse with an energy greater than 10 mJ at a repetition rate of up to 10 Hz. The linewidth of such pulses, as examined with a scanning interferometer (Tropel 240), was found to be Fourier transform limited. Frequency-tripled radiation corresponding to any wave-

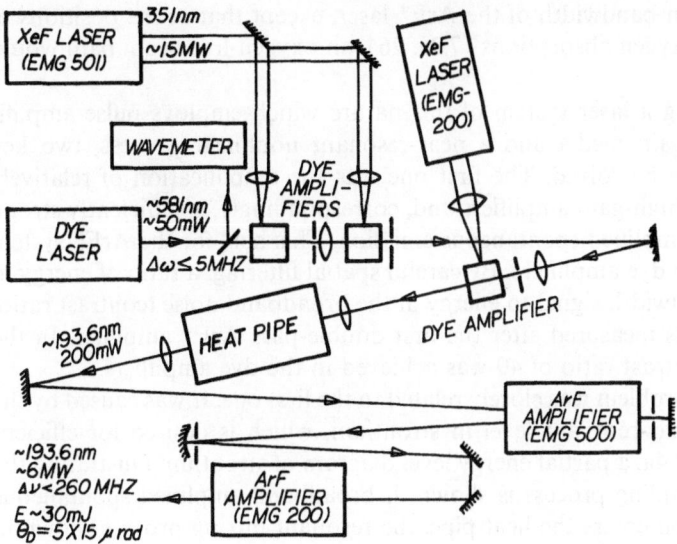

Fig. 7.3. ArF* laser system with cw dye laser, dye amplifier chain, strontium heat pipe, and ArF* amplifier

length within the ArF* gain profile is readily generated by focusing into a strontium heat pipe [7.24] producing ~5-ns third-harmonic pulses with peak power greater than 200 mW. This spectrally narrow third-harmonic radiation was subsequently amplified in a double and a single pass through two discharge-pumped ArF* amplifiers to produce output pulses ~5 ns in duration and energies up to 30 mJ.

In order to establish the linewidth of the source, a 1.5-GHz FSR Fabry-Perot interferometer was constructed with two dielectric-coated mirrors of 95 % reflectivity, resulting in a finesse of ~10. With this interferometer, the observed fringe width was one-fifth of the fringe separation, indicating a 300-MHz bandwidth. This bandwidth results from the convolution of the interferometer transmission ($\Delta\nu_{FP} = 150$ MHz) and the laser line. If the laser linewidth is assumed to be Fourier transform limited (180 MHz for a Gaussian 5-ns pulse), the total linewidth would be 235 MHz for Gaussian and 330 MHz for Lorentzian line shapes.

A study of the spatial properties of the 3×1-cm^2 output beam revealed it to have a diffraction-limited divergence of $\sim 5 \times 15$ µrad. This enables the production of focal intensities at 193 nm in excess of 10^{15} W/cm^2 with aspheric f2 optics.

Since the frequency of the cw dye laser can be electronically scanned over 1 cm^{-1}, continuously tunable coverage of the ultraviolet radiation over a 3-cm^{-1} interval is readily accomplished. In addition, the dye laser frequency is known to within 600 MHz by interferometric comparison with a stable HeNe laser source [7.20]. The 3-cm^{-1} interval can be set anywhere within the

~ 250-cm^{-1} gain bandwidth of the ArF* laser, except that at the positions of the molecular oxygen absorptions [7.25, 26], an oxygen-free beam path would be needed.

In developing a laser system of this nature which employs pulse amplification in high-gain media and a near-resonant nonlinear process, two key problems had to be solved. The first one was the amplification of relatively weak signals in high-gain amplifiers and, correspondingly, a sufficiently strong suppression of amplified spontaneous emission. This held for the ArF* system as well as for the dye amplifiers. By careful spatial filtering, a ratio of energy in the narrow-bandwidth signal to energy in the broadband noise (contrast ratio) of up to 300 was measured after the first double-pass ArF* amplifier. In the same way, a contrast ratio of 40 was achieved in the dye amplifiers.

The second problem was closely related to the first one. It was caused by the presence of a near-resonant level in strontium, which is desired for efficient tripling. In Fig. 7.4a, a partial energy level diagram of strontium illustrating the near-resonant tripling process is shown. If broadband amplified spontaneous dye laser emission enters the heat pipe, the resonant mixing process shown in Fig. 7.4b is also possible. For the tripling process to be dominant, the following condition must be fulfilled:

$$E_{580.2}/E_N > \Delta v^2/\Delta v_D \Delta v_L. \tag{7.1}$$

Here, $E_{580.2}$ is the energy in the narrow bandwidth at 580.2 nm and E_N is the total amplified spontaneous emission in the same spatial mode. With a detuning $\Delta v = 257$ cm^{-1}, a Doppler width $\Delta v_D \approx 0.1$ cm^{-1}, and a bandwidth

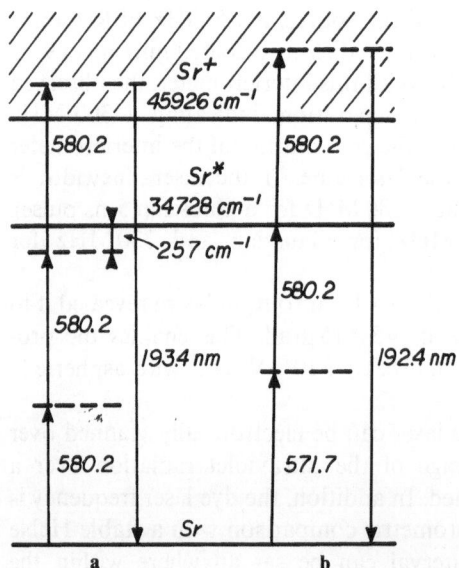

Fig. 7.4a, b. Near-resonant tripling process (a) in strontium and (b) competing resonant mixing

$\Delta v_L = 300 \, \text{cm}^{-1}$, of the spontaneous emission, the condition for the contrast ratio is given by

$$E_{580.2}/E_N > 2 \times 10^3. \tag{7.2}$$

In order to obtain such a high contrast ratio, a grating was put between the second and third amplifier stages. In this way the contrast ratio was increased markedly, so that any light produced at 192.4 nm was below the detection limit.

The results described above clearly demonstrated that an ArF* (193 nm) excimer system, tunable over the bandwidth available for amplification, and operating at the fundamental limits of spectral brightness, can be implemented. Since no scaling limit concerning output power was encountered, the generation of considerably higher powers is also implied in this case as with the KrF* (248 nm) system described in Sect. 7.2 above. Indeed, subsequent experiments involving an additional ArF* amplifier produced transform limited outputs of ~ 300 mJ. We note that tunable high brightness operation of ArF*, with a linewidth $\sim 1 \, \text{cm}^{-1}$, has been observed [7.27] with the use of anti-Stokes Raman scattering from molecular hydrogen.

7.4 Picosecond Rare-Gas-Halogen Sources

The discussion above in Sects. 7.2 and 3 indicated that the observed behavior of the high brightness KrF* (248 nm) and ArF* (193 nm) sources did not reveal, in the intensity range ~ 10 MW/cm^2, any limitation affecting scaling to higher intensity and power levels. Given the intrinsic bandwidth characteristic of rare-gas-halogen systems ($\sim 100 \, \text{cm}^{-1}$), considerable interest centers on the properties of picosecond amplification. Since the stimulated emission cross-sections σ [7.28] of these media are generally $\sigma > 10^{-16} \, \text{cm}^2$, saturation energies are in the range of $\sim 10^{-2} \, \text{J/cm}^2$, a very modest level. Operation with pulses of a duration ~ 1–10 ps then implies intensities on the scale of ~ 1–10 GW/cm^2, a magnitude sufficiently low that severe losses in window materials, optics, and the amplifying medium are not expected. Although the bandwidth ($\sim 100 \, \text{cm}^{-1}$) is sufficient to support a pulse width of ~ 0.1 ps, estimates [7.29] indicate that the presence of dispersion, although not representing a rigid fundamental limitation, will make practical operation with pulse lengths shorter than ~ 1 ps difficult.

High power picosecond operation has been achieved with both KrF* (248 nm) [7.30] and ArF* (193 nm) [7.31]. In both cases, the approach used was similar to that employed in the systems described in Sects. 7.2 and 3; an ultraviolet seed beam, generated by nonlinear scattering of high brightness visible radiation, is used to control the spectral and spatial properties of the amplified ultraviolet energy. The ArF* (193 nm) is described below; for the details of the corresponding KrF* (248 nm) system, the reader is referred to the literature [7.30].

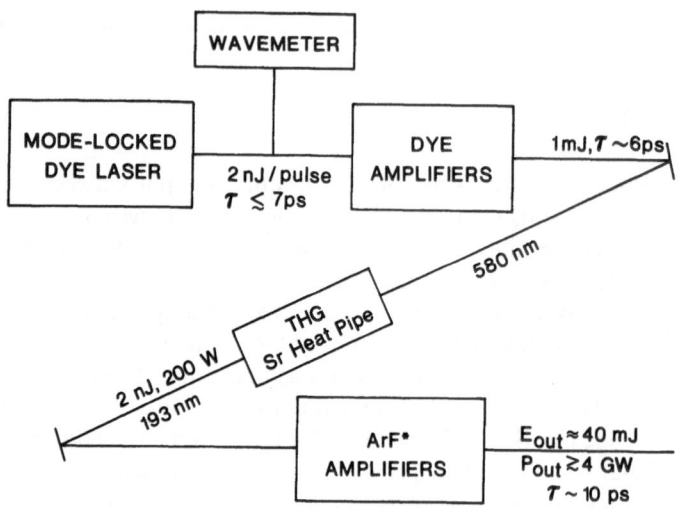

Fig. 7.5. Schematic diagram of 10-ps ArF* laser

A ~4-GW 193-nm ArF* excimer laser source with a characteristic output energy of ~30 ± 10 mJ, pulse duration ~10 ps, bandwidth ~5 cm^{-1}, and tunable over the full bandwidth of the ArF* gain profile is schematically shown in Fig. 7.5. In the configuration shown, the output of a synchronously pumped, mode-locked dye laser (Coherent Radiation 599–04, $\lambda = 580$ nm, pulse duration ~6 ps) was pulse amplified in a three-stage, XeF* excimer laser pumped dye amplifier. The amplified spontaneous emission arising from the dye amplifiers was suppressed by two 250-μm-thick saturable absorber cells which were installed between two consecutive amplifier stages. The most stable operation of the amplifier chain has been achieved with DQOCI (1×10^{-3} mol/l in methanol) as the saturable absorber. Besides DQOCI, DODCI and malachite green have been tested in different solvents. Under typical conditions, the 580-nm output pulse had an energy of ~1 mJ. The pulse duration of the visible pulse was measured with a streak camera (Hadland Photonics IMACON 500) and was determined to be ~5 ps.

The 1-mJ, 580-nm output pulse was focussed with a 35-cm focal length lens (f/50) into a strontium heat pipe, in which ~2 nJ of the third harmonic at 193 nm is produced. To obtain a spectrally clean output from the heat pipe, the broad-band background in the visible beam must be strongly suppressed. For this purpose a grating-pinhole combination with 12-cm^{-1} bandpass was installed between the second and the third dye amplifiers. With this filter installed, no spectral distortion of the third-harmonic output could be observed when monitored with a monochromator and optical multichannel analyzer, implying an upper limit for bandwidth and frequency jitter of the 193-nm pulse of 10 cm^{-1}. However, a temporal spread of the pulse is introduced by the inclusion of a grating [7.32]. In the present case, with a grating constant of

500 l/mm and a beam diameter of 3 mm, this spread is 3.5 ps and has been taken into account in the interpretation of the data from the autocorrelation measurements. By using a grating pair, this temporal elongation of the pulse can be eliminated and a reduction in the pulse width of 3.5 ps can be expected.

The 2-nJ, 193-nm output from the heat pipe is initially amplified in a double-pass ArF* amplifier. A spatial filter and a grating-pinhole combination with a 25-cm^{-1} bandpass is used to suppress amplified spontaneous ArF* emission. After further amplification in a single-pass amplifier, the energy in the short pulse is typically 5 mJ with an additional 5 mJ contained in amplified spontaneous emission. After final amplification in a second single-pass amplifier, 30 ± 10 mJ are typically measured in the short pulse and ~ 200 mJ in the amplified spontaneous emission. Short uv laser pulses have recently been generated by quenching of resonator transients [7.33].

It is important to note that the short-pulse energy observed represents a substantial fraction of the maximum energy available for extraction (~ 50 mJ) in a pulse shorter than the excimer lifetime. This level of extraction is direct evidence against the presence of a significant nonlinear loss mechanism affecting the amplification of the picosecond pulse up to an intensity of ~ 1 GW/cm^2. The duration of this final output pulse is determined to be ~ 10 ps, as shown in Fig. 7.6.

In addition to the temporal behavior of the ArF* pulse, the coherence properties of the beam were studied and compared to those of the 580-nm visible beam. In both cases, interference fringes were visible up to a maximum delay of ± 3 ps, indicating comparable coherence lengths for both the visible and ultraviolet pulses. Correspondingly, the bandwidth of the 193-nm pulse is estimated to be ~ 5 cm^{-1}.

The availability of spectrally bright picosecond rare-gas-halogen systems operating at the gigawatt level has a wide range of extremely important consequences for a wide range of scientific applications. Prominent among them are, naturally, nonlinear phenomena, particularly those associated with short wavelength generation. These aspects are examined in Chap. 8.

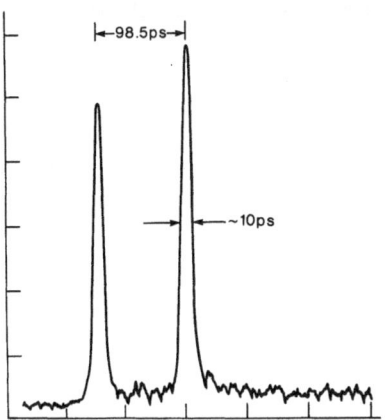

Fig. 7.6. Streak camera recording of ArF* laser emission after the final amplifier at high temporal resolution showing actual pulse duration

References

7.1   D.J.Bradley, D.R.Hull, M.H.R.Hutchinson, M.W.McGeach: Opt. Commun. **14**, 1 (1975)
7.2   W.G.Wrobel, H.Röhr, K.H.Steuer: Appl. Phys. Lett. **36**, 113 (1980)
7.3   T.R.Loree, K.B.Butterfield, D.L.Barker: Appl. Phys. Lett. **32**, 171 (1978)
7.4   J.Bokor, J.Zavelovich, C.K.Rhodes: Phys. Rev. A**21**, 1453 (1980)
7.5   J.Liegel, F.K.Tittel, W.L.Wilson, Jr., G.Marowsky: Appl. Phys. Lett. **39**, 369 (1981)
7.6   A.J.Andrews, A.J.Kearsley, M.C.Gowen, C.E.Webb: In *Digest of Topical Meeting on Excimer Lasers*, IEEE Cat. No. 79CH1470-4QEA (IEEE, New York 1979)
7.7   R.S.Hargrove, J.A.Paisner: In *Digest of Topical Meeting on Excimer Lasers*, IEEE Cat. No. 79CH1470-4QEA (IEEE, New York 1979)
7.8   T.Efthimiopoulos, J.Banic, B.P.Stoicheff: Can. J. Phys. **57**, 1437 (1979)
7.9   C.P.Christensen, L.W.Braverman, W.H.Steiner, C.Wittig: Appl. Phys. Lett. **29**, 424 (1976)
7.10 G.Reksten, T.Varghese, W.Margulis: Appl. Phys. Lett. **39**, 129 (1981)
7.11 T.J.Pacala, J.B.Laudenslager, C.P.Christensen: Appl. Phys. Lett. **37**, 366 (1980)
7.12 A.J.Mendelsohn, R.Normandin, S.E.Harris, J.F.Young: Appl. Phys. Lett. **38**, 603 (1981)
7.13 H.Egger, H.Pummer, C.K.Rhodes: Laser Focus **18**, 59 (1982)
7.14 J.Goldhar, J.R.Murray: Opt. Lett. **1**, 199 (1977)
7.15 J.R.Murray, J.Goldhar, A.Szöke: Appl. Phys. Lett. **32**, 551 (1978)
7.16 I.J.Bigio, M.Slatkine: Opt. Lett. **6**, 336 (1981)
7.17 I.J.Bigio, M.Slatkine: To be published
7.18 I.V.Tomov, R.Fedosejews, M.S.Richardson, W.J.Sargeant, A.J.Alcock, K.E.Leopold: Appl. Phys. Lett. **30**, 147 (1977); **31**, 747 (1977)
7.19 R.T.Hawkins, H.Egger, J.Bokor, C.K.Rhodes: Appl. Phys. Lett. **36**, 391 (1980)
7.20 F.V.Kowalski, R.T.Hawkins, A.L.Schawlow: J. Opt. Soc. Am. **66**, 965 (1976)
7.21 A.M.Hawryluk, J.A.Mangano, J.H.Jacob: Appl. Phys. Lett. **31**, 164 (1977)
7.22 I.I.Sobel'man: *Atomic Spectra and Radiative Transitions*, Springer Ser. Chem. Phys., Vol. 1 (Springer, Berlin, Heidelberg, New York 1979)
7.23 H.Egger, T.Srinivasan, K.Hohla, H.Scheingraber, C.R.Vidal, H.Pummer, C.K.Rhodes: Appl. Phys. Lett. **39**, 37 (1981)
7.24 C.R.Vidal: Appl. Opt. **19**, 3897 (1980)
7.25 M.Ackerman, F.Biaume: J. Mol. Spectrosc. **35**, 73 (1970)
7.26 K.Yoshino, D.E.Freeman, J.R.Esmond, W.H.Parkinson: Planetary and Space Science (in press)
7.27 J.C.White, J.Bokor, R.R.Freeman, D.Henderson: Opt. Lett. **6**, 293 (1981)
7.28 I.S.Lakoba, S.I.Yakovlenko: Sov. J. Quantum Electron. **10**, 389 (1980)
7.29 H.Egger: Private communication
7.30 P.H.Bucksbaum, J.Bokor, R.H.Storz, J.C.White: Opt. Lett. **7**, 399 (1982)
7.31 H.Egger, T.S.Luk, K.Boyer, D.F.Muller, H.Pummer, T.Srinivasan, C.K.Rhodes: Appl. Phys. Lett. **41**, 1032 (1982)
7.32 R.Wyatt, E.E.Marinero: Appl. Phys. Lett. **25**, 297 (1981)
7.33 F.P.Schäfer, Lee Wenchong, S.Szatmari: Appl. Phys. B**32**, 123 (1983)

8. Applications of Excimer Systems

K. Hohla, H. Pummer, and Ch. K. Rhodes

With 11 Figures

This chapter examines several areas of scientific application of excimer lasers, Since this is expected to be a rapidly developing field in coming years, the examples that have been selected should be regarded only as illustrative of this area rather than a comprehensive register.

8.1 Background

Depending upon the nature of the application and the particular system used, excimers can be utilized as efficient sources of either incoherent or coherent quanta. In the former role, they would serve basically as relatively narrow-band flashlamps. For rare-gas systems, such use has been made. For the latter, however, the radiation would be spatially coherent and possess a bandwidth typically much narrower than the spontaneous linewidth. The flashlamp or incoherent mode has certain technological advantages, since it is applicable to low gain or even weakly absorbing media and does not require optical elements of demanding specifications (e.g., high reflectivity and high damage threshold). On the other hand, a coherent source can generate radiation of far higher intensity and clearly can be spatially controlled with much greater precision. In the following discussion we shall concentrate on the laser mode.

Although presently the commercial applications of excimer systems are not fully developed, we should except that situation to be appreciably changed within the next three to five years. Indeed, we can reasonably anticipate a growth in the applications sector paralleling the past development of the molecular CO_2, CO, and HF infrared laser systems utilizing vibrational-rotational transitions and their associated technologies. The efficiencies of the infrared lasers may be as high as 30–50%, depending on the medium and its mode of operation. Although at the present time most excimer systems operate at efficiencies of $\sim 1\%$, HCl lasers with 3–4% efficiency are commercially available and efficiencies in the region of $\sim 5\%$ appear possible under appropriate circumstances.

An example in which excimer systems have a unique role is in the area of frequency upconversion, examples of which will be discussed in Sect. 8.2. Fields in which excimer lasers have already found extensive applications, or in which such applications can safely be predicted, include short-wavelength generation, laser pump sources, laser spectroscopy, laser chemistry, picosecond phenomena, material processing, and biology.

8.2 Wavelength Conversion of Excimer Lasers to the Vacuum and Extreme Ultraviolet Regions

A basic and long standing problem [8.1] in the field of coherent sources is that associated with the generation of coherent energy in the short wavelength range including the vacuum ultraviolet (vuv), the extreme ultraviolet (xuv), and the soft x-ray regions. Nominally, spectrally bright coherent sources of wavelengths $\lambda < 100$ nm are desired. Such a technology would serve a wide range [8.2] of important applications in both pure and applied scientific areas, including microlithography [8.3], solid-state surface characterization [8.4], trace element analysis [8.5], fusion plasma diagnostics [8.6], the analysis of thin films [8.7], and microholography [8.8]. In general, we conclude that a wide range of important applications would be benefited by the availability of coherent sources of short wavelength radiation.

It is generally understood [8.4, 9] that a high-power state-selective excitation source is the key requirement for successful generation of coherent high-spectral-brightness radiation in the xuv and x-ray ranges. The general lack of coherent sources below 100 nm is, therefore, directly attributable to the absence of the appropriate technology for excitation. Basically, a high-brightness technology is needed which can be configured in a manner appropriate for the excitation of matter radiating in the x-ray.

Recent experimental studies show that bright tunable radiation in the 30–200 nm spectral range can be generated with the use of high-brightness ultraviolet (uv) rare gas halogen (RGH) laser systems. Fundamentally, this approach is predicated on the extraordinary advances made in the spectral brightness of RGH media in recent times [8.10–13] with an experimentally demonstrated [8.12–14] enhancement in spectral brightness of approximately a factor of 10^{10}. The basic nature of these technical results was reviewed in Chap. 7. In addition, since nonlinear coupling is common to all processes involving frequency upconversion of RGH radiation, a premium value is placed on the ability to produce outputs of maximal power and intensity.

High-spectral-brightness sources can be used to generate coherent extreme ultraviolet radiation by either harmonic generation mechanisms or direct multiquantum excitation of appropriate gain media. Experiments over the past two years have indicated [8.15] that direct radiative excitation of an amplifying medium [8.16] by multiquantum processes probably represents the more efficient path for conversion to short wavelengths. It is a primary physical fact that a very strong scaling law favors the use of short pulses, in the picosecond range, for the excitation of amplifying media with nonlinear processes. Indeed, under common circumstances, the optical gain produced by an Nth order multiquantum excitation with a pulse of length τ scales as τ^{-N}, an extremely influential dependence that has enabled efficient up-conversion with a two quantum process in recent experiments [8.16]. Moreover, it has been shown in other recent studies [8.17, 18], under collision-free conditions, that multiquantum processes of unusually high order, one involving as many as 99 quanta [8.18] and an equivalent excitation energy of ~ 633 eV, can occur with unexpectedly high rates with the use of the high intensities achievable with high brightness picosecond RGH technology.

8.2.1 Harmonic Generation

Naturally, the high intensities available with the 10 ps output favor the observation of nonlinear processes. Therefore, harmonic generation in gaseous media serves as a simple and effective means for the production of radiation in the range below 100 nm. Preliminary experiments [8.19] examining the nonlinear scattering in various gases such as H_2, He, Ne, Ar, N_2, and CO indicate substantial production of both third (64 nm) and fifth (39 nm) harmonic radiation. Table 8.1 summarizes the results obtained in these early measurements. The maximum powers observed at the third harmonic (64 nm) and the fifth harmonic (39 nm) were ~ 20 kW and 200 W, respectively.

The density dependence of the nonlinear intensity is revealing. All neutral gaseous materials, with the exception of He and Ne, will present a loss at 64 nm arising from photoionization. The presence of this loss, for a given focal geometry, establishes a fundamental upper limit on the density of the nonlinear medium for optimum conversion. As shown in Fig. 8.1, N_2 and Ar both exhibit diminishing 64 nm output at densities above $\sim 1.5 \times 10^{18}$ cm^{-3} as a consequence of the absorption from photoionization. However, since the losses arising from photoionization at 64 nm in neon are totally absent, a different behavior is seen. Pressure-induced absorption [8.20] at 64 nm will occur at sufficiently high density in a manner similar to that observed in krypton, as well as many other materials, but this is not expected to be an appreciable effect for densities less than $\sim 2 \times 10^{20}$ cm^{-3}. With the use of a high-pressure pulsed valve for the

Table 8.1. Harmonic generation with 193 nm, 10 ps pulses

Gas	xuv power [W]	
	3rd harmonic	5th harmonic
H_2	2×10^4	~ 2
He	20	2
Ne	20	20
Ar	2×10^4	200
N_2	200	20
CO	20	0

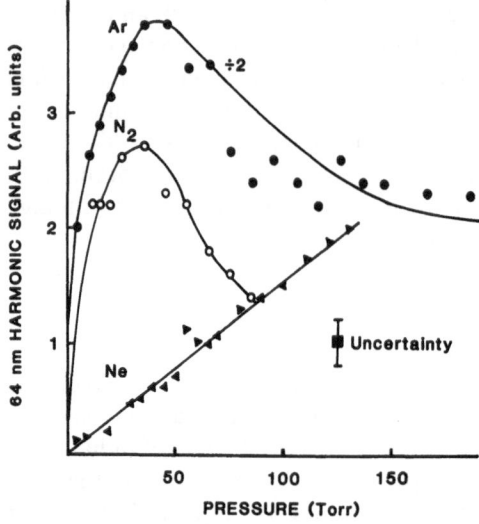

Fig. 8.1. Intensity of third-harmonic (64 nm) signal as a function of medium density using 10 ps 193 nm radiation

production of the nonlinear medium, the scaling behavior indicated for neon in Fig. 8.1 implies the generation of xuv power levels in the megawatt range. We note, indeed, that pulsed valves have been used to generate harmonic radiation [8.21] to produce wavelengths as short as 35.5 nm, the seventh harmonic of KrF* (248 nm). In addition, extended tunability has been demonstrated at 79 nm with frequency mixing techniques [8.22] as well as the capability for extremely high resolution ($\sim 0.1 \, \mathrm{cm}^{-1}$) in spectroscopic measurements [8.23] at 83 nm.

8.2.2 Stimulated Emission from Multiphoton Excited Inversions

Direct radiative excitation by multiquantum processes can be used to generate amplification in the vuv and xuv regions. Processes of this nature, which have previously been observed in the infrared [8.24], can now be readily extended to the shorter wavelength region with high spectral brightness picosecond RGH sources.

Molecular hydrogen represents the essential paradigm of this basic mechanism in the vuv range. The previously [8.25] studied $X^1\Sigma_g^+ \to E$, $F^1\Sigma_g^+ Q(2)$ transition in H_2 falls in a region in which exact resonance can be achieved for an allowed two-quantum excitation with 193 nm radiation. For irradiation with 10 ps pulses, estimates indicate that approximately 1 % of the H_2 ground state population can be transferred to the excited state at an intensity of $\sim 10^{11} \, \mathrm{W/cm}^2$ when proper account is made for losses due to photoionization of the $E, F^1\Sigma_g^+$ level at 193 nm. Therefore, at a medium density of $6 \times 10^{19} \, \mathrm{cm}^{-3}$, with an allowance for rotational partition, an inversion density of $\sim 10^{17} \, \mathrm{cm}^{-3}$ on the $E, F^1\Sigma_g^+ \to B^1\Sigma_g^+$ transition can be generated. Since the cross-section for the $E \to B$ transition $\sigma_{EB} \cong 6 \times 10^{-14} \, \mathrm{cm}^2$, an enormous gain

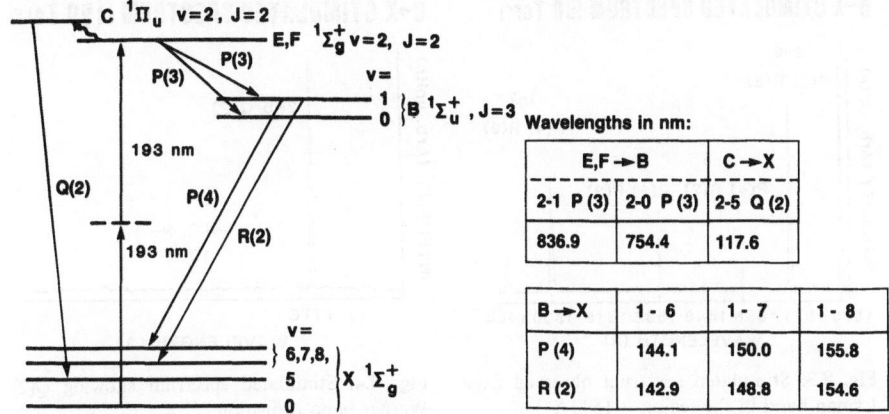

Fig. 8.2. Stimulated emission from H_2 following $E, F \leftarrow X (2-0) Q(2)$ two-photon excitation with 2 ArF* (193 nm) quanta. For the $E, F \rightarrow C$ transition, see text

constant $g_{EB} \sim 6 \times 10^3$ cm^{-1} is obtained, and strong stimulated emission on the relevant transitions in the near infrared from 750 to 920 nm is observed. As a direct consequence of the $E \rightarrow B$ emission, the $B \rightarrow X$ molecular Lyman transition develops an inversion of comparable magnitude which, together with the cross section for the $B \rightarrow X$ transitions, will exhibit a gain $g_{BX} \sim 10^3$ cm^{-1}. An amplification of this magnitude will lead to saturation in a distance of less than 1 mm. Naturally, this leads to the observation of intense stimulated emission [8.26] on the Lyman band in the 127–155 nm region. The shortest wavelength observed in this manner on the Lyman band was ~ 127 nm corresponding to the (0–3) transition [8.27]. The cascading stimulated transitions producing the Lyman $(1, v)$ bands are illustrated in Fig. 8.2. Most significantly, we note that the conversion efficiency from the 193 nm radiation to the up-converted light was determined to be $\sim 0.5\%$ for individual P and R branch transitions. Assuming that the vuv pulses occur on a time scale of ~ 10 ps, the signal strength observed indicated a peak power on the strongest transition of ~ 20 MW. In addition, strong stimulated emission has also been observed on the $Q(2)$ Werner band $C^1\Pi_u \rightarrow X^1\Sigma_g^+$ at 117.6 nm corresponding to the (2, 5) transition. Characteristic stimulated spectra illustrating the observed $B \rightarrow X$ bands ~ 154 nm and the 117 nm $C \rightarrow X$ transition are shown in Figs. 8.3 and 4, respectively. Table 8.2 contains values of various observed stimulated transitions.

As noted above, and illustrated in Figs. 8.2 and 4, strong stimulated emission is observed on the Werner band $Q(2)$ transition at 117 nm. As shown in Fig. 8.4, for the $Q(2)$ line, the transfer of population from the $E, F^1\Sigma_g^+$ to the $C^1\Pi_u$ state requires an *increase* of 22 cm^{-1} in the molecular energy, a fact clearly ruling out the role of stimulated emission as the mechanism for transfer of population from the C state. The probable energy transfer mechanism is, however, the collision of the excited molecules with electrons produced by the

B→X STIMULATED SPECTRUM (50 Torr)

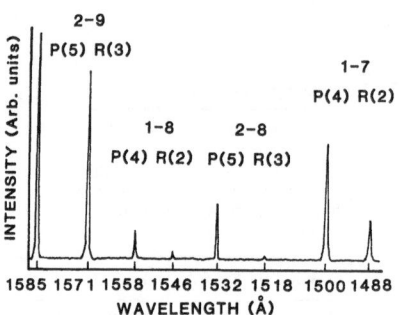

C→X STIMULATED SPECTRUM (150 Torr)

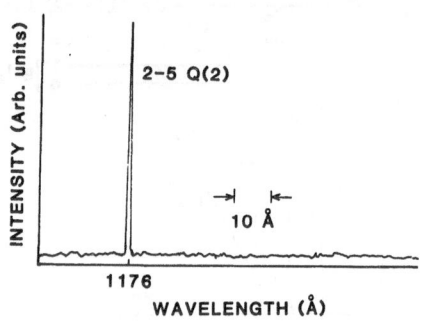

Fig. 8.3. Stimulated spectrum observed from Lyman band in the region ~1540 Å

Fig. 8.4. Stimulated spectrum showing $Q(2)$ Werner band transition

photoionization of the $E, F\,^1\Sigma_g^+$ state. Since the E $(2s\sigma)$ and C $(2p\pi)$ states strongly resemble their atomic counterparts, the electron collisional rate constant can be estimated [8.28, 29] to be $\sim 7 \times 10^{-6}\,cm^3\,s^{-1}$. Since electron densities of $\sim 10^{-16}\,cm^{-3}$ are expected to be produced by photoionization, a collisional rate of transfer of $\sim 7 \times 10^{10}\,s^{-1}$ is obtained, a value comparable to both the quenching rate of the E, F state by heavy body collisions [8.25, 30, 31]

Table 8.2. Transitions, corresponding wavelengths and the relative intensities of the observed stimulated emission in low pressure (20 Torr) H_2 on pumping the $Q(2)$ transition of the $X\,^1\Sigma_g^+ \to E,\ F\,^1\Sigma_g^+$ (0–2) band. Values in Column 3 are taken from the work of Dieke [8,32]

Transition	Wavelength [Å]		Observed relative strength
	Present work	Previous work	
Infrared			
$E \to B$			
2–1 $P(3)$	8370	8369.23	8
2–0 $P(3)$	7544.1	7544.06	0.8
vuv			
$B \to X$			
1–6 $P(4)$	1440.9	1440.7	3
1–6 $R(2)$	1428.8	1428.9	< 1
1–7 $P(4)$	1499.6	1499.6	33
1–7 $R(2)$	1487.6	1487.7	13
1–8 $P(4)$	1557.4	1557.6	6
1–8 $R(2)$	1545.4	1545.7	1
vuv			
$C \to X$			
2–5 $Q(2)$	1176.3	1176.8	< 1

and stimulated emission to the B state. A similar estimate [8.28, 29] for ion collisions indicates that the corresponding ion rate is approximately one order of magnitude less than the electron rate.

8.2.3 Raman Shifting

Near-resonant Raman shifting in metal vapor was the first efficient method to change the frequency of excimer lasers. *Djeu* et al. [8.33] reported the shifting of XeF* radiation to 585 nm in Ba vapor with near-unity photon-conversion efficiency. *Cotter* et al. [8.34] shifted XeCl radiation in Ba vapor to 475 nm and reported a conversion efficiency of 20 % and pulse energies >5 mJ. *Loree* et al. [8.35, 36] have investigated nonresonant Raman shifting of ArF, KrCl, KrF, and XeCl in H_2, D_2, CH_4, and liquid nitrogen. They find conversion efficiencies up to 50 % into the first Stokes and efficiencies of 1 % to >10 % into the second sixth Stokes and the first anti-Stokes lines. Raman shifting is thus a powerful tool for efficient wavelength conversion which will be of interest for many applications, including spectroscopy, photochemistry, and photolithography.

A discussion of backward Raman amplifiers which are investigated in connection with excimer lasers for fusion applications is beyond the scope of this short review. Detailed discussions of this topic can be found in the literature [8.37–40].

8.3 Studies of High-Lying Electronic States of Atoms and Molecules

Studies examining the collisional properties of high-lying electronic states of simple atoms and molecules can now be readily performed with the new excimer technology. This conclusion follows directly as a result of the discussion given above in Chap. 7 and Sect. 8.2.

In the implementation of these experimental analyses, two-quantum processes are becoming a powerful tool for the selective radiative coupling to atoms and molecules. Indeed, in the infrared, two quantum excitation has been successfully demonstrated in both up- and down-conversion of 10.6 µm radiation [8.41]. This represents an explicit example of frequency conversion of readily available radiation by carefully selected nonlinear processes. As is discussed below, similar mechanisms can be successfully used in the ultraviolet region for selective excitation of high lying states of atoms and small molecules.

Two photon coupling to molecular systems, including the Doppler-free aspects, was desribed by *Bischel* et al. [8.42], and *Bloembergen* and *Levenson* [8.43] who include references to earlier material. If we consider two waves with frequencies ω_1 and ω_2 and intensities I_1 and I_2, respectively, we may write the cross section $\sigma(\omega_2)$ from the absorption of the wave at frequency ω_2 as

$$\sigma(\omega_2) = \frac{(2\pi)^3}{\hbar c^2} I_1 v_2 |M_{fg}|^2 g(\omega_1, \omega_2). \tag{8.1}$$

Appearing in (8.1) are the lineshape factor $g(\omega_1, \omega_2)$, which contains the linewidth of the transition [8.44], and the two-quantum matrix element M_{fg}, which may be written in the form

$$M_{fg}(\omega_1, \omega_2) = \sum_k \left(\frac{\langle f|\hat{\varepsilon}_1 \cdot \mu_{op}|k\rangle \langle k|\hat{\varepsilon}_2 \cdot \mu_{op}|g\rangle}{E_{kg} - \hbar\omega_2} \right.$$
$$\left. + \frac{\langle f|\hat{\varepsilon}_2 \cdot \mu_{op}|k\rangle \langle k|\hat{\varepsilon}_1 \cdot \mu_{op}|g\rangle}{E_{kg} - \hbar\omega_1} \right). \tag{8.2}$$

In this expression ε_1 and ε_2 denote the polarization vectors of the optical waves, μ_{op} represents the electric dipole operator, and g, k, and f denote the ground, intermediate, and final states, respectively. The matrix elements contain the product of two transition moments over the appropriate energy denominator. Therefore, to obtain estimates for cases of interest we may scale from the measured quantity for NH_3 [8.42] provided that we have the relevant wavelengths, transition moments, and linewidths by application of (8.1, 2). The value of $\sigma(v_2)/I_1$ at 10.6 μm for NH_3 [8.45] is 3.78×10^{-20} cm^2 W^{-1} cm^2 for a linewidth factor $g(\omega_1, \omega_2)$ of 0.7×10^{-6} s, and dipole moments, in units of (debye)^2, given by $\mu_{gk}^2 = 0.026$ and $\mu_{fk}^2 = 0.030$, and an energy denominator $\Delta\omega_{kg}$ of 0.165 cm^{-1}. From this example we may now provide estimates for the two-photon excitation of electronic levels in several interesting cases.

Table 8.3 illustrates the results of estimates for several atomic and molecular systems. It should be noted that the range in susceptibility spans five orders of magnitude. Experiments have been successfully performed confirming the estimates for the production of $H_2(E, F^1\Sigma_g^+)$, $Kr(4p^56p)$, $Xe(5p^56p)$, and $Xe^+(5p^5)$.

Since the most quantitative studies have been made on H_2, we shall present those results below to serve as a quantitative example of this method. In the experiments dealing with molecular hydrogen [8.46] the following data were obtained: (i) an experimental determination of the coupling parameter for two-photon excitation for hydrogen at 193 nm, (ii) an assessment of the $E \rightarrow B$ radiative lifetime, (iii) an experimental measurement of the collisional properties of electronically excited $(E, F^1\Sigma_g^+)$ hydrogen molecules with H_2 and He, and (iv) the establishment of a bound on the photoionization cross-section at 193 nm of the $E, F^1\Sigma_g^+$ level. These results constitute a demonstration of a new general experimental technique enabling selective excitation of atomic and molecular states whose energies lie above ~ 12 eV. This aspect is clear in the case of H_2, since the principal limiting factor in the experimental analysis of the collisional behavior of electronically excited states of hydrogen has been the

Table 8.3

System	Wavelength λ [Å]	Excited product	Energy denominator $\Delta\omega$ [cm^{-1}]	Line width [MHz]	μ_1^2 [debye]^2	μ_2^2	σ/I [cm^4/W]	Remarks
H$_2$	1930	$E, F^1\Sigma_g^{+\,(c)}$ ($v=2, j=2$)	45×10^3	10^{6a}	1^b	1	1.8×10^{-31}	$B^1\Sigma_u^+$ Intermediate state
CO	1930	$F^1\Sigma^+$	$13 \times 10^3 (A^1\Pi)$ $10 (a^3\Pi)$	10^{6a}	1 10^{-5}	0.01 $\sim 10^{-3}$	1.6×10^{-32}	$A^1\Pi$ and $a^3\Pi$ Intermediate state^d
N$_2$	1930	$E^3\Sigma_g^+$	50×10^3	10^{6a}	1	0.01	1.1×10^{33}	$b'^{\,1}\Sigma_u^+$ Intermediate state
Kr	1930	Kr* $4p^56p$ (103,762 cm^{-1})	30×10^3	10^{6a}	1	1	3×10^{-31}	$4p^55s$ Intermediate state
I$_2$	1930$^{(e)}$	0442 $^1\Sigma_g^+$ ($^1S + {}^1S$)	4×10^3		1	0.1	1.7×10^{-30}	$D^1\Sigma_u^{+\,(f)}$ Intermediate state
Xe	2484	Xe* ($5p^56p$)	23×10^3	10^{6a}	1	1	3×10^{-31}	$5p^56s$ Intermediate state
	1930	Xe$^+$	16×10^3	10^3	1	1	1×10^{-34}	
N$_2$O	2484	N$_2$ + O*	28×10^3	$\sim 10^{9g}$	~ 1	$\sim 10^{-3}$	$\sim 3 \times 10^{-35}$	Weak intermediate state amplitude

[a] Estimated linewidth

[b] J. E. Hesser: J. Chem. Phys. **48**, 2518 (1968)

[c] R. M. Crosswhite: *The Hydrogen Molecule Wavelength Tables of Gerhard Heinrich Dieke* (Wiley-Interscience, New York 1972)

[d] P. H. Krupenie: The Band Spectrum of Carbon Monoxide, NSRDS-NBS No. 5 (USGPO, Washington, DC 1966)

[e] R. S. Mulliken: J. Chem. Phys. **55**, 288 (1971)

[f] J. A. Myer, J. A. Samson: J. Chem. Phys. **52**, 716 (1970)

[g] Some interfering absorption present from Cordes bands. See J. A. Nyer: J. Chem. Phys. **52**, 716 (1970)

absence of an effective means for selective excitation of the molecular states. Optical pumping techniques [8.47], with the exception of the $B\,^1\Sigma_u^+$ state, are made extremely difficult by the lack of a good window material below 1000 Å. Naturally, even states could not be excited by conventional absorption from the $X\,^1\Sigma_g^+$ level, even without this limitation. Discharge excitation, on the other hand, lacks sufficient specificity to be useful in a wide range of circumstances. For detailed collisional studies of hydrogen, a selective means of excitation unhampered by the lack of a transparent dielectric below 1000 Å is needed, a condition achieved by this technique.

Equally significant was the verification of the theoretical estimate for the two-photon coupling parameter in H_2. As the experiments demonstrated [8.46], the experimentally established and estimated magnitudes agreed to within 10 %, a remarkably close figure in light of the approximations used in the original calculations.

An important conclusion resting on this finding concerns the density of excited species that can be generated. In these experiments, excited state densities of $\sim 2.5 \times 10^{11}\,\mathrm{cm}^{-3}$ were generated with a laser linewidth of $\sim 10^2\,\mathrm{cm}^{-1}$. Since linewidths considerably less than $10^{-1}\,\mathrm{cm}^{-1}$ have been observed, as discussed in Chap. 7, for rare-gas-halogen laser systems at pulse energies *greater* than those used in these studies, the excitation rate can be increased more than a thousandfold, allowing the production of excited state species at densities greater than $10^{14}\,\mathrm{cm}^{-3}$. Number of densities of this scale will apply generally to a wide class of systems.

The implications of this conclusion regarding the achievable excitation density for fundamental studies of collisional phenomena and frequency conversion to short wavelengths are manifest. For example, at excitation densities on the order of $10^{14}\,\mathrm{cm}^{-3}$, assuming that the dominant collisional pathway for the $H_2\,(E, F\,^1\Sigma_g^+)$ level is

$$H_2(E, F\,^1\Sigma_g^+) + H_2(X\,^1\Sigma_g^+) \rightarrow H_2(C\,^1\Pi_u) + H_2(X\,^1\Sigma_g^+), \tag{8.3}$$

as indicated by the experiments [8.46], stimulated emission is expected [8.48] on the $C\,^1\Pi_u \rightarrow X^1\Sigma_g^+$ bands at $\sim 155\,\mathrm{nm}$. Since the cross section [8.26] for the C-X transition is $\sim 1.1 \times 10^{-13}\,\mathrm{cm}^2$, gain coefficients of $\sim 11\,\mathrm{cm}^{-1}$ appear feasible. This value is sufficient to generate strong coherent emission in a 1-cm region of excited material without the use of an optical resonator. Naturally, we would also expect that excited state-excited state collisions should also be readily observable at excited molecular densities on the order of $10^{14}\,\mathrm{cm}^{-3}$. It is evident that this general technique of excitation can be applied to a wide variety of systems and experimental conditions.

Indeed, as the discussion above in Sect. 8.2.2 clearly demonstrates, actual stimulated emission of the C-X Werner band transitions has been observed [8.27], in the manner suggested, since the first edition of this volume. An important modification, however, must be noted. The population of the $H_2(C\,^1\Pi_u)$ level appears to be collisionally generated by electron collisions in the process

$$H_2(E, F{}^1\Sigma_g^+) + e^- \rightarrow H_2(C{}^1\Pi_u) + e^- \tag{8.4}$$

rather than by the heavy body collision indicated in (8.3).

8.3.1 Nonlinear Processes in Molecules

The rare-gas-halogen excimer lasers have emerged in the last few years as a valuable tool of research in the areas of ultraviolet physics and chemistry. These systems have been used to dissociate a variety of molecules following single-photon and multiphoton absorption, both in unimolecular processes [8.49–56] and in processes involving collisions arising from buffer gases [8.57]. Because of the ultraviolet wavelengths utilized, excimer lasers commonly ionize molecules [8.58] and can also interact with the photofragments in a secondary photo-excitation process [8.59].

Recent studies have involved the analysis of N_2O [8.60], $OCCl_2$ [8.61], $NOCl$ [8.62], and SO_2 [8.63], the latter at 248 nm and the remainder at 193 nm. Normally, multiphoton dissociation results in a wide variety of processes and product distributions. Indeed, excited radical species, such as $S(3p^34s)^3S_1$ and $SO(B{}^3\Sigma^-)$, can be produced at substantial densities.

A particularly interesting case, which illustrates the versatility of the molecular responses to intense ultraviolet radiation, is that of the nonlinear photolysis of $OCCl_2$ at 193 nm. Under irradiation at 193 nm at an intensity of $\sim 50\,MW/cm^2$, strong fluorescent emission at 258 nm from Cl_2^* is observed [8.61]. Analysis of the data leads to the conclusion that the primary mechanism of Cl_2^* formation is a direct two quantum absorption in $OCCl_2$ of the type

$$\gamma(193\,nm) + \gamma(193\,nm) + OCCl_2 \rightarrow CO + Cl_2^*. \tag{8.5}$$

Since two 193 nm quanta represent an energy greater than the ionization potential of $OCCl_2$, however, ionic fragments are expected to dominate with very little direct neutral fragment production. Contrarily, the neutral channel appears very strong.

The first ionization limit [8.64] of $OCCl_2$ is at 11.8 eV. The second and third limits are nearly degenerate at [8.50] ~ 12.6 eV, an energy less than 0.25 eV below that corresponding to the energy of two 193 nm quanta. These limits correspond to the ionization of electrons from the chlorine nonbonding orbitals. Recalling the rather weak $CO–Cl_2$ bond [8.65] of ~ 1.1 eV, to a reasonable first approximation, the $OCCl_2$ molecule may be viewed as two weakly connected diatomics CO and Cl_2. In this picture, under 193 nm excitation, the Cl_2 molecule is ionized, with ~ 0.2 eV kinetic energy imparted to the free electron. We note that the potential energy curve of the $Cl_2^* + e^-$ system

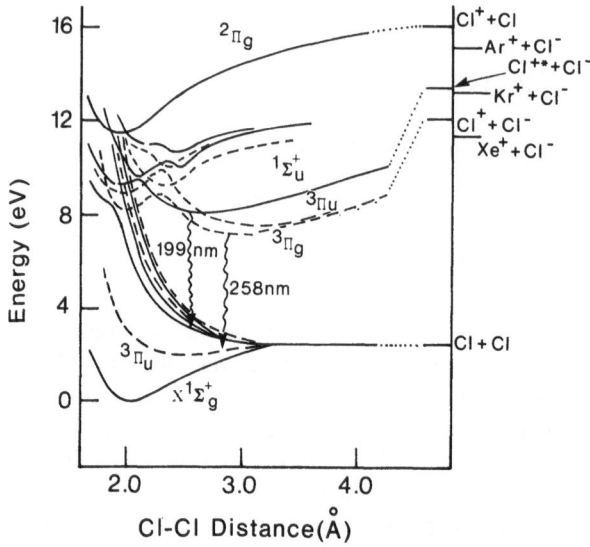

Fig. 8.5. Some potential curves of Cl_2 pertiment to the present studies [8.66]. The solid lines represent singlet states and broken lines represent triplet states. The states connected by the 199 and 258 nm fluorescence are also indicated. The asymptotic values and the atomic energy levels are based on [8.67]. All atoms and ions are in their ground states, with the exception of Cl^{+*}, which denotes $Cl^*(^1S)$

is crossed, in this energy range, by a number of curves of the neutral Cl_2 molecule, representing both singlet and triplet states [8.58], as shown in Fig. 8.5, and that some curves have slopes as steep as $\sim 40\,eV/Å$. Viewed from the standpoint of the curve-crossing mechanism commonly operative in dissociative recombination, the separation between the capture point [8.59] R_c and the stabilization (crossing) point R_s can be as small as $10^{-2}\,Å$ for the Cl_2 system under these circumstances. The associated stabilization time [8.68] that is the time needed for the system to move from R_c and R_s is then on the order of 10^{-14}–10^{-15} s, a value comparable to the orbital period of the electron in the field of the positive ion. Thus, the fraction of systems reaching the crossing point R_s (the survival factor) is expected to be close to unity, so that the probability for curve crossing depends mainly on the strength of the configuration interaction matrix element connecting the two states. The magnitude of such an interaction can be high enough that a large fraction of the $OCCl_2$ molecules excited above their ionization limit will indeed dissociate into neutral fragments by a process which, in this model, is equivalent to internal dissociative recombination.

SO$_2$ exhibits an extremely complicated response under irradiation at 248 nm [8.63]. In this case, two strong channels of molecular fragmentation were found to occur from an electronically excited configuration of SO_2. One channel produced O atoms while the other generated S atoms which underwent additional electronic excitation. Figure 8.6 illustrates the energy pathways

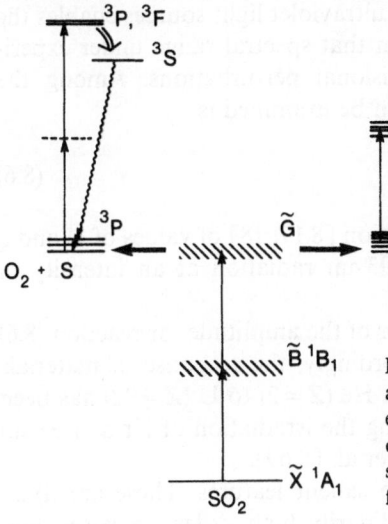

Fig. 8.6. A schematic diagram of the various excitation and dissociation processes reported in this study. The energy levels are not drawn to scale. Vertical arrows denote photoexcitation, horizontal arrows denote dissociation, and the wiggly arrows represent the observed fluorescence

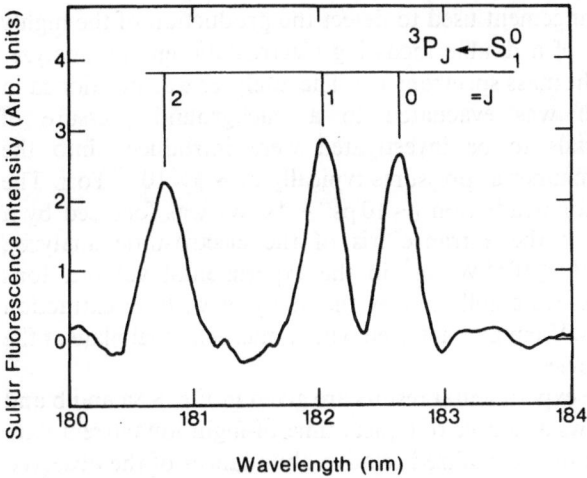

Fig. 8.7. The atomic fluorescence from the excited $(3p^3 4s)^3 S_1$ state to the three levels of the $(3p^4)^3 P_J$ ground state in sulfur. The SO_2 pressure was 0.4 Torr, and the laser fluence at the focus was $\sim 10\,J/cm^2$

observed. Strong atomic emissions were present at $\sim 180\,nm$ from the excited sulfur atoms, as shown in Fig. 8.7.

8.3.2 Nonlinear Processes in Atoms

The high spectral brightness sources described in Chap. 7 clearly enable the observation of high-order electronic nonlinearities in atoms. Indeed, the availability of spectrally bright picosecond ultraviolet light sources enables the study of nonlinear coupling mechanisms in that spectral range under experimental circumstances unaffected by collisional perturbations. Among the simplest general physical processes that can be examined is

$$N\gamma + X \rightarrow X^{q+} + qe^- . \tag{8.6}$$

Preliminary studies have led to the observation [8.17, 18] of values of N and q as high as 99 and 10, respectively, for 193 nm radiation at an intensity of $\sim 10^{14}$ W/cm^2.

Of particular significance is the behavior of the amplitude for reaction (8.6) as a function of atomic number (Z). Accordingly, the response of materials spanning the range in atomic number from He ($Z=2$) to U ($Z=92$) has been measured [8.18]. Similar processes involving the irradiation of Kr at 1.06 μm have recently been described by L'Huillier et al. [8.69].

The experimental results exhibited two salient features. These are (i) an unexpectedly strong coupling for extraordinarily high order processes, and (ii) a coupling strength which is dramatically enhanced at higher Z-values.

The experimental arrangement used to detect the production of the highly ionized species consisted of a double focussing electrostatic energy analyzer operated as a time-of-flight mass spectrometer. The analyzer was positioned in a vacuum vessel which was evacuated to a background pressure of $\sim 10^{-7}$ Torr. The materials to be investigated were introduced into the chamber in a controlled manner at pressures typically of $\sim 3 \times 10^{-7}$ Torr. The 193 nm ArF* laser used for irradiation (~ 10 ps, ~ 4 GW) was focussed by a $f = 50$ cm lens in front of the entrance iris of the electrostatic analyzer, producing an intensity of $\lesssim 10^{14}$ W/cm^2 in the experimental volume. Ions formed in the focal region were collected by the analyzer with an extraction field in the range of 50–500 V/cm and detected with a microchannel plate at the exit of the electrostatic device.

Representations of the experimental results are given in Fig. 8.8a and b and Table 8.4. Figure 8.8a shows a sample of typical time-of-flight ion current data for Xe. Table 8.4 contains the normalized relative abundances of the observed ion charge states for Xe, derived from Fig. 8.8a and uncorrected for detector sensitivity. Experiments indicate that the detector is about four times as sensitive for the high xenon ion states than for Xe$^+$. Similar data have been recorded for He, Ne, Ar, Kr, I, Hg, and U. In Fig. 8.8b, the observed ions and the total energies required for their generation in the electronic ground state are given.

A remarkable feature of the data is the magnitude of the total energy which can be communicated to the atomic systems, especially for high Z materials. The total energy investment [8.70–73] of ~ 633 eV, a value equivalent to 99 quanta, needed to generate U^{10+} from the neutral atom with neglect to the small contribution associated with molecular binding [8.74] in the experimental material UF$_6$, represents the highest energy value reported for a collision-

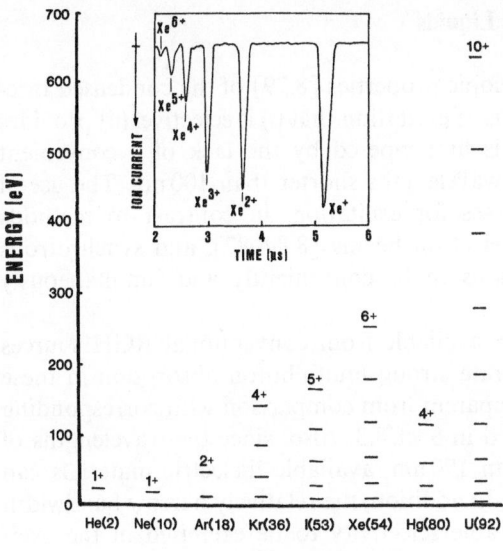

Fig. 8.8a, b. Data concerning multiple ionization of atoms for 193 nm irradiation at $\sim 10^{14}\,\text{W/cm}^2$. (**a**) Inset: Typical time-of-flight ion current signal for xenon. (**b**) Plot of total ionization energies of the observed charge states as a function of atomic number (Z)

Table 8.4. Charge state distribution of xenon derived from Fig. 8.8a

Charge state	1+	2+	3+	4+	5+	6+
Relative abundance	44	26	20	7	5	1

free nonlinear process. The removal of the tenth electron from uranium, which requires [8.71] $\sim 133\,\text{eV}$ if viewed as an independent process, requires a minimum of 21 quanta.

Although a full understanding of these phenomena has not been presented, it is clear that the coupling strength implied by the observed scale of energy transfer at an intensity of $\sim 10^{14}\,\text{W/cm}^2$ very substantially exceeds that anticipated from conventional theoretical formulations describing multiquantum ionization [8.75, 76]. It appears possible [8.18] that a collective atomic response [8.71], viewed essentially as an atomic plasmon [8.78], may be a major factor in the enhanced coupling in heavy materials.

8.4 Studies of Condensed Matter

Research on condensed matter can now be conducted in a variety of new areas involving ultraviolet radiation. In this section, we discuss three examples involving the spectroscopy of liquids, photoemission studies of surfaces, and laser annealing of semiconductors.

8.4.1 Spectroscopy of Cryogenic Liquids

Ideally, the study of the spectroscopic properties [8.79] of the condensed rare-gas mixtures would utilize a means of excitation that (i) is selective, (ii) provides a high excitation rate, and (iii) is unhampered by the lack of a convenient transparent window material at wavelengths shorter than 100 nm. The use of ultraviolet multiquantum processes for excitation, in contrast to methods involving α particles [8.80–83], electron beams [8.84–87], and synchrotrons [8.88], enables all three conditions to be conveniently and simultaneously satisfied.

The optical intensities readily available from conventional RGH sources are more than sufficient to generate strong multiphoton absorption in these condensed materials, an aspect apparent from comparison with corresponding gas-phase measurements discussed in Sect. 8.3. Also, since the wavelengths of RGH sources are all greater than 190 nm, available dielectric materials can conveniently be used as windows. In addition, the relatively narrow bandwidth $(\Delta \tilde{v} \leq 100 \, \mathrm{cm}^{-1})$ enables considerable selectivity to be exercised in the excitation of certain mixtures.

Relatively simple apparatus, as shown in Fig. 8.9, can be used to conduct these studies. A commercial He refrigeration system was used to cool a stainlesssteel sample cell, the temperature of which was controlled with a heating coil and monitored using a Si temperature-sensing diode. The cell, a 3.8-cm cube, was fitted with LiF windows 2.5 cm in diameter, sealed in place with indium gaskets. A copper radiation shield surrounded the cell, and the entire assembly was enclosed in a vacuum jacket.

The excimer laser used was of standard configuration with a pulse duration of ∼10 ns. It was focussed into the center of the cell with $f/10$ optics, yielding a

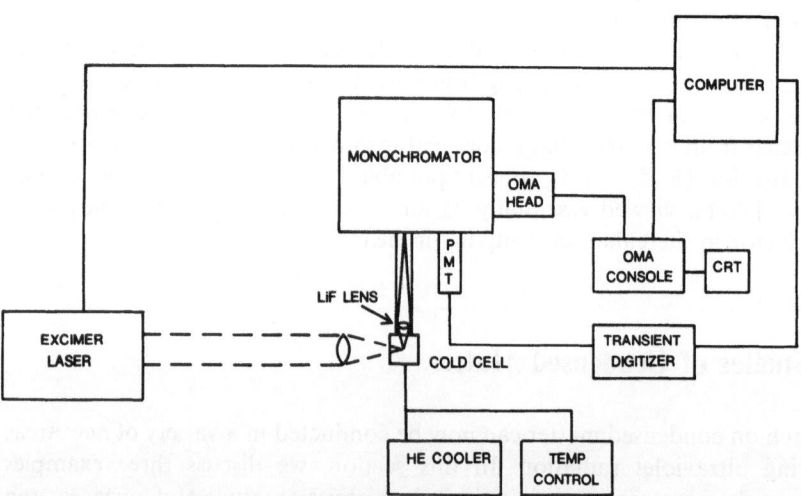

Fig. 8.9. Block diagram of apparatus used in liquid rare-gas experiments

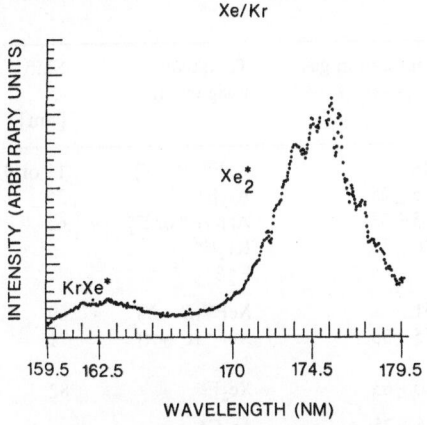

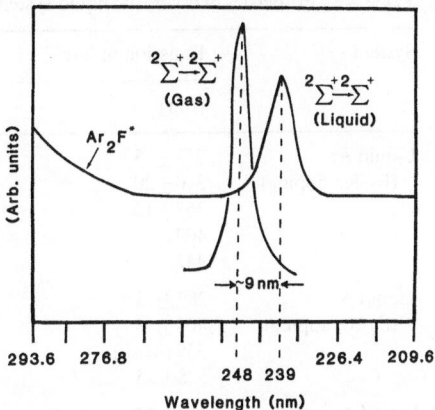

Fig. 8.10. Fluorescence spectrum observed from Xe–Kr mixture (100 ppm Xe) irradiated at 193 nm. The modulation observed on the Xe_2^* emission is generated by the response of the optical multichannel analyzer

Fig. 8.11. KrF ($B{\to}X$) emission in argon gas host and in liquid argon host (100 ppm F_2, 220 ppm Kr in liquid Ar; irradiation at 193 nm)

range of intensities used in these studies [8.89] between 10^7 and 10^9 W/cm². Normally, the fluorescence was collected by a simple optical system at right angles with respect to the direction of propagation of the laser beam used for excitation.

A typical result obtained from an analysis of Xe/Kr mixtures is shown in Fig. 8.10. The experimental findings provided by these studies represent the first examples of ultraviolet multiphoton excitation of the liquid rare gases and the first use of narrow-bandwidth sources of excitation in that process. From the intensity and concentration dependence of the resulting vuv fluorescence, it has been possible to construct a qualitative scheme for the excitation pathways. Of particular interest are (i) the apparent ability to excite the bound ground-state dimers directly by a simple two-quantum amplitude and (ii) the observation of efficient liquid-phase molecule-molecule ($Kr_2^* {\to} Xe_2^*$) energy transfer in the Xe/Kr system. Finally, these data provided a positive identification of the 162.5-nm emission band characteristic of Xe/Kr media as arising from KrXe* species, settling a subject of recent controversy [8.80].

Very similar experiments [8.90] have been recently performed in liquid argon and liquid nitrogen doped with rare-gases and fluorine to examine the behavior of rare gas halogen species in the environment of a liquid host. Table 8.5 indicates the range of species tentatively identified in these studies along with the wavelengths characteristic of their emission. A prominent feature is the relatively large shifts in the emission frequencies observed in the liquids as compared to the corresponding gas phase spectra. Figure 8.11 illustrates this comparison for KrF* observed in liquid argon doped with Kr and F_2. Significantly, the spectral shift of ~9 nm is considerably greater than the widths associated with the emissions.

Table 8.5. Properties of cryogenic liquid systems

System	Emission in liquid [nm]	Emission in gas	Tentative assignment	Shift [nm]
Liquid Ar (F$_2$/Kr doping)	237 ± 3	248	KrF* ($B \to X$)	11 or 9
	310 ± 20	285 ± 25	Ar$_2$F*	25
	367 ± 10	305 ± 33	ArKrF* or F$_2^*$	62
	409	420 ± 35	Kr$_2$F*	
	442	?	F$_2^*$?	
Liquid Ar (F$_2$/Xe doping)	282 ± 3	351	XeF* ($B \to X$)	69
	404 ± 3	475 ± 35	XeF* ($C \to A$)	71
	339 (weak)	?	?	
	528 ± 25	610 ± 65	Xe$_2$F*	82
Liquid Ar (F$_2$ doping)	310 ± 20	285 ± 25	Ar$_2$F*	
	367 (weak)			
	$\left. \begin{matrix} 409 \\ 456 \end{matrix} \right\}$ overlap	F$_2^*\left\{ \begin{matrix} ? \\ ? \end{matrix} \right.$	F$_2\left\{ \begin{matrix} ? \\ ? \end{matrix} \right.$	
Liquid N$_2$ (F$_2$ doping)	One very broad band at 443 ± 30 nm			

8.4.2 Photoemission Studies of Condensed Matter Surfaces

Certainly, the detailed study of the properties of semiconductor materials ranks among the most important areas of research in condensed matter. In order to pursue optimally the development of semiconductor materials and the methods for efficiently processing these substances into appropriate devices, microscopic physical models representing the material growth and processing are needed. Viewed very generally, the ability to fashion the detailed atomic configuration of semiconducting materials is desired.

A microscopic atomic scale model of semiconductor growth and processing is of little value without a corresponding technique of measurement valid at the atomic level. The use of xuv and x-ray radiation figures prominently in several experimental techniques that provide detailed structural data on a spatial scale given by atomic dimensions. Among the experimental methods applicable to this type of analysis are photoemission [8.91, 92], extended x-ray absorption fine structure (EXAFS) [8.93, 94], and trace-element analysis utilizing fluorescence [8.95]. Ideally, it would be desirable to obtain direct information on the detailed atomic configurations on a time scale with sufficient resolution to record dynamically the alterations of the structure arising from the laser-induced process.

Laser processing of semiconductor materials, such as the annealing of surfaces described in the following Sect. 8.4.3, has become a widespread practice in recent years which holds great promise for the fabrication of superior microelectronic components. Several diagnostic techniques have been used to study the properties of many laser-processed systems. These methods include time-of-flight analysis [8.96], Raman scattering [8.97, 98], photoacoustic studies [8.99], reflection [8.100], positron annihilation [8.101], photoemission [8.102], low-energy electron diffraction (LEED) [8.103], and transmission electron microscopy (TEM) [8.104]. Although some measurement techniques provide atomic-scale structural data, such as photoemission [8.102], and others provide time resolution in the nanosecond range, such as the pulsed Raman studies [8.101], no currently practiced technique combines the high measurement rate required to obtain the short-time resolution *and* spatial resolution of the scale of atomic dimensions. For example, if the measurements could be performed sufficiently rapidly, photoemission and EXAFS studies can, in principle, provide the dynamical high-resolution data needed for correspondence with a microscopic model of laser-processed semiconductor materials.

8.4.3 Laser Annealing of Semiconductors

Laser processing of materials has become an active area of research in recent years [8.105]. Particularly, the annealing of semiconductors to induce the regrowth of crystalline films is a process currently under intense study. Early work in this area, conducted by *Gibbons* and coworkers on ion-implanted silicon [8.106], examined the use of continuous wave lasers and electron beams to induce the annealing action. Among the interesting properties found were nearly perfect recrystallization and essentially complete electrical activity for impurity concentrations exceeding the solubility limit [8.107].

Excimer lasers have properties well suited to requirements of surface annealing [8.108, 109]. Among these properties are optical beam uniformity, the availability of high average powers, and a frequency of operation corresponding to a high absorption coefficient in the material undergoing annealing.

Silicon serves as an excellent example of the potential for laser annealing processes. The energy density required is $\sim 1.5\,\mathrm{J/cm^2}$. When used on B and As ion-implanted silicon, essentially one hundred percent free carrier recovery was obtained at the energy density [8.108]. In addition, the production of smooth flat surfaces free from damage was also reported [8.109].

Other lasers processing techniques are also developing. One important area involves laser-induced photochemical reactions to produce localized metallic deposition on semiconductors [8.110]. Spatial resolutions on the order of $1\,\mu\mathrm{m}$ were obtained

8.5 Dye Laser Excitation

Pulsed dye lasers are usually pumped by flashlamps, N_2 lasers, higher harmonics of Nd lasers, or excimer lasers. While flashlamps offer a low-cost capability to produce quite substantial average powers, they cannot generate

Table 8.6. Characteristics of commercially available excimer lasers

Laser medium	ArF	KrCl	KrF	XeCl	XeF
Wavelength [nm]	193	222	248	308	351
Output energy [J]	0.5	0.05	> 1.0	0.5	0.4
Average power [W]	10	1	> 20	10	6
Bandwidth		$\sim 100\,cm^{-1}$		Multi-line	
Pulse duration			Typically 10–20 ns		
Beam divergence			$\sim 10\,m\,rad$		

high pulse powers because energy storage times of organic dyes are short (ns range). The matching between the flashlamp spectra and the absorption bands of dyes is poor, which leads to high thermal loads and low overall efficiency. N_2 lasers have been used extensively to generate short pulse duration – high repetition rate pulses; they cannot, however, be scaled to high pulse energies. The excitation of high power, high energy dye lasers, therefore, tends to be an exclusive domain of Nd and excimer lasers. A comparison of the two systems shows that available pump energies are comparable (1 J range). In the case of Ne, the second, third, and fourth harmonic of the fundamental at 1.06 μm are used as pump radiation. For excimer pumping, the range of available pump wavelengths is given in Table 8.6 and extends to somewhat shorter wavelengths with 248 nm being the shortest wavelength used to date. The excimer laser seems to be simpler in its set-up because the radiation can be used without frequency shifting, and because the active medium is gaseous, thermal problems which are associated with high repetition rate pumping will be less severe than in the case of Nd lasers. When used as a dye laser pump, the very large number of modes oscillating in the excimer laser can be of an advantage because this makes the energy deposition over the beam profile more homogeneous. Overall, it can be stated that excimer lasers are extremely well suited for pumping of dye lasers.

Excimer lasers have been used as dye laser pumps since 1976 [8.111]. The most commonly used excimers are XeCl, KrF, and XeF, as their emission coincides with the absorption bands of several dyes. Research in this field has centered on the extension of dye-laser radiation towards shorter wavelengths, investigation of laser efficiency, studies of dye stability under high pump power conditions and pump induced processes [8.112].

The shortest dye laser wavelengths have been reported as 311.2 nm [8.113] and 318 nm [8.114] with a KrF laser providing the excitation. These authors reported tripling of this radiation to 106 nm, indicating that the third harmonic of dye laser radiation can cover the spectrum down to the transmission limit of crystalline materials. Dye laser efficiencies η can be substantial with $\eta = 28\,\%$ [8.111, 115], 18 % [8.116, 117], 14 % [8.113], and > 40 % [8.118] being reported by different groups. Especially in the ultraviolet region, the efficiencies can be high. *Tomin* et al. [8.119] reported a KrF* pumped dye laser system operating between 348 and 366 nm with an overall efficiency of 0.5 %. *Cassand* et al. [8.120] described a XeCl* pumped dye laser with $\eta > 25 \%$, an average power of 1 W and 100 Hz repetition rate. Dye stability under KrF* and XeF irradiation has been investigated by *Stappaerts* [8.121], and *Cassand* et al. [8.122] reported pump induced absorption processes which might influence laser efficiency. Finally, in a related area, various materials suitable as saturable absorbers at 248 nm [8.123] and 193 nm [8.124] have been examined.

8.6 Remote Sensing

In laser based remote sensing, atmospheric constituents are identified and their concentration is determined through observation of the characteristic interaction of the laser radiation with specific atmospheric constituents. Techniques such as resonance fluorescence and resonance Raman scattering involve tuning the laser to a frequency characteristic of the sample and observing the radiation from the sample at the same or a slightly different frequency. The temporal behavior of the observed signal contains the information on the spatial distribution of the sample and the intensity indicates its concentration. For general application of this method, the laser radiation must be tunable over a wide range [8.125].

Remote sensing and detection is of importance for a basic understanding of atmospheric processes, meteorology, and pollution control. In all laser based techniques, a light pulse which interacts in a specific and characteristic manner with one or more atmospheric compounds is propagated through the atmosphere. Part of the laser radiation is redirected by Rayleigh, Mie, or Raman scattering and can be detected by a ground based station which usually contains both laser and detector. As light which is scattered from a more distant position will reach the detector at a later time, the temporal behavior of the back scattered signal contains spatial information about the atmospheric layer through which the laser pulse propagates. Although, in the most fundamental application of the LIDAR technique [8.125], the time dependent intensity of the back scattered signal contains the density distribution of particles and molecules as a function of distance, the type of molecules cannot be identified from nonresonant scattering.

To give the reader a feeling for the signals involved, the following simple estimate may be used. A more detailed discussion is given by *Uchino* et al. [8.126]. We assume that a laser pulse of 1 J energy and 600 nm wavelength is used to probe the atmosphere at a height h of 50 km. The number of counts in the detection system which is caused by light in Rayleigh backscattered from a layer with thickness $\varDelta$ at height h can be written as

$$N(h) = \eta N_0 A \frac{\beta(h)\varDelta}{h^2} T^2 . \tag{8.7}$$

In (8.7), η is the quantum efficiency of the detection system, A is the aperture of the collecting telescope, N_0 is the number of transmitted photons, $\beta(h)$ is the volume backscattering coefficient, and T is the transmission of the atmosphere. For $\lambda = 600$ nm, we find a total Rayleigh scattering cross-section of $\sigma_R \sim 3 \times 10^{-27}$ cm². The transmission of a 50 km high column is therefore $T \sim 1$. Choosing $\eta = 10^{-1}$, $A = 1$ m², $\beta(50\,\mathrm{km}) = 2 \times 10^{-14}$, and $\Delta = 5 \times 10^3$ m and neglecting any effects of Mie scattering we find $N(h) \sim 3 \times 10^{-3}$ counts per pulse.

It is obvious that high pulse energies at high repetition rates and short wavelengths are needed for LIDAR techniques. The choice of short wavelength lasers is favorable as the Rayleigh cross-section scales with ω^4 while Mie scattering (dependent on the particle size) scales with a lower power dependence (typically ω^1). The main advantage of excimer lasers in this field, however, is their scalability to repetition rates of $\sim 10^3$ Hz and pulse energies of several hundred mJ. This certainly will greatly facilitate the routine application of ground based LIDAR systems.

If an absorbing species, e.g. O_3, is present in the atmosphere and the total backscattering coefficient is known as a function of distance, the spatially resolved concentration of the absorber can be measured. In this way, *Uchino* et al. [8.126, 127] have measured the stratospheric O_3 distribution using a XeCl* laser. They pointed out that the OH radical absorption coincides with the XeCl laser wavelength. Furthermore, the detection of metallic atoms and ions should be possible at low concentrations (10–10^3 cm^{-3}) using resonant scattering.

Another technique for remote sensing which was used by *Halpern* et al. [8.128] is photodissociation followed by detection of fluorescence from an excited product. NH_3 was dissociated at 193 nm and the fluorescence from NH was observed. *Rodgers* et al. [8.129] described photofragmentation of atmospheric trace gases followed by laser-induced fluorescence as a technique to detect molecules which do not have bonding-excited states that fluoresce and therefore cannot directly be detected. NO_2, NO_3, and HNO_2 are given as examples.

We can conclude in general that tunable radiation of high average power and narrow bandwidth can be obtained at any wavelength in the visible and ultraviolet region from ~ 800 to ~ 193 nm with excimer lasers. Depending on the specific requirements of the remote measurements, different techniques can be used for optimum performance in either region. With the enhanced output of excimer lasers, the output characteristics of the dye and modified excimer lasers, and in particular the energy and average power, are expected to improve substantially, greatly facilitating a wide range of LIDAR measurements. In addition, the new higher range of intensities that are now possible with the picosecond systems discussed in Sect. 7.5 may open up further techniques, such as observation of nonlinear processes, for remote sensing [8.130].

8.7 Solid-State and Photodissociation Laser Excitation

Excimer lasers enjoy wide application as sources of excitation for a wide range of laser media including solid-state systems and gas phase photodissociative media.

8.7.1 Solid-State Lasers

The first excimer pumped solid-state lasers were described at the Laser '78, Orlando, FL and at the Topical Meeting on Excimer Lasers in 1979 [8.131, 132]. *Baer* et al. reported a XeF* pumped Tm:YLF laser emitting at 453 nm. These studies were aimed at the identification of a suitable storage medium which allows deposition of optical excimer laser energy, followed by controlled extraction at a shifted wavelength. *Ehrlich* et al. have investigated a KrF* pumped Ce^{3+}:YLF laser emitting at 325 and 309 nm. They concluded from the fluorescence band that this laser should be tunable between 305 and 335 nm and should be scalable to high powers [8.133]. In an extension of this work, laser radiation at 286 nm has been observed from Ce^{3+} using LaF_3 as a host [8.134]. The laser can be pumped by ArF*, KrF*, and doubled Ar^+ (257 nm) and the tuning range is 275–315 nm. This is so far the shortest wavelength obtained from a solid laser material. The wavelength is beyond the uv cut-off of organic dye lasers and could find important applications for remote sensing of atmospheric species like O_3, SO_2, and OH.

8.7.2 Photodissociation Lasers

Photodissociation of molecular compounds which can lead to electronically excited fragments has been successfully used to generate inverted media. A prominent member of this class of lasers is the atomic iodine laser which is pumped by photolytic production of $I(^2P^0_{1/2})$ at 2400–3100 Å from iodides like CF_3I and C_3F_7I.

It has been found that for a large number of molecules ultraviolet absorption can represent a direct and selective channel for the formation of electronically excited fragments. Historically, Xe flashlamps and other plasma sources, such as exploding wires, have been used as excitation sources for these lasers. It is obvious that the use of an ultraviolet laser as the photolytic source allows a substantially more sophisticated control of experimental conditions. The large number of excimer wavelengths together with their Raman shifted components makes a considerable number of molecules accessible to this pump technique. Furthermore, it has been shown that vibrational excitation of molecular species can shift or broaden their absorption bands into coincidence with fixed frequency excitation sources [8.135].

However, the comparatively low pump rates achievable with such sources has limited their applications to systems with long upper state lifetimes. Spontaneous as well as laser emission from various excimer laser media have been used to generate inversion and stimulated emission in a substantial number of elements. Molecular species which can be used as starting substances are chosen for a number of features. These are: (i) a sufficient volatility at practical temperatures (e.g., vapor pressures of ~ 0.1–1 Torr at $\lesssim$ a few hundred °C), (ii) a sufficient absorption at one of the excimer emissions (e.g., absorption lengths of no more than a few cm at the above pressures) to guarantee efficient energy coupling, (iii) the bound free absorption has to be sufficiently selective with respect to population of the upper laser level versus the lower laser level, and (iv) the medium has to be transparent at the laser wavelength. The maximum obtainable energy of the laser photon equals the energy of the pump photon minus the dissociation energy of the molecular ground state. If the dissociative pump transition ends in a strongly repulsive region of the upper state surface, the available photon energy is further reduced and the velocity distribution of the dissociation products can become strongly superthermal. These effects, their influence on laser lineshape and kinetic processes relevant to photolytically pumped lasers were discussed in a recent article by *Ehrlich* et al. [8.136].

One of the first type of media to be investigated as laser candidates were the group VI atoms O, S, and Se. Starting from several compounds (e.g., N_2O, CO_2, OCS, and OCSe), it was known that dissociative channels exist into the atomic ^1S state which have quantum efficiencies of $\sim 90\%$. Lasing can then occur on the electric dipole forbidden transitions $^1S - {}^1D$ and $^1S - {}^3P$. Due to the large excited state lifetimes, these media are of interest for high-energy-density optical-storage techniques and laser emission as suggested by *Murray* et al. [8.137] from Se [8.138] and S [8.139] has been demonstrated.

A second important group of photodissociation lasers which operates on the resonance line of group I and IIIA elements was first proposed in 1965 by *Zare* et al. [8.140]. However, due to the lack of intense uv light sources these lasers could not be realized at that time. With the arrival of powerful ArF* and KrF* lasers this obstacle has been removed and laser action has been reported by several groups. Using alkali iodides and bromides as parent molecules, *Ehrlich* et al. [8.141] have demonstrated lasing on the resonance lines of Na, K, Rb, and Cs at power levels of 1–10 kW. Emission from atomic In at 451 nm was reported by *Burnham* [8.142], from Ga, In, Al, and Bi by *Deutsch* et al. [8.143], from Tl by *Ehrlich* et al. [8.144] and *Burkhard* et al. [8.145, 146], and from Ga by *Hemmati* et al. [8.147]. Because these lasers are automatically locked to the atomic resonance lines, they are of practical interest for trace detection and remote sensing.

Further metal-atom laser include Pb at 405.8, 368.3, and 364.0 nm [8.148] and the Fe laser at 360, 385, 395, 540, 558, and 563 nm starting from iron pentacarbonyl [8.149, 150]. The atomic iodine laser has been pumped with XeBr* [8.151] and with XeCl* and KrF* radiation [8.152]. In the XeCl* and

KrF* pumping experiments, the main interest was directed at the high degree of flexibility in pulse duration and resonator design which the excimer pump makes possible. An example of a process in which a molecular dissociation fragment appears inverted is given by the photodissociation of HgBr$_2$ at 193 nm [8.153]. The product HgBr* undergoes stimulated emission between 502 and 505 nm as discussed in Sect. 5.7.

References

8.1   A.L.Schawlow, C.H.Townes: Phys. Rev. **112**, 1940 (1958)
8.2   S.Jorna, N.A.Bailey, J.Hirth, R.Mueller, R.Meyerott, S.Schneider, D.A.Shirley, W.Smith, R.Spitzer, P.A.Sullivan, J.K.Thomas, G.T.Trammel: "X-Ray Laser Applications Study", Report PD-LJ-159, Physical Dynamics, Inc., La Jolla (July 1977)
8.3   A.R.Neureuther: In *Synchrotron Radiation Research*, ed. by H.Winick, S.Doniach (Plenum Press, New York 1980) p. 223
8.4   C.K.Rhodes: In *Novel Materials and Techniques in Condensed Matter, ed. by* G.W.Crabtree, P.Vashishta (Elsevier, New York 1982) p. 151
8.5   J.Jaklovic, J.A.Kirby, M.P.Klein, A.S.Robertson, G.S.Brown, P.Eisenberger: Solid State Commun. **23**, 679 (1977)
8.6   C.K.Rhodes, R.L.Carman: Appl. Opt. **21**, 3799 (1982)
8.7   L.L.Warren, T.H.Geballe: Mater. Sci. Eng. **50**, 149 (1981)
8.8   J.C.Solem, G.C.Baldwin: Science **218**, 229 (1982)
8.9   R.C.Elton, R.W.Waynant, R.A.Andrews, M.A.Bailey: "X-Ray and Vacuum uv Lasers", Naval Research Laboratory report 7412 (May 1972)
      R.W.Waynant, R.C.Elton: Proc. IEEE **64**, 1059 (1976)
8.10 J.R.Murray, J.Goldhar, A.Szöke: Appl. Phys. Lett. **32**, 551 (1977)
8.11 R.T.Hawkins, H.Egger, J.Bokor, C.K.Rhodes: Appl. Phys. Lett. **36**, 391 (1980)
8.12 H.Egger, T.Srinivasan, K.Hohla, H.Scheingraber, C.R.Vidal, H.Pummer, C.K.Rhodes: Appl. Phys. Lett. **39**, 37 (1981)
8.13 H.Egger, T.S.Luk, K.Boyer, D.F.Muller, H.Pummer, T.Srinivasan, C.K.Rhodes: Appl. Phys. Lett. **41**, 1032 (1982)
8.14 H.Egger, H.Pummer, C.K.Rhodes: Laser Focus **18**, 59 (1982)
8.15 C.K.Rhodes: "Generation of vacuum ultraviolet and extreme ultraviolet radiation by nonlinear processes with excimer lasers", in *Proc. 6th Intern. Workshop on Laser Interaction and Related Plasma Phenomena*, ed. by H.Hora, G.H.Miley (Plenum Press, New York to be published)
      H.Pummer, H.Egger, T.S.Luk, T.Srinivasan, C.K.Rhodes: "Vacuum ultraviolet stimulated emission from two-photon excited molecular hydrogen" (submitted to Phys. Rev. A)
8.16 H.Pummer, H.Egger, T.S.Luk, T.Srinivasan, C.K.Rhodes: "High-power vuv stimulated emission from two-photon excited H$_2$". *AIP Conf. Proc.* **100**, (AIP, New York, 1983)
8.17 T.S.Luk, H.Pummer, K.Boyer, M.Shahidi, H.Egger, C.K.Rhodes: "Collision-free generation of highly ionized atomic species with 193 nm radiation", *AIP Conf. Proc.* **100** (AIP, New York 1983)
8.18 T.S.Luk, H.Pummer, K.Boyer, M.Shahidi, H.Egger, C.K.Rhodes: "Anomalous collision-free multiple ionization of atoms with intense picosecond ultraviolet radiation" (to be published)
8.19 T.Srinivasan, K.Boyen, H.Egger, T.S.Luk, D.F.Muller, H.Pummer, C.K.Rhodes: In *Picosecond Phenomena III*, ed. by K.B.Eisenthal, R.M.Hochstrasser, W.Kaiser, A.Laubereau, Springer Ser. Chem. Phys., Vol. 23 (Springer, Berlin, Heidelberg, New York 1982) p. 19

8.20   J.P.Colpa: In *Physics of High-Pressures and Condensed Phases*, ed. by A. van Itterbeek (North-Holland, Amsterdam 1965) p. 490
       M.Rothschild, W.Gornik, J.Zavelovich, C.K.Rhodes: J. Chem. Phys. **75**, 3794 (1981)

8.21   J.Bokor, L.Eichner, R.H.Storz, P.H.Bucksbaum, R.R.Freeman: In *AIP Conf. Proc.* **100**, 143 (AIP, New York 1983)

8.22   T.Srinivasan, H.Egger, H.Pummer, C.K.Rhodes: "Generation of extreme ultraviolet radiation at 79 nm by sum frequency mixing", IEEE J. Quantum Electron. (to be published)

8.23   M.Rothschild, H.Egger, R.T.Hawkins, J.Bokor, H.Pummer, C.K.Rhodes: Phys. Rev. A**23**, 206 (1981)

8.24   J.Eggelston, J.Dallarosa, W.K.Bischel, J.Bokor, C.K.Rhodes: J. Appl. Phys. **50**, 3867 (1979)
       C.K.Rhodes: In *High Power Lasers and Applications*, ed. by K.L.Kompa, H.Walter, Springer Ser. Opt. Sci., Vol. 9 (Springer, Berlin, Heidelberg, New York 1978) p. 163

8.25   D.J.Kligler, J.Bokor, C.K.Rhodes: Phys. Rev. A**21**, 607 (1980)

8.26   R.W.Dreyfus, R.T.Hodgson: Phys. Rev. A**9**, 2635 (1974)

8.27   H.Pummer, H.Egger, T.S.Luk, T.Srinivasan, C.K.Rhodes: Phys. Rev. A (to be published)

8.28   L.Spitzer, J.L.Greenstein: Astrophys. J. **114**, 407 (1951)

8.29   E.M.Purcell: Astrophys. J. **116**, 457 (1952)

8.30   D.J.Kligler, C.K.Rhodes: Phys. Rev. Lett. **40**, 309 (1978)

8.31   H.Schmoranzer, J.Imschweiler: In *Proc. 8th Intern. Conf. on Atomic Physics*, ed. by I.Lindgren, A.Rosén, S.Svanberg (Wallin & Dalholm Boktr. AB, Lund 1982) p. A40

8.32   G.H.Dieke: J. Mol. Spectrosc. **2**, 494 (1958)

8.33   N.Djeu, R.Burnham: Appl. Phys. Lett. **30**, 473 (1977)

8.34   D.Cotter, W.Zapka: Opt. Commun. **26**, 251 (1978)

8.35   T.R.Loree, R.C.Sze, D.L.Barker: Appl. Phys. Lett **31**, 37 (1977)

8.36   T.R.Loree, R.C.Sze, D.L.Barker, P.B.Scott: IEEE J. QE-**15**, 337 (1979)

8.37   T.R.Murray, J.Goldhar, D.Eimerl, A.Szöke: Appl. Phys. Lett. **33**, 399 (1978)

8.38   J.R.Murray, J.Goldhar, D.Eimerl, A.Szöke: IEEE J. QE-**15**, 342 (1979)

8.39   N.Djeu: Appl. Phys. Lett. **35**, 663 (1979) ·

8.40   R.R.Jacobs, J.Goldhar, D.Eimerl, S.B.Brown, J.R.Murray: Appl. Phys. Lett. **37**, 264 (1980)

8.41   R.R.Jacobs, D.Prosnitz, W.K.Bischel, C.K.Rhodes: Appl. Phys. Lett. **29**, 710 (1976)
       H.Pummer, W.K.Bischel, C.K.Rhodes: J. Appl. Phys. **49**, 976 (1978)

8.42   W.K.Bischel, P.J.Kelly, C.K.Rhodes: Phys. Rev. A**13**, 1829 (1976)

8.43   N.Bloembergen, M.D.Levenson: In *High-Resolution Laser Spectroscopy*, ed. by K.Shimoda, Topics Appl. Phys., Vol. 13 (Springer, Berlin, Heidelberg, New York 1976) p. 315

8.44   J.E.Bjorkholm, P.F.Liao: Phys. Rev. A**14**, 751 (1976)

8.45   W.K.Bischel, R.R.Jacobs, C.K.Rhodes: Phys. Rev. A**14**, 1294 (1977)
       H.Pummer, W.K.Bischel, C.K.Rhodes: J. Appl. Phys. **49**, 976 (1978)

8.46   D.J.Kligler, C.K.Rhodes: Phys. Rev. Lett. **40**, 309 (1978)

8.47   D.L.Atkins, E.H.Fink, C.Bradley Moore: J. Chem. Phys. **52**, 1604 (1970)
       E.H.Fink, D.L.Atkins, C.Bradley Moore: J. Chem. Phys. **59**, 900 (1972)

8.48   R.T.Hodgson, R.W.Dreyfus: Phys. Rev. Lett. **28**, 536 (1972)
       C.R.Jeppeson: Phys. Rev. **44**, 165 (1933)

8.49   W.M.Jackson, J.B.Halpern, C.-S.Lin: Chem. Phys. Lett. **55**, 254 (1978)

8.50   A.P.Baronawski, J.R.McDonald: Chem. Phys. Lett. **56**, 369 (1978)

8.51   Z.Karny, R.Naaman, R.N.Zare: Chem. Phys. Lett. **59**, 33 (1978)

8.52   C.Fotakis, M.Martin, K.P.Lawley, R.J.Donovan: Chem. Phys. Lett. **67**, 1 (1979)

8.53   H.K.Haak, F.Stuhl: Chem. Phys. Lett. **68**, 399 (1979)

8.54   J.J.Tiee, F.B.Wampler, W.W.Rice: Chem. Phys. Lett. **68**, 403 (1979)

8.55   W.W.Rice, F.B.Wampler, R.C.Oldenborg, W.B.Lewis, J.J.Tiee, R.T.Pack: J. Chem. Phys. **72**, 2948 (1980)

8.56   W.G.Hawkins, P.L.Houston: J. Chem. Phys. **73**, 297 (1980)

8.57   H.Hemmati, G.J.Collins: Chem. Phys. Lett. **67**, 5 (1979)

8.58   S.Rockwood, J.P.Reilly, K.Hohla, K.L.Kompa: Opt. Commun. **28**, 175 (1979)

8.59   J.R.McDonald, A.P.Baronavski, V.M.Donnelly: Chem. Phys. **33**, 161 (1978)

8.60   J.Zavelovich, M.Rothschild, W.Gornik, C.K.Rhodes: J. Chem. Phys. **74**, 6787 (1981)

8.61   M.W.Wilson, M.Rothschild, C.K.Rhodes: J. Chem. Phys. **78**, 3779 (1983)

8.62   M.W.Wilson, M.Rothschild, D.F.Muller, C.K.Rhodes: J. Chem. Phys. **76**, 4452 (1982)

8.63   M.W.Wilson, M.Rothschild, D.F.Muller, C.K.Rhodes: J. Chem. Phys. **77**, 1837 (1982)

8.64   D.Chadwick: Can. J. Chem. **50**, 737 (1972)

8.65   H.Okabe: *Photochemistry of Small Molecules* (Wiley, New York 1978)

8.66   S.D.Peyerimhoff, R.J.Buenker: Chem. Phys. **57**, 279 (1981)

8.67   C.E.Moore: *Atomic Energy Levels*, Natl. Stand. Ref. Data Ser. Natl. Bur. Stand. 35 (1971) Vol. I–III

8.68   J.N.Bardsley: J. Phys. B**1**, 365 (1968)
       J.N.Bardsley, M.A.Biondi: In *Advances in Atomic and Molecular Physics*, ed. by D.R.Bates (Academic Press, New York 1970) p. 1

8.69   A.L'Huillier, L.A.Lompre, G.Mainfray, C.Manus: Phys. Rev. Lett. **48**, 1814 (1982)

8.70   I.I.Sobelman: *Atomic Spectra and Radiative Transitions*, Springer Ser. Chem. Phys., Vol. 1 (Springer, Berlin, Heidelberg, New York 1979)
       R.D.Cowan: *The Theory of Atomic Structure and Spectra* (University of California Press, Berkeley 1981)

8.71   T.A.Carlson, C.W.Nestor, Jr., N.Wasserman, J.D.McDowell: Atomic Data **2**, 63 (1970)

8.72   F.T.Porter, M.S.Freedman: J. Phys. Chem. Ref. Data **7**, 1267 (1978)

8.73   J.Bearden, A.F.Burr: Rev. Mod. Phys. **39**, 125 (1967)

8.74   G.L.DePoorter, C.K.Rofer-Depoorter: Spectrosc. Lett. **8**, 521 (1975)

8.75   Y.Gontier, M.Trahin: Phys. Rev. **172**, 83 (1968)

8.76   H.R.Reiss: Phys. Rev. A**1**, 803 (1970)
       H.R.Reiss: Phys. Rev. Lett. **25**, 1149 (1970)
       H.R.Reiss: Phys. Rev. D**4**; 3533 (1971)
       H.R.Reiss: Phys. Rev. A**6**, 817 (1972)

8.77   G.Wendin: J. Phys. B**3**, 455 and 466 (1970); B**4**, 1080 (1971); B**5**, 110 (1972); B**6**, 42 (1973)

8.78   F.Bloch: Z. Phys. **81**, 363 (1933)

8.79   I.Ya.Fugol': Adv. Phys. **27**, 1 (1978)

8.80   J.Jortner, L.Meyer, S.A.Rice, E.G.Wilson: J. Chem. Phys. **42**, 4250 (1965)

8.81   Y.Salamero, A.Birot, H.Brunet, H.Dijols, J.Galy, P.Millet, J.P.Mantagne: J. Chem. Phys. **74**, 288 (1981)

8.82   A.Gedanken, B.Raz, J.Jortner: J. Chem. Phys. **59**, 5471 (1973)

8.83   O.Cheshnovsky, B.Raz, J.Jortner: J. Chem. Phys. **59**, 5554 (1973)

8.84   N.G.Basov, E.M.Balashov, O.V.Bogdankevitch, V.A.Danilychev, G.N.Kashnikov, N.P.Lantshov, D.D.Khodkevitch: J. Luminescence **1/2**, 834 (1970)

8.85   N.G.Basov, V.A.Danilychev, Yu.M.Popov: Sov. J. Quantum Electron. **1**, 29 (1971)

8.86   D.A.Emmons, R.M.Lerner: Opt. Commun. **11**, 155 (1974)

8.87   J.C.Hill, O.Hebey, G.K.Walters: Phys. Rev. Lett. **26**, 1213 (1971)

8.88   B.Sonntag: In *Rare Gas Solids*, Vol. II, ed. by M.L.Klein, J.A.Venables (Academic Press, New York 1977) p. 1012

8.89   D.F.Muller, M.W.Wilson, M.Rothschild, C.K.Rhodes: Phys. Rev. A**25**, 1004 (1982)

8.90   H.Jara: Private communication

8.91   M.Cardona, L.Ley (eds.): *Photoemission in Solids I and II*, Topics Appl. Phys., Vols. 26 and 27 (Springer, Berlin, Heidelberg, New York 1978)

8.92   H.Sommer, G.Leonhardt, A.Meisel, D.Hirsch: Jpn. J. Appl. Phys. **17**, Suppl. 17-2, 278 (1978)

8.93   H.Winick, A.Bienenstock: Ann. Rev. Nucl. Part. Phys. **28**, 33 (1978)
       W.Gudat, C.Kunz: In *Synchrotron Radiation*, ed. by C.Kunz, Topics Current Phys., Vol. 10 (Springer, Berlin, Heidelberg, New York 1979) p. 55

8.94   A.Bianconi, L.Incoccia, S.Stipcich (eds.): EXAFS *and Near Edge Structures*, Springer Ser. Chem. Phys., Vol. 27 (Springer, Berlin, Heidelberg, New York, Tokyo 1983)

8.95   J.Jaklovic, J.A.Kirby, M.P.Klein, A.S.Robertson, G.S.Brown, P.Eisenberger: Solid State Commun. **23**, 679 (1977)

8.96   B.Stritzker, A.Pospieszczyk, J.A.Tagle: Phys. Rev. Lett. **47**, 356 (1981)

8.97   H.W.Lo, A.Compaan: Appl. Phys. Lett. **38**, 179 (1981)

8.98   H.W.Lo, A.Compaan: Phys. Rev. Lett. **44**, 1604 (1980)

8.99   R.A.McFarlane, H.L.Dunlop, L.D.Hess, G.L.Olson: In *Laser and Electron Beam Processing of Materials*, ed. by C.W.White, P.S.Peercy (Academic Press, New York 1980) p. 215

8.100   D.H.Auston, J.A.Golovchenko, A.L.Simons, C.M.Surko, T.N.Venkatesan: Appl. Phys. Lett. **34**, 777 (1979)

8.101   F.H.Hsu, C.M.White: In *Laser and Electron Beam Processing of Materials*, ed. by C.W.White, P.S.Peercy (Academic Press, New York 1980) p. 461; *Proc. 5th Intern. Conf. on Positron Annihilation*, ed. by R.S.Hasiguti, K.Fujiwara (Japan Institute of Metals, Sendai, 1979)

8.102   Y.J.Chabal, J.E.Rowe, D.A.Zwemer: Phys. Rev. Lett. **46**, 600 (1981)

8.103   Y.J.Chabal, R.J.Culbertson, L.C.Feldman, J.E.Rowe: J. Vac. Sci. Tech. **18**, 880 (1981)

8.104   M.Tamura, N.Natsuaki, T.Tokuyama: In *Laser and Electron Beam Processing of Materials*, ed. by C.W.White, P.S.Peercy (Academic Press, New York 1980) p. 247
        L.Reimer: *Transmission Electron Microscopy*, Springer Ser. Opt. Sci., Vol. 36 (Springer, Berlin, Heidelberg, New York, Tokyo 1984)

8.105   R.Aussenegg, A.Leitner, M.E.Lippitsch (eds.): *Surface Studies with Lasers*, Springer Ser. Chem. Phys., Vol. 33 (Springer, Berlin, Heidelberg, New York, Tokyo 1983)
        S.U.Campisano: Appl. Phys. A**30**, 195 (1983)
        C.J.Chen, R.M.Osgood: Appl. Phys. A**31**, 171 (1983)

8.106   A.Gat, J.F.Gibbons: Appl. Phys. Lett. **32**, 142 (1978)
        A.Gat, J.F.Gibbons, T.J.Magee, J.Pony, V.R.Deline, P.Williams, C.A.Evans, Jr.: Appl. Phys. Lett. **32**, 276 (1978)

8.107   J.F.Gibbons: In *Proceedings of the Symposium on Laser and Electron Beam Processing of Electronic Materials*, ed. by C.L.Anderson, G.K.Celler, G.A.Rozgonyi (Electrochemical Society, Inc., Princeton, NJ 1980) p. 1

8.108   R.T.Young, G.A.van der Leeder, J.Narayan, W.H.Christie, R.F.Wood, P.E.Rothe, J.I.Levatter: IEEE EDL-**3**, 280 (1982)
        R.T.Young, R.F.Wood: Ann. Rev. Mater. Sci. **12**, 323 (1982)

8.109   H.Ryssel, H.Glawischnig (eds.): *Ion Implantation Techniques*, Springer Ser. Electrophys., Vol. 10 (Springer, Berlin, Heidelberg, New York 1982)

8.110   D.J.Ehrlich, T.F.Deutsch, R.M.Osgood, Jr.: In *Laser and Electron Beam Processing of Materials*, ed. by C.W.White, P.S.Peercy (Academic Press, New York 1980) p. 671

8.111   B.Godard, O.de Witte: Opt. Commun. **19**, 325 (1976)
        D.G.Sutton, G.A.Capelle: Appl. Phys. Lett. **29**, 563 (1976)

8.112   V.S.Antonov, K.L.Hohla: Appl. Phys. B**30**, 109 (1983); B**32**, 9 (1983)

8.113   W.Zapka: Appl. Phys. **20**, 283 (1979)

8.114   W.Zapka, D.Cotter: Opt. Commun. **36**, 79 (1981)

8.115   V.I.Tomin, A.J.Alcock, W.J.Sarjeant, K.E.Leopold: Opt. Commun. **26**, 396 (1978)

8.116   W.Buffer, R.Schieder, H.Telle, R.Rave, W.Brinkwerth: Opt. Commun. **33**, 85 (1980)

8.117   H.Telle, W.Huffer, D.Basting: Opt. Commun. **38**, 402 (1981)

8.118   O.Uchimo, T.Mizunami, M.Maeda, Y.Miyazoe: Appl. Phys. **19**, 35 (1979)

8.119   V.I.Tomin, A.J.Alcock, W.J.Sarjeant, K.E.Leopold: Opt. Commun. **28**, 336 (1978)

8.120   P.Cassand, P.B.Corkum, A.J.Alcock: Appl. Phys. **25**, 17 (1981)

8.121   E.A.Stappaerts: Appl. Opt. **16**, 3079 (1977)

8.122   P.Cassand, R.S.Taylor, P.B.Corkum, A.J.Alcock: Opt. Commun. **38**, 131 (1981)

8.123   W.B.Lewis: "Saturable Absorbers at 248nm", Los Alamos National Laboratory Memorandum (November 30, 1981)

8.124   D.F.Muller, M.Rothschild, K.Boyer, C.K.Rhodes: IEEE J. QE-**18**, 1865 (1982)

8.125   E.D.Hinkley (ed.): *Laser Monitoring of the Atmosphere*, Topics Appl. Phys., Vol. 14 (Springer, Berlin, Heidelberg, New York 1976)

8.126   O.Uchino, M.Maeda, M.Hirano: IEEE J. QE-**15**, 1094 (1979)

8.127   O.Uchino, M.Maeda, T.Shibata, M.Hirano, M.Fujiwara: Appl. Opt. **24**, 4175 (1980)
        O.Uchino, M.Madea: Appl. Phys. Lett. **33**, 807 (1978)

8.128   J.B.Halpern, W.M.Jackson, V.McGrary: Appl. Opt. **18**, 590 (1979)

8.129 M.O.Rodgers, K.Asai, D.D.Davis: Appl. Opt. **19**, 3597 (1980)
8.130 T.Srinivasan, H.Egger, T.S.Luk, H.Pummer, C.K.Rhodes: In *Laser and Optical Remote Sensing*, ed. by D.Killinger, A.Mooradian, Springer Ser. Opt. Sci., Vol. 39 (Springer, Berlin, Heidelberg, New York,Tokyo 1983)
8.131 D.J.Ehrlich, P.F.Moulton, R.M.Osgood, Jr.: Digest of Topical Meeting on Excimer Lasers, IEEE Cat. No. 79CH1470-4QEA (IEEE, New York 1979)
8.132 J.W.Baer, M.G.Knights, E.P.Chicklis: Digest of Topical Meeting on Excimer Lasers, IEEE Cat. No. 79CH1470-4QEA (IEEE, New York 1979)
8.133 D.J.Ehrlich, P.F.Moulton, R.M.Osgood, Jr.: Opt. Lett. **4**, 184 (1979)
8.134 D.J.Ehrlich, P.F.Moulton, R.M.Osgood, Jr.: Opt. Lett. **5**, 339 (1980)
8.135 H.Pummer, J.Eggleston, W.K.Bischel, C.K.Rhodes: Appl. Phys. Lett. **32**, 427 (1978)
8.136 D.J.Ehrlich, R.M.Osgood, Jr.: IEEE J. QE-**16**, 257 (1980)
8.137 J.R.Murray, C.K.Rhodes: J. Appl. Phys. **47**, 5041 (1976)
8.138 H.T.Powell, A.U.Hazi: Chem. Phys. Lett. **59**, 71 (1978)
8.139 H.T.Powell, D.Prasnitz, B.A.Schleicher: Appl. Phys. Lett. **34**, 571 (1979)
8.140 R.N.Zare, D.R.Herschbach: "Atomic and Molecular Fluorescence Excited by Photodissociation", ed. by K.E.Shuler, W.R.Bennett, Jr., Appl. Opt. Suppl. No. 2 (Chemical Lasers) pp. 193–200 (1965)
8.141 D.J.Ehrlich, R.M.Osgood, Jr.: Appl. Phys. Lett. **34**, 655 (1979)
8.142 R.Burnham: Appl. Phys. Lett. **30**, 132 (1977)
8.143 T.F.Deutsch, D.J.Ehrlich, R.M.Osgood, Jr.: Opt. Lett. **4**, 378 (1979)
8.144 D.J.Ehrlich, J.Maya, R.M.Osgood, Jr.: Appl. Phys. Lett. **33**, 931 (1978)
8.145 P.Burkhard, W.Lüthy, T.Gerber: Opt. Lett. **5**, 522 (1980)
8.146 P.Burkhard, W.Lüthy, T.Gerber: Opt. Commun. **34**, 451 (1980)
8.147 H.Hemmati, G.J.Collins: Appl. Phys. Lett. **34**, 844 (1979)
8.148 H.Hemmati, G.J.Collins: IEEE J. QE-**16**, 594 (1980)
8.149 D.W.Trainor, S.A.Mani: Appl. Phys. Lett. **33**, 31 (1978)
8.150 D.W.Trainor, S.A.Mani: J. Chem. Phys. **68**, 5481 (1978)
8.151 J.C.Swingle, C.E.Turner, Jr., J.R.Murray, E.V.George, W.F.Krupke: Appl. Phys. Lett. **28**, 387 (1976)
8.152 E.E.Fill, W.Skrlac, K.J.Witte: Opt. Commun. **37**, 123 (1981)
8.153 E.J.Schimitschek, J.E.Celto, J.A.Trias: Appl. Phys. Lett. **31**, 608 (1977)

List of Figures

List of Tables

Subject Index

Applied Physics B
Photophysics and Laser Chemistry

Fields and Editors:

Laser Physics and Spectroscopy

High-Resolution Laser Spectroscopy:
V.P.Chebotayev, Novosibirsk
Laser Spectroscopy: **T.W.Hänsch,** Stanford U.
Quantum Electronics: **A.Javan,** MIT
Ultrafast Phenomena: **W.Kaiser,** TU München
Laser Physics and Applications:
H.Walther, U.München

Chemistry with Lasers

Chemical Dynamics and Structure: **K.L.Kompa,** MPI Garching
Laser-Induced Processes: **V.S.Letokhov,** Moscow
Dye Laser and Photophysical Chemistry:
F.P.Schäfer, MPI Göttingen
Laser Chemistry: **R.N.Zare,** Stanford U.

Photophysics

Optics: **W.T.Welford,** Imperial College
Nonlinear Optics and Nonlinear Spectroscopy:
T.Yajima, Tokyo U.

Editor: **H.K.V.Lotsch,**
Springer-Verlag, P.O.Box 105280,
D-6900 Heidelberg 1, Federal Republic of Germany

Special Features:
- rapid publication (3–4 months)
- no page charges for concise reports
- 50 offprints free of charge

Subscription information and/or **sample** copies are available from your bookseller or directly from Springer-Verlag, Journal Promotion Dept., P.O.Box 105280, D-6900 Heidelberg, FRG

Springer-Verlag
Berlin
Heidelberg
New York
Tokyo

W. Demtröder

Laser Spectroscopy
Basic Concepts and Instrumentation
2nd corrected printing. 1982. 431 figures. XIII, 696 pages
(Springer Series in Chemical Physics, Volume 5)
ISBN 3-540-10343-0

Contents: Introduction. – Absorption and Emission of Light. –
Widths and Profiles of Spectral Lines. – Spectroscopic Instrumentation. – Fundamental Principles of Lasers. – Lasers as Spectroscopic
Light Sources. – Tunable Coherent Light Sources. – Doppler-Limit-
ed Absorption and Fluorescence Spectroscopy with Lasers. – Laser
Raman Spectroscopy. – High-Resolution Sub-Doppler Laser Spectroscopy. – Time-Resolved Laser Spectroscopy. – Laser Spectroscopy of Collision Processes. – The Ultimate Resolution Limit. –
Applications of Laser Spectroscopy. – References. – Subject Index.

Lasers and Chemical Change
By A. Ben-Shaul, Y. Haas, K. L. Kompa, R. D. Levine
1981. 245 figures. XII, 497 pages
(Springer Series in Chemical Physics, Volume 10)
ISBN 3-540-10379-1

Contents: Lasers and Chemical Change. – Disequilibrium. –
Photons, Molecules, and Lasers. – Chemical Lasers. – Laser
Chemistry. – References. – Author Index. – Subject Index.

High-Power Lasers and Applications
Proceedings of the Fourth Colloquium on Electronic Transition
Lasers in Munich, June 20–22, 1977

Editors: **K.-L. Kompa, H. Walther**
2nd printing. 1979. 142 figures, 27 tables. IX, 228 pages
(Springer Series in Optical Sciences, Volume 9)
ISBN 3-540-08641-2

Contents: Excimer Lasers. – Chemical Lasers. – Other Laser
Systems. – Frequency Conversion. – Applications.

G. Brederlow, E. Fill, K. J. Witte

The High-Power Iodine Laser
1983. 46 figures. IX, 182 pages
(Springer Series in Optical Sciences, Volume 34)
ISBN 3-540-11792-X

Contents: Introduction. – Basic Features. – Principles of High-Power
Operation. – Beam Quality and Losses. – Design and Layout of an
Iodine Laser System. – The ASTERIX III System. – Scalability and
Prospect of the Iodine Laser. – Conclusion. – References. – Subject
Index.

Springer-Verlag
Berlin
Heidelberg
New York
Tokyo